TK 1005 .K495 2008

Khan, Shoaib.

Industrial power systems

INDUSTRIAL POWER SYSTEMS

INDUSTRIAL POWER SYSTEMS

Shoaib Khan

NEW ENGLAND INSTITUTE OF TECHNOLOGY
LIBRARY

CRC Press
Taylor & Francis Group
Boca Raton London New York

CRC Press is an imprint of the
Taylor & Francis Group, an **informa** business

CRC Press
Taylor & Francis Group
6000 Broken Sound Parkway NW, Suite 300
Boca Raton, FL 33487-2742

© 2008 by Taylor & Francis Group, LLC
CRC Press is an imprint of Taylor & Francis Group, an Informa business

No claim to original U.S. Government works
Printed in the United States of America on acid-free paper
10 9 8 7 6 5 4 3 2

International Standard Book Number-13: 978-0-8247-2443-6 (Hardcover)

This book contains information obtained from authentic and highly regarded sources. Reprinted material is quoted with permission, and sources are indicated. A wide variety of references are listed. Reasonable efforts have been made to publish reliable data and information, but the author and the publisher cannot assume responsibility for the validity of all materials or for the consequences of their use.

No part of this book may be reprinted, reproduced, transmitted, or utilized in any form by any electronic, mechanical, or other means, now known or hereafter invented, including photocopying, microfilming, and recording, or in any information storage or retrieval system, without written permission from the publishers.

For permission to photocopy or use material electronically from this work, please access www.copyright.com (http://www.copyright.com/) or contact the Copyright Clearance Center, Inc. (CCC) 222 Rosewood Drive, Danvers, MA 01923, 978-750-8400. CCC is a not-for-profit organization that provides licenses and registration for a variety of users. For organizations that have been granted a photocopy license by the CCC, a separate system of payment has been arranged.

Trademark Notice: Product or corporate names may be trademarks or registered trademarks, and are used only for identification and explanation without intent to infringe.

Library of Congress Cataloging-in-Publication Data

Khan, Shoaib.
   Industrial power systems / Shoaib Khan, Sheeba Khan, Ghariani Ahmed.
     p. cm.
   Includes bibliographical references and index.
   ISBN 978-0-8247-2443-6 (alk. paper)
   1. Electric power systems. I. Khan, Sheeba. II. Ahmed, Ghariani. III. Title.

TK1005.K495 2007
621.319--dc22
                                                    2007032836

Visit the Taylor & Francis Web site at
http://www.taylorandfrancis.com

and the CRC Press Web site at
http://www.crcpress.com

Dedication

My father dedicated the draft version of this book to my mother, his wife Shamim, whose love, patience and support inspired him to pursue knowledge and share it with others.

The publication of this book was my father's dream, as it will allow the knowledge he acquired over the years to be shared with others on a larger scale. I'm grateful for having had the opportunity to play a small role in making his dream come true. I would like to dedicate the final version to him.

Sheeba Khan

Contents

Acknowledgments xxv
Shoaib A. Khan xxvii

Chapter 1 Introduction 1

Chapter 2 System Planning 3

2.1 Introduction 3
2.2 Basic Design Considerations 3
    2.2.1 Safety of Life 3
    2.2.2 Reliability of Utility Power Supply 4
    2.2.3 Reliability of Plant Distribution System 4
    2.2.4 Simplicity of Operation and Maintenance 5
    2.2.5 Voltage Regulation and Flicker 5
    2.2.6 Cost (Last Priority) 5
2.3 Plant Distribution Systems 5
    2.3.1 An Overview 5
    2.3.2 Plant Main Substation 5
    2.3.3 Primary Distribution System 6
    2.3.4 Secondary Distribution System 7
    2.3.5 In-Plant Generation 7
    2.3.6 Emergency Power Supply 7
    2.3.7 Power Supply for Monitoring and Control Systems 8
    2.3.8 DC Power Supply for Protection and Control 9
2.4. Distribution Types 9
    2.4.1 Simple Radial 9
    2.4.2 Expanded Radial 9
    2.4.3 Primary Selective 10
    2.4.4 Secondary Selective 10
    2.4.5 Sparing Transformer 10
2.5. Plant Power Demand and Load Estimate 11
    2.5.1 Estimate of Power Demand and Energy Cost 11
        2.5.1.1 Energy Consumption or Average Kilowatt Hours Required for One Unit of Production 11
        2.5.1.2 Typical Demand Factors for Utilization Equipment 12
        2.5.1.3 Estimate of Power Demand from Equipment List 15
    2.5.2 Factors Used for Load Estimate and Power Demand 15
2.6 Voltage Considerations 15
    2.6.1 Voltages Used in North America 15
    2.6.2 Voltages Used in Europe 16

        2.6.3   Voltage and Frequency Used in Other Countries 16
        2.6.4   Voltage Drop and Flicker.. 17
                2.6.4.1   Steady-State Voltage Drop 17
                2.6.4.2   Voltage Flicker... 20
                2.6.4.3   Voltage Drop Due to Motor Starting.................................. 24
                2.6.4.4   Voltage Drop/Rise Due to Switching 24
        2.6.5   Voltage Sag and Threshold Voltage.. 24
References... 25
Bibliography... 26

Chapter 3   Power System Studies.. 27

3.1   Introduction and Overview 27
3.2   Useful Formulae.. 27
3.3   Load Flow 28
        3.3.1   Overview... 28
        3.3.2   Load-Flow Runs.. 29
3.4   Short Circuits 29
        3.4.1   Contribution from Different Sources... 29
        3.4.2   Steps Required for a Short-Circuit Study 31
        3.4.3   Medium- and High-Voltage Circuit Breakers................................... 31
                3.4.3.1   Contact Parting or Interrupting Duty.................................. 32
                3.4.3.2   Close-and-Latch Duty 33
        3.4.4   Interrupting Duty of Low-Voltage Circuit Breakers (<1000 V) 35
                3.4.4.1   Metal-Enclosed Circuit Breakers [S6, S7] 35
                3.4.4.2   Molded-Case Circuit Breakers 36
        3.4.5   Interrupting Duty of Medium- and High-Voltage Fuses [S5]............ 36
        3.4.6   For Time-Delayed Relaying Devices... 36
        3.4.7   Rotating Machine Impedance Multipliers [S9] 36
        3.4.8   Short-Circuit Current Multiplying Factors for Higher X/R Ratio 37
        3.4.9   X/R Ratio of System Component 38
        3.4.10 System and Equipment Data... 39
        3.4.11 Example: Short-Circuit Calculation for Equipment Rating............... 45
                3.4.11.1   Hand Calculations 45
                3.4.11.2   Computer Solution... 52
3.5   Protective Device Coordination 53
        3.5.1   Introduction... 53
        3.5.2   Data Required for a Study 54
        3.5.3   Time Interval for Coordination of Overcurrent Elements 54
                3.5.3.1   Between Relays Controlling Circuit Breakers.................... 54
                3.5.3.2   Between Upstream and Downstream Fuses 55
                3.5.3.3   Upstream Fuse and Downstream Circuit Breaker............... 55
                3.5.3.4   Static Trip Units of Low-Voltage Circuit Breakers 55
        3.5.4   Computer Programs for Overcurrent Coordination........................... 55
3.6   Arc-Flash Hazard Calculations... 55
        3.6.1   Introduction... 55

		3.6.2	Steps Required for the Analysis	57
3.7	Harmonic Analysis			57
	3.7.1	Introduction		57
	3.7.2	Methods and Techniques		57
	3.7.3	Necessary Steps for the Study		58
3.8	Power System Stability			59
	3.8.1	An Overview		59
	3.8.2	Transient Behavior of Synchronous Generators and Motors		59
	3.8.3	Stability Study for an Ammonia Plant		60
3.9	Other Common Studies and Calculations			62
	3.9.1	Switching Transients		62
	3.9.2	Cable Ampacity Calculations		63
	3.9.3	DC Power Supply		63
	3.9.4	Motor Starting		63
	3.9.5	Ground Grid		64

References 64
Bibliography 64

Chapter 4 System Neutral Grounding 67

4.1	Introduction and Overview	67
4.2	Ungrounded System	68
	4.2.1 Normal Operation	68
	4.2.2 Ground Fault on Phase "a"	68
	4.2.3 Derivation of Equivalent Circuit	69
	4.2.4 Overvoltage Due to Ground Fault through an Impedance	69
	4.2.5 Conclusion and Recommendation	70
4.3	High-Resistance Grounded System	71
	4.3.1 Normal Operation	71
	4.3.2 Ground Fault on Phase "a"	71
	4.3.3 Example of a High-Resistance Grounding Calculation	72
	4.3.4 HR Grounding for Delta-Connected Systems (System Neutral not Available)	74
	4.3.5 HR Grounding for Low-Voltage (480 and 600 V) Systems	75
	4.3.6 Measurement of System Charging Current	75
	4.3.7 Conclusions	75
	4.3.8 Recommendations	76
4.4	Low-Resistance (LR) Grounded System	76
	4.4.1 LR Grounding for Wye-Connected System	76
	4.4.2 LR Grounding for Delta-Connected System	77
	4.4.3 Conclusions	77
	4.4.4 Recommendations	78
4.5	Solidly Grounded Neutral System	79
	4.5.1 Solidly Grounded Neutral	79
	4.5.2 Power System Considerations	79
	4.5.3 Conclusions	79

	4.5.4	Recommendations 80
4.6	Generator Neutral Grounding 80
	4.6.1	Solidly Grounded Neutral 80
	4.6.2	High-Resistance Grounding 81
		4.6.2.1 Resistor Directly Connected to Generator Neutral 81
		4.6.2.2 Distribution Transformer-Resistor Combination 82
		4.6.2.3 Example: Sizing of Generator Neutral Grounding Equipment 82
	4.6.3	Low-Resistance Grounding 84
	4.6.4	Generator HR Grounded and System LR Grounded 85
	4.6.5	Hybrid High-Resistance Grounding 85
	4.6.6	Low-Reactance Grounding 85
	4.6.7	Resonant Grounding 85
4.7	Grounding of Mine Power System 86
	4.7.1	An Overview 86
	4.7.2	Safety Considerations in HV Substation 86
	4.7.3	Factors Involved in Shock Hazard 86
	4.7.4	Neutral Grounding Resistor 87
	4.7.5	Ground-Fault Relaying and Ground-Conductor Monitoring 88
	4.7.6	Low-Voltage (600 V and 480 V) Systems 88
4.8	Neutral Grounding Equipment 89
	4.8.1	HR Grounding for Wye-Connected MV Systems 89
	4.8.2	HR Grounding for Wye-Connected LV Systems 89
	4.8.3	LR Grounding for Wye-Connected MV Systems 90
	4.8.4	HR Grounding for Wye-Connected Generator Neutral 90
	4.8.5	Grounding Transformer 90
	4.8.6	Zig-Zag Grounding Transformer 91
	4.8.7	Wye-Delta Grounding Transformer 92
4.9	System Capacitance Data 93
	4.9.1	Surge Capacitors for Rotating Machines 93
	4.9.2	Overhead Transmission/Distribution Lines 93
	4.9.3	Power Cable 94
	4.9.4	Power Transformers 95
	4.9.5	Outdoor Apparatus Bushings 95
	4.9.6	Isolated Phase, Segregated, and Nonsegregated Bus 95
	4.9.7	Synchronous Motors with Class B Insulation System 95
	4.9.8	Air-Cooled Generators 96
	4.9.9	Induction Motors 96
References 97
Bibliography 97

Chapter 5	Power Transformers and Reactors 99
5.1	General 99
	5.1.1	Basic Construction 99
	5.1.2	Normal (Usual) Service Conditions 99

Contents

	5.1.3	Unusual Service Conditions	100
		5.1.3.1 Altitude Correction Factors	100
		5.1.3.2 Unusual Cooling Air and Water Temperature	101
	5.1.4	Transformer Categories and Short-Circuit Withstand	102
	5.1.5	Short-Circuit Impedance or Impedance Voltage (Z_T)	103
	5.1.6	Vector Group, Angular Displacement, and Terminal Markings	103
	5.1.7	Taps or Tap Changer	105
		5.1.7.1 Off-Circuit Taps	106
		5.1.7.2 Load Tap Changer	106
	5.1.8	Parallel Operation	106
5.2	Oil-Filled (Immersed) Transformers		107
	5.2.1	Classification of Mineral-Oil-Immersed Transformers	107
	5.2.2	Voltage, Power Rating, and Temperature Rise	107
	5.2.3	Impedance	108
	5.2.4	Cooling Designation	108
	5.2.5	Oil or Liquid Preservation System	110
		5.2.5.1 Sealed-Tank System	110
		5.2.5.2 Gas-Oil Seal System	110
		5.2.5.3 Conservator Design	111
		5.2.5.4 Conservator Diaphragm	112
	5.2.6	Winding Insulation Levels	112
	5.2.7	Tap Changers	113
		5.2.7.1 Tap Changer Design	113
		5.2.7.2 Basic Construction Features	113
		5.2.7.3 Controls and Indication	113
	5.2.8	Transformer Bushings	114
	5.2.9	Bushing Current Transformers	114
	5.2.10	Accessories	115
	5.2.11	Cooling Equipment	116
	5.2.12	Nameplate Information	117
	5.2.13	Core Grounding	117
	5.2.14	Transformer Gas Analysis: Dissolved Gases in Transformers	117
5.3	Nonflammable Liquid-Filled Transformers		118
	5.3.1	Overview	118
	5.3.2	Askeral-Immersed Transformers	118
	5.3.3	Silicone-Immersed Transformers	119
	5.3.4	Other Available Transformer Types	119
5.4	Dry-Type Transformers		119
	5.4.1	Type and Rating	119
	5.4.2	Insulation Class and Temperature Rise	120
	5.4.3	Primary System Fault Level for Short-Circuit Withstand	120
	5.4.4	Insulation Level	121
	5.4.5	Impedance	121
	5.4.6	Tap Changer or Taps	121
	5.4.7	Accessories	122
	5.4.8	Surge Arresters for Dry-Type Transformers	122

5.5	Transformers for Nonlinear Loads	122
	5.5.1 Standard Design	122
	5.5.2 Transformer Losses	122
	5.5.3 K-Rated Transformers	123
5.6	Generator Step-Up and Other Special Transformers	124
5.7	Installation of Oil-Filled Transformers and Reactors	126
	5.7.1 General	126
	5.7.2 Indoor (Inside Buildings)	126
	5.7.3 Outdoor Installation	127
5.8	Reactors	127
5.9	Inspection and Testing	128
	5.9.1 Factory Routine and Type Tests	128
	5.9.2 Test of Transformer	130
	5.9.3 Routine Power Transformer Inspection	131
	5.9.4 Internal Inspection of Transformer	132
References		132
Bibliography		133

Chapter 6	Instrument Transformers	135
6.1	Introduction and Overview	135
6.2	Current Transformers	135
	6.2.1 Bushing Type	135
	6.2.1.1 Bar Type	135
	6.2.1.2 Window Type	135
	6.2.1.3 Wound Type	136
	6.2.1.4 Post Type	136
	6.2.1.5 Auxiliary CT	136
	6.2.1.6 Linear Couplers (Air-Core Mutual Reactors)	136
	6.2.1.7 Optical or Digital CT	136
	6.2.2 Equivalent Circuit	137
	6.2.3 Excitation Characteristics	138
	6.2.4 Accuracy Class and Burden	139
	6.2.5 Current Transformers Saturation	141
	6.2.5.1 Saturation with Symmetrical Fault Current (No DC Component)	141
	6.2.5.2 Saturation with Asymmetrical Fault Current (with DC Component)	141
	6.2.5.3 Example	141
	6.2.5.4 Recommendations	142
	6.2.5.5 Saturation Factor and Time to Saturate	143
	6.2.6 Remnant Flux	144
	6.2.7 Accuracy Voltage of Bushing Current Transformers	144
	6.2.8 Accuracy Voltage of Current Transformers Used in Switchgear	144
	6.2.9 Current Transformers in Series	145
	6.2.10 Polarity of CT: Polarity Marking	145

Contents

- 6.2.11 Auxiliary Current Transformers 146
- 6.2.12 Metering Current Transformers 146
- 6.3 Voltage Transformers 146
  - 6.3.1 Inductive Voltage Transformer 146
  - 6.3.2 Capacitor Voltage Transformers 147
  - 6.3.3 Standard Accuracy Class and Burden 148
  - 6.3.4 Ratings and Characteristics 148
  - 6.3.5 Protection of Voltage Transformers 149
  - 6.3.6 Ferro-Resonance in Voltage Transformer 150
- 6.4 Grounding of Secondary 150
- 6.5 European Standards 151
  - 6.5.1 Current Transformers 152
    - 6.5.1.1 Protection-Class Current Transformers 152
    - 6.5.1.2 Example 152
    - 6.5.1.3 Measuring or Metering Current Transformers 152
  - 6.5.2 Voltage Transformers 152
    - 6.5.2.1 Standard Accuracy Class 152
    - 6.5.2.2 Voltage Factor 153
- References 153
- Bibliography 154

Chapter 7 Switchgear, Circuit Breakers, and Motor Control Center 155

- 7.1 Low Voltage 155
  - 7.1.1 Short Circuit 155
  - 7.1.2 Important Features to Be Specified 155
- 7.2 Medium Voltage 156
- 7.3 Load-Interrupter Switchgear 157
- 7.4 Power Fuse 166
  - 7.4.1 Advantages 166
  - 7.4.2 Disadvantages 166
  - 7.4.3 Fuse Size for Motor Branch Circuit 167
  - 7.4.4 Application of Fuse Let-Through Charts 167
- 7.5 Medium- and High-Voltage Circuit Breaker 167
  - 7.5.1 Overview of Medium- and High-Voltage Circuit-Breaker Standards 168
  - 7.5.2 Basic Performance Requirements: Medium- and High-Voltage Circuit-Breaker Standards 169
  - 7.5.3 Circuit-Breaker Types 170
    - 7.5.3.1 Air-Magnetic Air Quenching 170
    - 7.5.3.2 Oil Circuit Breakers 170
    - 7.5.3.3 Air-Blast Breakers 171
    - 7.5.3.4 SF_6 Gas Circuit Breakers 171
    - 7.5.3.5 Vacuum Circuit Breakers 172
  - 7.5.4 Circuit-Breaker Short-Circuit Rating and E/X Method 173
- 7.6 SF_6 Gas-Insulated Switchgear 174

	7.6.1	Main Features 175
	7.6.2	Design Concepts 175
	7.6.3	Specifications 176
7.7	Low- and Medium-Voltage Motor Control Centers 177	
	7.7.1	Low-Voltage Motor Control Center 177
	7.7.2	Medium-Voltage Motor Control Center 178
	7.7.3	Short-Circuit Capacity 178
	7.7.4	Coordinated Protection in Medium-Voltage Fused Starters 179
	7.7.5	Main and Vertical Bus Bars 179
	7.7.6	Minimum Size 179
	7.7.7	IEC-Rated Contactors 179

References 180

Chapter 8   Station Battery 181

8.1	Introduction 181
8.2	System Types 181
8.3	DC Distribution Systems 182
8.4	Types of Battery 185
8.5	Battery Chargers 188
8.6	Application Criteria 189

	8.6.1	Application Voltage Levels 189
	8.6.2	Rating 190
	8.6.3	Load Classification 190
8.7	Battery Sizing 191	
	8.7.1	System Voltage 191
	8.7.2	Correction Factor 193
	8.7.3	Duty Cycle 193

References 193
Bibliography 194

Chapter 9   Application and Protection of Medium-Voltage Motors 195

9.1	Introduction and Overview 195
9.2	Load Characteristics 195

	9.2.1	Load Categories 195	
		9.2.1.1	Constant-Torque Drives 195
		9.2.1.2	Torque as a Function of Speed Drives 196
	9.2.2	Steady-State Power 196	
	9.2.3	Average Starting Torque and Acceleration 197	
		9.2.3.1	Acceleration Torque 197
		9.2.3.2	Inertia 198
		9.2.3.3	Friction 198
		9.2.3.4	Frequency of Starting 198
		9.2.3.5	Induction-Motor Equipment Network 199
9.3	Squirrel-Cage Induction Motors 199		

	9.3.1	Torque Characteristics 199
	9.3.2	NEMA Design Letters 200
	9.3.3	NEMA Code Letter 201
	9.3.4	Inertia.. 201
	9.3.5	Service Factor 202
	9.3.6	Insulation Systems and Temperature Rise................................203
	9.3.7	Surge-Withstand Capability..203
	9.3.8	Power Supply Voltage and Frequency 203
	9.3.9	Starting or Locked-Rotor Power Factor..................................204
	9.3.10	Enclosures 204
	9.3.11	Motors for Classified (Hazardous) Areas 205
	9.3.12	Bearing and Shaft Currents 206
		9.3.12.1 Sleeve Bearings 206
		9.3.12.2 Antifriction Bearing 207
		9.3.12.3 Thrust Bearing 207
		9.3.12.4 Bearing Protection Against Failures 207
		9.3.12.5 Supply Voltage Unbalance......................................207
		9.3.12.6 Shaft Current 207
	9.3.13	Accessories 208
	9.3.14	Motor Terminal Box 208
9.4	Wound Rotor (Slip Ring) Induction Motors...209
9.5	Synchronous Motors 211
	9.5.1	An Overview 211
	9.5.2	Power Supply Voltage and Frequency 211
	9.5.3	Insulation System and Temperature Rise 212
	9.5.4	Torque Characteristics 212
	9.5.5	Inertia (Wk^2) 213
	9.5.6	Excitation System... 213
	9.5.7	Rotor Protection and Monitoring.. 214
	9.5.8	System Power Factor Improvement 214
	9.5.9	Torque Pulsations during Acceleration................................. 215
	9.5.10	Current Pulsations 216
	9.5.11	Application... 217
9.6	Electric Motors for Variable Frequency Drives................................... 217
9.7	Voltage Drop and Acceleration Time.. 218
	9.7.1	Guidelines 218
	9.7.2	Voltage Drop Using Hand Calculations..............................220
		9.7.2.1 Short-Circuit Method 220
		9.7.2.2 Per-Unit Method 220
		9.7.2.3 Motor Representation 221
		9.7.2.4 Voltage Drop Caused by Running Load.............221
		9.7.2.5 Commonly Used Formulae and Equations........221
		9.7.2.6 Example Voltage Drop Due to Motor Starting...................222
	9.7.3	Voltage Dip When a Sudden Load Is Applied to a Small Generator 224
	9.7.4	Voltage Drop and Acceleration Time Using Computer Software ... 225

	9.7.5	Estimate of Acceleration Time Using Hand Calculations	226
	9.7.6	Estimate of Deceleration Time	227
	9.7.7	Simplified Method for Estimating the Acceleration Time of Centrifugal Drives (Fans, Blowers)	227

9.8 Motor Controllers and Starting Methods 228
    9.8.1 Fused Starter 228
    9.8.2 Switchgear-Type Controllers 230
    9.8.3 Starting Methods 230
        9.8.3.1 Direct-on Line Start 230
        9.8.3.2 Reactor Start 230
        9.8.3.3 Autotransformer Start 230
        9.8.3.4 Capacitor Start 232
        9.8.3.5 Reactor-Capacitor Start 232
        9.8.3.6 Captive Transformer Start 233
        9.8.3.7 Variable-Voltage Start 233
        9.8.3.8 Variable-Frequency Start 233
        9.8.3.9 Part Winding Start 234
References 234
Bibliography 234

Chapter 10 Power and Control Cables 237

10.1 Introduction and Overview 237
10.2 Cable Selection Criteria 237
    10.2.1 Load Current Criteria 238
    10.2.2 Voltage-Drop Limitations 238
    10.2.3 Fault-Current Criteria 239
    10.2.4 Insulation Levels and System Grounding 239
10.3 Cable Shielding 239
    10.3.1 Inner Shield 239
    10.3.2 Outer Shield 240
    10.3.3 Outer or Overall Shield Size 240
    10.3.4 Shield and Outer Sheath Grounding 241
    10.3.5 Recommended Practice for Industrial Cable System 241
10.4 Additional Application Considerations 243
    10.4.1 Short-Circuit Conductor Heating 243
    10.4.2 Application of Power Cables for Variable-Frequency Drives 244
        10.4.2.1 Feeder Cable to the Drive Cubicle 244
        10.4.2.2 Feeder from the Drive Cubicle to the Motor 245
10.5 Cable Insulation 246
    10.5.1 Standards 246
    10.5.2 Insulation Characteristics 247
    10.5.3 Insulation Levels 247
    10.5.4 Insulation Reliability 248
10.6 Testing 249

Contents xvii

    10.6.1 High-Voltage DC Testing (DC Hi Pot) of Medium-Voltage Cable Insulation ... 249
    10.6.2 EPR Insulation Testing ... 249
10.7 Control Cables ... 250
References ... 251
Bibliography ... 252

Chapter 11 Protection ... 253

11.1 Introduction ... 253
11.2 Protection and Coordination Principles ... 254
    11.2.1 Protection Schemes and Relay Selection ... 254
        11.2.1.1 Current Transformer Connections ... 254
        11.2.1.2 Power System Device Function Numbers ... 254
        11.2.1.3 Selection Criteria ... 255
    11.2.2 Necessary Steps to Carry Out Relay-Setting and Time–Current Coordination ... 255
11.3 Transformer Protection ... 257
    11.3.1 Protective Devices and Features ... 257
        11.3.1.1 Devices Provided with the Oil-Immersed Transformers ... 257
        11.3.1.2 Primary Overcurrent Protection ... 259
        11.3.1.3 Secondary Overcurrent Protection ... 259
        11.3.1.4 Differential Protection ... 260
        11.3.1.5 Time Settings ... 260
        11.3.1.6 Typical Settings ... 260
    11.3.2 Transformer Protection with a Primary Fuse ... 260
        11.3.2.1 Minimum Fuse Rating ... 260
        11.3.2.2 Limitation of Fuse for Transformer Protection ... 262
    11.3.3 Protection of Low-Voltage Unit Substation Transformers ... 263
        11.3.3.1 Secondary Circuit Breaker with Static Trip ... 263
        11.3.3.2 Primary Protection with Circuit Breaker ... 264
        11.3.3.3 Primary Protection with Fuse ... 264
    11.3.4 Electrical Code Rules for Transformer Protection ... 264
    11.3.5 Transformer Through-Fault Current Damage Curves ... 266
        11.3.5.1 Category and kVA Range ... 266
        11.3.5.2 Short-Circuit Impedance ... 266
        11.3.5.3 Example ... 268
    11.3.6 Differential Protection of Transformers ... 268
        11.3.6.1 Phase Differential Scheme ... 268
        11.3.6.2 Ground Differential Scheme ... 269
        11.3.6.3 Guidelines for Phase Differential ... 270
    11.3.7 Protection of Grounding Transformers ... 271
    11.3.8 Overexcitation Protection ... 272
    11.3.9 Surge Protection ... 273
11.4 Motor Protection ... 274

11.4.1 Data Required for Relay Selection and Setting 274
    11.4.1.1 Data from Motor Manufacturer.. 274
    11.4.1.2 Data from Mechanical Equipment Vendor......................... 275
    11.4.1.3 Data from Power Utility Company..................................... 275
  11.4.2 Motor-Protection Schemes.. 276
    11.4.2.1 Selection of Current Transformers 276
    11.4.2.2 Relay Characteristics and Modeling 276
    11.4.2.3 Motor Acceleration Torque... 276
    11.4.2.4 Protection Redundancy 277
    11.4.2.5 The Use of RTD Bias into Motor Current.......................... 278
    11.4.2.6 Motors Driving High-Inertia or High-Torque Load 278
    11.4.2.7 NEMA E2 Motor Controllers.. 278
    11.4.2.8 Electrical Code Requirements... 278
  11.4.3 Recommended Protection... 279
    11.4.3.1 Protection against Abnormal Power Supply Conditions ... 279
    11.4.3.2 Protection against Faults in the Motor or Motor Feeder ... 279
    11.4.3.3 Protection against Abnormal Conditions Caused by
             the Driven Load... 279
    11.4.3.4 Protection against Abnormal Conditions Caused by
             Environment 279
    11.4.3.5 Recommended Protection Schemes 280
  11.4.4 Protective Devices and Suggested Settings 281
    11.4.4.1 Undervoltage Protection (Device 27).................................. 281
    11.4.4.2 Unbalanced Voltage/Current (Device 46) 282
    11.4.4.3 Underfrequency (Device 81)... 285
    11.4.4.4 Overload Protection (Device 49) 285
    11.4.4.5 Winding Overtemperature (Device 49W).......................... 285
    11.4.4.6 Locked Rotor or Fail to Accelerate (Device 51S).............. 285
    11.4.4.7 Phase Overcurrent or Short Circuit (Device 50) 287
    11.4.4.8 Ground Fault (Device 51G) 287
    11.4.4.9 Differential (Device 87M).. 288
    11.4.4.10 Loss of Field Protection for Synchronous Motors,
             Device 40... 288
    11.4.4.11 Out-of-Step Protection (for Synchronous Motors),
             Device 55 or 78 288
    11.4.4.12 Rotor Protection and Monitoring for Synchronous
             Motors.. 289
    11.4.4.13 Slip-Ring Flashover Protection for Wound Rotor
             Induction Motors 289
    11.4.4.14 Out-of-Phase Reenergizing Protection 290
    11.4.4.15 Surge Protection 291
    11.4.4.16 Variable-Frequency Drive Protection................................ 291
  11.4.5 Electrical Code Requirements 292
    11.4.5.1 Canadian Electrical Code, CSA C22.1-2002 292
    11.4.5.2 National Electrical Code, NEC-2002 294
  11.4.6 Typical Coordination Curves 295

Contents   xix

        11.4.6.1 5000 hp Compressor Time–Current Coordination Curve 295
        11.4.6.2 1750 hp Time–Current Coordination Curve 295
11.5 Generator Protection 296
    11.5.1 Data Required for Relay Selection and Settings 297
    11.5.2 Recommended Protection Schemes 298
        11.5.2.1 High-Resistance Grounded Generator 298
        11.5.2.2 Unit-Connected Generator 298
        11.5.2.3 Resistance-Grounded Generator Connected to LV-Grounded System 301
    11.5.3 Protective Devices and Suggested Settings 301
        11.5.3.1 Accidental or Inadvertent Energizing 301
        11.5.3.2 Volts per Hertz or Overfluxing Protection (Device 24) 301
        11.5.3.3 Reverse Power or Antimotoring (Device 32) 303
        11.5.3.4 Loss of Field/Excitation (Device 40) 304
        11.5.3.5 Negative-Sequence Current (Device 46) 305
        11.5.3.6 Generator Thermal Overload Protection (Device 49) 305
        11.5.3.7 Winding Overtemperature Protection (Device 49W) 305
        11.5.3.8 Voltage-Restrained Overcurrent Protection (Device 51V) 306
        11.5.3.9 Overvoltage Protection (Device 59) 307
        11.5.3.10 Out-of-Step Protection (Device 78) 307
        11.5.3.11 Frequency — Under or Over (Device 81 U/O) 307
        11.5.3.12 Differential Phase (Device 87G) 307
        11.5.3.13 Differential, Ground (Device 87GN) 308
        11.5.3.14 Stator Ground-Fault Protection 309
        11.5.3.15 Field Ground Protection 310
        11.5.3.16 Generator Breaker Fail Protection (Device 52BF) 310
    11.5.4 Diesel Generator Connected to a Low-Voltage Bus 312
    11.5.5 Generator Tripping Schemes 312
        11.5.5.1 Sequential Tripping 313
11.6 Feeder Protection 313
11.7 Capacitor Protection 314
    11.7.1 Fuse for Individual Capacitor Unit Protection 314
    11.7.2 Relaying for Capacitor-Bank Protection 315
        11.7.2.1 Overcurrent Relay for Capacitor-Bank Protection 315
        11.7.2.2 Under- and Overvoltage Protection 316
        11.7.2.3 Current- or Voltage-Unbalance Protection 316
11.8 Reactor Protection 317
11.9 Bus Protection 317
    11.9.1 Current-Transformer Saturation 317
    11.9.2 Bus-Differential Schemes 318
        11.9.2.1 Nondirectional Overcurrent Relays 318
        11.9.2.2 Stabilizing Resistor in Series with the Overcurrent Relay 319
        11.9.2.3 High-Impedance Voltage Differential Scheme 320

　　　　　　11.9.2.4 Moderately High-Impedance Bus-Differential Scheme 320
　　　　　　11.9.2.5 Linear Coupler Bus-Differential Scheme 320
References 321
Bibliography 322

Chapter 12 High-Voltage Substation Design Considerations 325

12.1　Introduction 325
12.2　Necessary Information for the Design of Substations 325
　　　12.2.1 Considerations for the Location of the Substations 325
　　　12.2.2 Basic Information Required 327
12.3　Standards and Design Principles for Substation Design 327
　　　12.3.1 System of Units 327
　　　12.3.2 Applicable Standards 328
　　　12.3.3 Standard Transformer Connections and System Phasing 328
12.4　Substation Electrical Configuration 330
　　　12.4.1 Switching Configuration 330
　　　12.4.2 Commonly Used Substation Configurations 331
　　　12.4.3 Recommended Configurations 334
12.5　Secondary System Aspects of Substation Design 335
　　　12.5.1 Substation Protection 335
　　　12.5.2 Local and Remote Control and Metering and Indication
　　　　　　 Requirements 336
　　　12.5.3 Relay Building Location 338
　　　12.5.4 Utility Telecommunication Requirements 338
　　　12.5.5 Current and Voltage Transformers Requirements 339
12.6　High-Altitude Considerations 340
　　　12.6.1 Effect of Altitude on Insulation Levels 340
　　　12.6.2 Effect of Altitude on Conductor Size 340
　　　12.6.3 Definition of Altitude Zones for Substation Design 341
12.7　Substation Design Considerations 342
　　　12.7.1 Layout and Conductor Spacing 342
　　　12.7.2 General Considerations for Bus Design 342
　　　12.7.3 Rigid Bus Design 345
　　　12.7.4 Strain Bus Design 346
12.8　Grounding Design and Lightning Protection 347
　　　12.8.1 Introduction 347
　　　12.8.2 Substation Area 347
　　　12.8.3 General Surge Protection 348
　　　12.8.4 Equipment Surge Protection 349
　　　12.8.5 Direct-Stroke Shielding Principles 349
　　　　　　12.8.5.1 Overhead Shield Wires 349
　　　　　　12.8.5.2 Masts 351
　　　12.8.6 Lightning Protection Shielding Calculation Methods 351
12.9　Medium- and Low-Voltage Substations and Switchgear 352
12.10 Site Testing and Commissioning 352

Contents   xxi

        12.10.1 Scope of Work 352
        12.10.2 Objectives 353
        12.10.3 Responsibilities... 353
        12.10.4 Safety Procedures 353
        12.10.5 Training of the Employer's Staff 354
        12.10.6 Maintenance Aspects 354
References... 354
Bibliography... 355

Chapter 13 Substation Grounding Design Considerations 361

13.1   Introduction and Overview 361
13.2   Soil-Resistivity Measurements.. 361
        13.2.1 Soil Resistivity 361
                13.2.1.1 Rectangular Volume 362
                13.2.1.2 Soil Characteristics... 362
        13.2.2 Soil Model and Its Impact on Grid Design............................. 362
                13.2.2.1 Uniform Earth 362
                13.2.2.2 Two-Layer Earth... 362
        13.2.3 Measurement Techniques 364
13.3   Permissible Potential Difference... 365
        13.3.1 Step Voltage 365
        13.3.2 Touch Voltage 366
        13.3.3 Mesh Voltage 367
        13.3.4 Transferred Voltage.. 367
        13.3.5 Typical Shock Situation for Gas-Insulated Switchgear 367
13.4   Maximum Ground Current 368
        13.4.1 Worst-Case Symmetrical Fault Current (I_f or $3I_0$) 368
        13.4.2 Symmetrical Grid Current 369
        13.4.3 Maximum Grid Current for Design.. 370
        13.4.4 Effect of Station Ground Resistance....................................... 370
        13.4.5 Effect of Overhead Ground Wires and Neutral Conductors........... 371
        13.4.6 Effect of Direct Buried Pipes and Cables............................... 371
        13.4.7 Computation of Current Division Factor 371
        13.4.8 Grid Current Returning through the Earth............................. 372
        13.4.9 Decrement Factor... 373
13.5   Selection of Conductors and Joints 373
        13.5.1 Basic Requirement 373
        13.5.2 Data Required to Design.. 373
                13.5.2.1 Type of Connector 374
                13.5.2.2 Maximum Allowable Temperature 374
        13.5.3 Minimum Size (for Hard-Drawn Copper Wire) 374
13.6   Design of Grounding System 374
        13.6.1 Evaluation of Ground Resistance... 375
        13.6.2 Design Procedure... 376
        13.6.3 Maximum Mesh and Step Voltage... 378

13.7	Treatment of Substation Fence	380
	13.7.1 Fence-Grounding System Design	381
	13.7.1.1 Station Fence Connected to the Main Grid and also to Its Own Ground Electrodes	381
	13.7.1.2 Station Fence Is Not Connected to the Station Ground Grid but Has Its Own Grounding System	381
	13.7.2 Station Fence-Grounding Details	382
13.8	Case Study (Ground Grid Design)	382
	13.8.1 Design Parameters	382
	13.8.2 Grid Design	384
13.9	Conclusion	386
References		386
Bibliography		386

Chapter 14 Electrical Aspects of Power Generation 387

14.1	Introduction and Overview	387
14.2	Generator Rating and Parameters	387
14.3	Excitation System	390
	14.3.1 Brushless Excitation System	391
	14.3.2 Static Excitation System	392
	14.3.3 Voltage-Regulating System	392
	14.3.4 Power System Stabilizer	393
14.4	Synchronizing	393
14.5	Integration into the Power System	395
	14.5.1 General	395
	14.5.2 Higher Reactance	396
	14.5.3 Current-Limiting Reactor	396
	14.5.4 Duplex Reactor	396
	14.5.5 Unit-Connected System	396
	14.5.6 Fault-Current Limiters	397
14.6	Plant Power System and Utility Interface	398
14.7	System Disturbances and Islanding	399
	14.7.1 Faults on HV (Utility) System	399
	14.7.2 Performance Criteria for Plant Electrical System and Equipment	400
	14.7.3 Steps to Improve Plant Electrical System Ride-Through Capability	401
14.8	Induction Generator	403
	14.8.1 General	403
	14.8.2 Characteristics	403
	14.8.3 Terminal Voltage	403
	14.8.4 Excitation	404
	14.8.5 Protection	404
	14.8.6 Controls	405
	14.8.7 Advantages and Disadvantages	405

		14.8.8 Differences with an Induction Motor	406
14.9	Station Auxiliaries		406
	14.9.1	Auxiliary System Load	406
	14.9.2	Power Supply Arrangement	406
		14.9.2.1 Power Plants	406
		14.9.2.2 Industrial Power Systems	407
	14.9.3	Design Criteria	408
		14.9.3.1 System Design and Equipment	408
		14.9.3.2 Voltage Considerations	409
		14.9.3.3 Short-Circuit Considerations	409
		14.9.3.4 Auxiliary Bus Capacity	409
References			410
Bibliography			411

Chapter 15 Application of Capacitors 413

15.1	Introduction	413
15.2	Capacitor Application	413
	15.2.1 Capacitor Construction	413
	15.2.2 Individual Capacitors	414
	15.2.3 Motor Capacitors	414
	15.2.4 Switched Capacitors	415
	15.2.5 Switching Device Rating	416
15.3	Large-Bank Construction	417
	15.3.1 Bank Construction	417
	15.3.2 Capacitor Unit Rating	418
	15.3.3 Bank Monitoring Criteria	419
	15.3.4 Bus Insulation Systems	420
	15.3.5 Interlocking and Safety	420
15.4	Control and Monitoring Systems	420
15.5	Capacitor Protection and Monitoring	421
	15.5.1 Capacitor Fusing	421
	15.5.2 Protective Control and Monitoring	424
15.6	Capacitor-Bank Resonances	426
	15.6.1 Introduction	426
	15.6.2 Example of a Resonance Risk Calculation	427
References		428
Bibliography		428

Chapter 16 Impact of Nonlinear Loads on Power System and Equipment 429

16.1	Introduction	429
16.2	Harmonics and Resonance	429
	16.2.1 What are Harmonics?	429
	16.2.2 Harmonic Current Generated by Nonlinear Loads	431
	16.2.3 Effects of Harmonics	431

		16.2.3.1 Motors and Generators 432
		16.2.3.2 Transformers... 432
		16.2.3.3 Capacitors 433
		16.2.3.4 Power Cables 433
		16.2.3.5 Electronic Equipment 433
		16.2.3.6 Switchgear and Relaying 434
		16.2.3.7 Fuses 434
		16.2.3.8 Telephone Interference 434
	16.2.4	Harmonic Resonance 435
		16.2.4.1 Parallel Resonance 435
		16.2.4.2 Series Resonance 437
	16.2.5	IEEE Standards for Harmonic Distribution Levels 438
16.3	Variable-Frequency Drives 440
16.4	System Neutral Grounding 441
16.5	Design Recommendations for Mitigation of Harmonics 442
	16.5.1 Capacitor Bank Applications 442
	16.5.2 Variable-Frequency-Drive Application.. 443
References 445
Bibliography 445

Index... 447

Acknowledgments

This book was inspired by power systems courses developed and taught by my father for IEEE Continuing Education (Montreal) over several years. It was intended to respond to what he felt was a need in the field for an engineering/design guide for all working engineers, particularly younger ones who are not exposed to power systems during their undergraduate studies and who may not get guidance or training at their work place. It was an idea which was enthusiastically supported and encouraged by members of IEEE Continuing Education (Montreal), the not-for-profit engineering education organization my father collaborated with for over thirty years.

When he passed away unexpectedly, the book was incomplete. IEEE members approached me and asked that I continue where my father had left off. I was apprehensive about taking on such a project without an engineering background, but IEEE was insistent that it was needed, and offered to assist in its completion. I knew my father had been very committed to the project, and finally, upon reflection, I decided to take it on, despite my concerns. Fortunately, my father's former colleagues, as well as members of IEEE came to my aid whenever requested, and provided the guidance and assistance needed to complete my father's book. Michael Care, Gabor Furst, and Vijay Sood provided much-needed expertise and help in editing, and Maurice Huneault also assisted. Finally, this project might not have been completed if not for the dedication of Ahmed Ghariani, my father's former assistant and mentee, who worked tirelessly to compile materials, draw diagrams and provide constant assistance whenever needed. This book would not have been possible without the contributions and dedication of these dear friends and colleagues.

Sheeba Khan

SHOAIB A. KHAN
May 4, 1936 - February 10, 2005

Shoaib Khan was an electrical engineer who worked for over 40 years in power systems engineering, application, and protective relaying for power (hydro, thermal, nuclear), pulp and paper, mining and metals, and chemical plants in North America and overseas. Born in India, he received a bachelor of science degree in electrical engineering with major in electric power from Banares Hindu University in 1957. He worked with Kaiser Engineers and M.N. Dastur & Co in India until 1965, when he immigrated to Canada with his wife and young daughter. In Canada, he was responsible for the engineering, construction and operation of steel, aluminium and power plants. He soon became known as an expert in power systems and protection, and organized, developed and taught courses in those areas. He wrote numerous technical papers on power systems, application engineering and protection, some of which have been published in IAS/PES transactions and *Pulp & Paper Canada*.

Shoaib was a senior member of the Institute of Electrical and Electronic Engineers (IEEE), and was very active with the IEEE Education Committee (Montreal Section) for over thirty years. He received the IEEE Centennial and Third Millennium Medals for achievement in education, the Honoured Engineers Medal and the Meritorious Engineering Award from the IEEE Pulp and Paper Industry Committee in 2001. He was also an auxiliary professor with the Department of Electrical Engineering and a guest lecturer with the Department of Chemical Engineering at McGill University, Montreal, where he organized, developed, and taught courses in power systems and protection.

Shoaib will be remembered by all his colleagues and friends for his dedication to the advancement of applied engineering, and for his hard work and dedication to education in his field of endeavor. He was not only an outstanding engineer, but also a true friend to his colleagues and a committed educator and supporter of young engineers.

1 Introduction

Based on the author's 40 years of hands-on experience in many industries, *Industrial Power Systems* provides the practicing engineer with modern, wide-ranging, and practical information, from the planning and design of electrical supply to electrical installations for industrial power systems. Using materials from IEEE courses developed for practicing engineers (under the banner of the IEEE Montreal Section), this comprehensive book provides a wealth of practical experience in a readable tone.

The book illustrates the importance of power systems, which is sometimes overlooked in practice. A power system is very much like a culture, varying from one type of industry to another in the same manner as traditions vary from one ethnic group to another. Each industry tries to maintain its own traditional practices which were evolved just after World War II. Those practices were influenced by economic constraints, process needs, technology available at that time, and fast growth. Some industries still try to maintain the same traditions when it comes to their power system but keep modernizing in the areas of controls and communications.

The main reason for the present situation is a lack of awareness for a reliable system on the part of power engineers and senior managers. The power system is still treated as a service only and not considered as an essential element to maintain product quality and continuity of operations. Also, the plant power distribution system generally represents a relatively small portion of the entire plant cost (5% to 10%), yet the production and output of the other 90% to 95% of plant investment is dependent on the service delivered by that investment in the power distribution system. The investment will return a profit only if electric power is continuously available in the quantities and of the quality desired.

This book covers the salient engineering features and design procedures, including power system studies, grounding, instrument transformers, medium-voltage meters, and many more topics. Chapters are easy to use and sufficiently detailed to address plant design engineering problems. An exhaustive list of standards and technical papers is also provided for further study.

Long overdue, *Industrial Power Systems* is a must-have for anyone involved in power engineering, including students, instructors, engineers, and senior management involved in the design and maintenance of power distribution systems.

2 System Planning

2.1 INTRODUCTION

The power system plays an important role in a process plant. The plant operations and production depend on a safe and reliable power system. Each plant power system design, whether new or an expansion to an existing system, must be analyzed to ensure that it is safe, reliable, meets the present objective, and permits expansion for future needs. This chapter outlines the system planning considerations and provides guidelines. References [S1] and [S2] provide guidelines for planning and designing a reliable power system.

2.2 BASIC DESIGN CONSIDERATIONS

The industrial power system planning and design must include the following considerations at the initial and planning or design stage.

2.2.1 Safety of Life

Human life is of utmost importance, and safety shall never be compromised. Electrical codes prescribe minimum installation practices, such as working space, clearance from live parts, minimum protection against overcurrent, etc. Major applicable electrical codes for North America are: National Electrical Code (NEC) [S17], Canadian Electrical Code (CEC) [S18], National Electrical Safety Code, etc. Sometimes states, provinces, or municipalities institute local codes in addition to NEC and CSA. Insurance underwriters such as Factory Mutual institute their standards and guidelines, which are over and above the relevant electrical codes and standards.

Electrical equipment minimum quality standards are prescribed in the standards and guides of the Institute of Electrical and Electronics Engineers (IEEE), the National Electrical Manufacturer's Association (NEMA), Underwriters Laboratories (UL), and the International Electrotechnical Commission (IEC). UL maintains a continuing service in testing and certifying the products of electrical manufacturers, principally those to be used for industrial and commercial applications.

The system planning and design shall include the following to ensure the safety of personnel and preservation of plant property:

- Equipment and installation shall conform to relevant codes and standards.
- Provide adequate working space and safe clearances around the electrical equipment, dead-front equipment for low- and medium-voltage systems, insulated bus and connections for metal-enclosed equipment, and adequate system and equipment grounding.

- Design the system to permit maintenance of equipment and circuits in a de-energized state without plant shutdown.
- Provide fully rated and protected equipment to withstand maximum short-circuit and load currents.
- Provide personnel protective equipment (PPE) such as insulated gloves, fire-retardant or fireproof clothing, and warning signs.
- Provide operations and maintenance instructions, such as built-in wiring and interlocking diagrams.
- Install emergency lighting for the safety and safe exit of personnel during power outage.

2.2.2 Reliability of Utility Power Supply

Ensure that the quantity, quality, and reliability of the utility power supply meets the plant power requirements. A dedicated line is more reliable and is subject to fewer power interruptions than a shared line. Distribution lines may have interruption rates of about 12 times per 100 km when compared with a 120 kV transmission line of the same length. Hence power supply at transmission voltage such as 120 kV or 230 kV is recommended.

Utility power systems, including transmission and distribution lines, are subject to disturbances such as lightning strokes and ground faults that cause voltage sags. These voltage sags may cause the undervoltage and control devices to trip and motors to stall, resulting in plant shutdown.

2.2.3 Reliability of Plant Distribution System

Plant power distribution system design must be considered during the planning and conceptual design stage. The IEEE Gold Book [S8] provides guidelines for the design of a reliable power system as well as surveys carried out on the failure of equipment and systems. The key steps to increase the reliability of the plant power system are:

- Select modern, standard, and reliable equipment. Apply good installation and preventive maintenance practice.
- Use a minimum of two circuits or feeds, each from a different bus, to major and critical load centers. Do not run both circuits in the same cable tray or duct bank.
- Do not use bare conductor overhead lines within the plant boundary. Run the distribution feeders above ground wherever possible; the failure rate of a directly buried cable is considerably higher.
- System neutral grounding reduces transient overvoltage on single line-to-ground faults, thus minimizing insulation failures.
- A coordinated short-circuit and overcurrent protection isolates the faulted circuit, protects the equipment, confines the power outage to the protected zone, makes it easier to locate the fault, and prevents fires.

System Planning

2.2.4 Simplicity of Operation and Maintenance

The majority of faults in utility distribution networks are caused by environmental reasons, such as a tree branch falling on the bare conductor overhead line, lighting strikes, etc. For this reason, the distribution networks are interconnected to facilitate alternative power supply routes. However, the majority of the faults in industrial systems are caused by insulation failure and sometimes by inadvertent or accidental contacts. The design of an industrial power distribution system shall be simple, utilizing radial feeds. This approach simplifies the interlocking and the maintenance, thus increasing safety.

2.2.5 Voltage Regulation and Flicker

This subject is covered in section 2.6.

2.2.6 Cost (Last Priority)

The cost of an electric power system is small compared with the total cost. Safety, reliability, voltage regulation, maintenance, and provision for future expansion shall be given priority.

2.3 PLANT DISTRIBUTION SYSTEMS

2.3.1 An Overview

A typical process plant distribution system in a single (one)-line diagram form is shown in fig. 2.1. This is composed of a main substation, primary distribution, secondary distribution, and in-plant generation. Electric utility companies supply power at high voltages (HV) via transmission lines or sometimes insulated power cable. HV power is stepped down to medium voltage for primary distribution to different plant facilities or load centers.

Each industry has implemented some specific and unique features in its plant electrical systems that have evolved with time and are based on experience. System planners and designers are advised to study a few typical plant distribution systems of similar installations before developing their own. IEEE guides are an excellent help for everyone involved with this task.

2.3.2 Plant Main Substation

High-voltage power from the utility is stepped down in the plant main substation for the primary distribution system. The electrical and nonelectrical items include:

- *Nonelectrical*: Gantry for terminating incoming and outgoing transmission lines, support structures, equipment foundation, perimeter fence, control building, grounding, underground oil containment, etc.
- *Electrical items*: Power transformers, circuit breakers, disconnect switches, instrument transformers, reactors, capacitors, power and control cable, protection and control, communications, lighting, heating, etc.

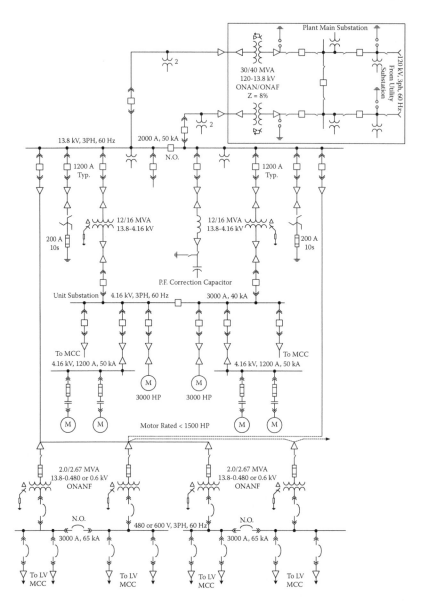

FIGURE 2.1 Typical one-line (single line) diagram

2.3.3 Primary Distribution System

This includes distribution from the plant main substation or generating station to the primary load centers or switchgear located in different plant facilities. A radial system with two feeders to each area is used for greater reliability. The following voltage levels can be considered as a guide:

- 4.16 kV for small and medium-sized plants (load up to 15 MVA)

System Planning

- 13.8 kV for medium and large plants (load exceeding 20 MVA)
- 34.5 kV for large plants where individual facilities are remote from each other

2.3.4 SECONDARY DISTRIBUTION SYSTEM

This includes distribution from the primary load centers to secondary load centers, unit substations, low-voltage switchgear, and utilization equipment such as motor control centers, motors, heating, and lighting. For greater reliability, a secondary selective arrangement is recommended for the load centers.

2.3.5 IN-PLANT GENERATION

In-plant generation is used when one of the following conditions is present:

- Steam is available at a suitable pressure and temperature. (It is economical to build a power plant using electric generators driven by steam turbines.)
- Power is not available from a utility company.
- Purchased power from the utility company is unreliable or the cost is very high.

2.3.6 EMERGENCY POWER SUPPLY

Emergency power is required where an outage of normal source will be detrimental to process and equipment. This may include motors, valves, emergency lighting, controls, etc. Electric generators driven by diesel engines are used. The unit is started automatically after about a 20-s delay upon loss of power. The diesel generator is started periodically to ensure that it is in a proper operating condition.

The diesel engine kW rating is selected to meet the active power demand. However, the generator kVA rating, reactance (X_d' and X_d''), and excitation system shall be selected to suit the following:

- Continuous load is the kVA vector sum of active and reactive power

$$\left(\sqrt{kW^2 + kVAR^2}\right).$$

- Motor starting kVA is at a very low power factor.
- Nonlinear loads such as a variable-frequency drive (VFD) are present. Since the generator is designed for a sinusoidal load with a 5% harmonic distortion, the kVA rating needs to be derated or the design altered for higher distortion factors.
- Apply resistance grounding for the generator neutral. (This subject is covered in chapter 4.)

2.3.7 Power Supply for Monitoring and Control Systems

Power supply for monitoring and control systems shall be reliable, unaffected by voltage dip or sag and transients, and shall meet the requirements of load characteristics including voltage, frequency, and harmonics.

An uninterruptible power supply (UPS) is required for electronic, analogue, and digital control and monitoring systems. This consists of a rectifier to convert AC source to DC, battery, inverter, and static transfer switch. Normal AC power for control is through the inverter, which is filtered, regulated, and isolated from system disturbances. Typical nonredundant and redundant UPS systems are shown in figs. 2.2 and 2.3, respectively.

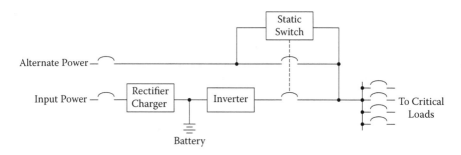

FIGURE 2.2 Nonredundant UPS

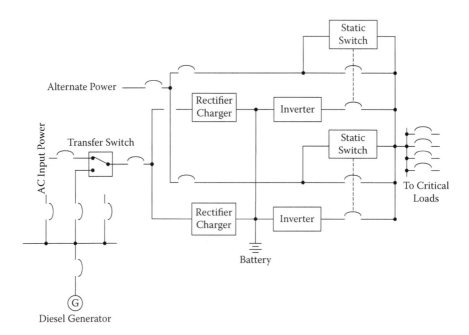

FIGURE 2.3 Redundant UPS

System Planning

In a nonredundant UPS, one charger and one inverter is used. In a redundant UPS, two chargers, two inverters, and an alternated AC source from an emergency or standby diesel generator are used to maintain the power supply under multiple contingencies.

2.3.8 DC Power Supply for Protection and Control

Power supply at 125 V DC is used for HV circuit breaker, medium-voltage switchgear, protection, control, monitoring, communications, emergency lighting, and emergency backup loads such as DC lube oil pump, etc. DC power supply consists of station battery, battery charger, and DC distribution. Selection and application of a station battery is covered in chapter 8.

2.4. DISTRIBUTION TYPES

Distribution systems commonly used in industrial plants are briefly described here.

2.4.1 Simple Radial

The simple radial system (fig. 2.4) has no redundancy and can be used where loss of power for an extended period is not detrimental to the process.

2.4.2 Expanded Radial

The expanded radial scheme (fig. 2.5) is an expansion of the above and has been used in many industries. The industrial power system at medium voltage (2.4 to 34.5 kV) is low-resistance (100–400 A) grounded, and power cable (insulated) is used for

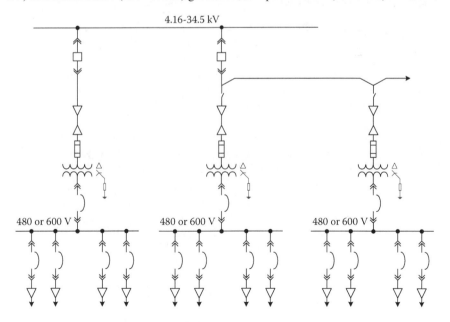

FIGURE 2.4 Simple radial   **FIGURE 2.5** Expanded radial

distribution. The majority of the faults with cable systems start as a line-to-ground fault and escalate to a three-phase fault if not cleared within a reasonable time. On a line-to-ground fault, the fuse will either not see the fault current or will take a long time to melt. However, the ground-fault relay at the switchgear set at about 5% will trip the feeder breaker. The power is lost in a large area, and a longer time is required to detect and isolate the fault.

2.4.3 Primary Selective

In the primary selective arrangement (fig. 2.6), two primary feeders are brought to each substation transformer. Half of the transformers are connected to each of the two feeders. Each primary feeder is designed to carry the entire load. Though the problem caused by the line-to-ground faults is the same as with the expanded radial system, the power to the load centers can be restored quickly by transferring to the alternate feeder.

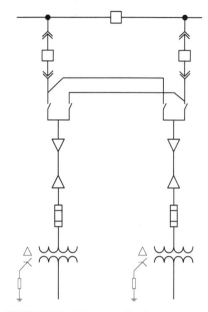

FIGURE 2.6 Primary selective

2.4.4 Secondary Selective

The secondary selective arrangement (fig. 2.7) can be achieved between two single transformer stations or double-ended stations by using a tie-circuit breaker. For low-voltage systems, the tie breaker is normally kept in open position, and an interlock between the main and the tie prevents paralleling of the transformers. An electrically operated manual transfer scheme can be used to close the tie breaker for a few cycles before tripping the selected main breaker. This scheme permits a planned shutdown of one transformer or primary feeder without dropping any load. However, the fault level may exceed the equipment rating during this brief period. Because both low-voltage buses are in an energized state, the possibility of developing a fault during this period is remote.

For medium-voltage systems, the tie breaker may operate in normally open or normally closed position. However, the relaying becomes more complex with parallel operation. The preferred and a commonly used arrangement is to operate the system with the tie breaker open and provide an auto- or manual-transfer scheme. The application of an auto-transfer scheme is common in thermal power plants.

2.4.5 Sparing Transformer

The sparing transformer scheme (fig. 2.8) is used where several low-voltage transformer stations of the same capacity are installed in one area. The load is transferred to the spare transformer upon the loss of any unit in the system. This minimizes the spare transformer capacity.

System Planning

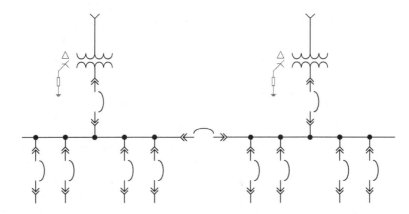

FIGURE 2.7 Secondary selective

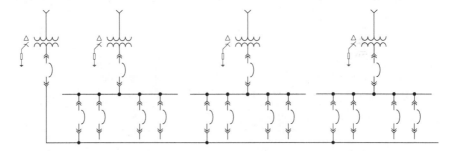

FIGURE 2.8 Sparing transformer

2.5. PLANT POWER DEMAND AND LOAD ESTIMATE

2.5.1 Estimate of Power Demand and Energy Cost

Estimates of power consumption, power demand, and total power cost are carried out during the project feasibility stage and during the contract negotiations with the power utility. The results of the power cost study, the power consumption of utilization equipment, and the approximate location of equipment is required for system planning and sizing the electrical equipment.

The plant power demand, power cost, electrical load estimate in different areas, and selection of major electrical equipment can be made using the following approach.

2.5.1.1 Energy Consumption or Average Kilowatt Hours Required for One Unit of Production

Historical data on energy consumption or average kilowatt hours required to produce one unit of product can be used for estimating:

- Energy consumption based on the plant output and operating hours
- Average power for the period
- Power demand for each facility and net demand for the plant

- Monthly and yearly power cost consisting of energy cost and demand charge:

$$\text{Avg. Power} = \frac{\text{kWh/unit} \times \text{Total Production}}{\text{Operating Hours}}$$

$$\text{Maximum Demand} = \frac{\text{Avg. Power}}{\text{Load Factor}}$$

An example illustrating the method for estimating the power requirement and power demand for an alloy steel plant producing 60,000 tons/year of finished product is given in tables 2.1 and 2.2, respectively. A plant producing Fe Cr (ferrochromium) operates 24 hours per day and 365 days per year, whereas the operating period for other plant facilities depends on the production requirement. These tables provide the plant operating period for each facility, including the number of shifts per day and number of days per week. Load factors and coincidence factors are based on past experience with similar plants.

Table 2.1 has been used to estimate the energy consumption and maximum power demand for each facility and total plant. Table 2.2 shows the plant facilities arranged in groups based on their operating period and shifts, which helps in organizing the plant operations and estimating plant manpower requirements.

Tariff as agreed with the power utility:

- $6.00/kW of maximum power demand
- $6.50/kW in excess or less than 80% of contract demand
- $0.04/kW for energy consumed.

Estimate of power cost:
  Power demand (based on a 30-min period) as derived from table 2.1 = 39,200 MW
  Energy consumption per year = 166.3×10^6 kWh
  Energy consumption per month = $13,860 \times 10^3$ kWh (based on an average 730 h)
  Energy cost per month = $\$0.04 \times 13,860 \times 10^3 = \$554,400$
  Demand charge per month = $\$6.0 \times 39,200 = \$235,200$
  Total power cost per month = $789, 600

2.5.1.2 Typical Demand Factors for Utilization Equipment

Typical demand factors for utilization equipment are given in table 2.3.

$$\text{Total Power Demand} = \Sigma \, (\text{Equipment Rating} \times \text{Demand Factor})$$

added space between equipment and rating.

$$\text{Net Power Demand} = \frac{\text{Total Power Demand}}{\text{Coincidence Factor}}$$

System Planning

TABLE 2.1
Estimate of Power Requirement (Basis — 60,000 ton/yr)

Department/Facility	Operating	Period, h/yr	Production, ton/yr 10^3	Energy, kWh/ton	Consumption, kWh/yr 10^6	Avg. Power, kW	Load Factor	Max. Demand (30 min)
Fe Cr plant	Continuous	8,760	18	4,000	72	8,220	0.9	9,130
Arc furnace	330 d/yr	7,920	80	500	40	5,050	0.3	17,000
V.O.D.	330 d/yr	7,920	80	10	0.8	101	0.2	505
Continuous casting	330 d/yr	7,920	77	15	1.2	152	0.4	380
Slab grinding	3 shifts/d 5 d/week	6,240	73	28	2.1	340	0.5	680
Hot mill	16 h/week	832	72	65	4.7	5,650	0.65*	8,700
Coil preparation	3 shifts/d5 d/wk	6,240	55	27	1.5	240	0.5	480
Cold mill	330 d/yr	7,920	55	320	17.6	2,230	0.6	3,720
Hot annealing and pickling	3 shifts/d5 d/wk	6,240	54	55	3	480	0.6	800
Skin pass mill	2 shifts/d 5 d/wk	4,160	40	140	5.6	1,350	0.65	2,080
Slitting	2 shifts/d 5 d/wk	4,160	40	35	1.4	340	0.6	570
Leveling and cutting	5 d/wk	4,160	28	65	1.8	440	0.6	730
Strip grinding	3 shifts/d 5 d/wk	6,240	16.5	160	2.6	420	0.6	700
Services	...	7,920	60	130	7.9	1,000	0.8	1,250
Total	...	...	...	...	**166.3**	**26,673**		**47,605**
Overall annual operation	...	8,760	60	2,772	166.3	18,990	0.484	39,200
Excluding Fe Cr plant	...	8,760	60	1,572	94.3	...	...	...

Note: Overall coincidence factor = 0.823 (diversity factor = 1.215). Overall coincidence factor is based on assuming 1.0 for Fe Cr plant and 0.781 for other loads.

TABLE 2.2
Estimate of Power Demand (Basis — 60,000 ton/yr)

Department/Facility	Operating Period	Production, ton/yr 10^6	kWh/ton	Max. Demand, kW	Coincidence Factor for Group	Max. Demand for Group	Group No.	Operation Shift [a]
Fe Cr plant	Continuous	18	4000	9130	1	9130	1	1, 2, 3
Arc furnace	330 d/yr	80	500	17000	1	17885	2	1, 2, 3
VOD	330 d/yr	80	10	505				
Cont. casting	330 d/yr	77	15	380				
Cold mill	330 d/yr	55	320	3720	1	3720	3	1, 2, 3
Hot mill	16 h/wk	72	65	8700	1	8700	4	3, 1
Slab grinding	3 shifts/d, 5 d/wk	73	28	68	1.15	3080	5	1, 2, 3
Coil preparation	3 shifts/d, 5 d/wk	55	27	04				
Hot annealing and pickling	3 shifts/d, 5 d/wk	54	55	80				
Cold annealing and pickling	3 shifts/d, 5 d/wk	70	58	08				
Strip grinding	3 shifts/d, 5 d/wk	16.5	160	700				
Skin pass mill	3 shifts/d, 5 d/wk	40	140	2080	1.15	2900	6	3, 1
Slitting	2 shifts/d, 5 d/wk	40	35	570				
Leveling and cutting		28	65	700				
Services		...	...	1250	1	1250	7	1, 2, 3

[a] Times for shift nos.: shift 1, 7:00 a.m.–3:00 p.m.; shift 2, 3:00 p.m.–11:00 p.m.; shift 3, 11:00 p.m.–7:00 a.m.

System Planning

TABLE 2.3
Demand Factors for Utilization Equipment

Equipment	Demand Factor, %	Equipment	Demand Factor, %
Arc furnace	90–100	Paper mills	55–70
Compressors	30–90	Refineries	55–70
Conveyors, blowers/fans	80–90	Welders, resistance	10–40
Lighting	100	Heating equipment	85–95
Cranes	30–50	Induction furnace	85–95
Welders, arc	30–50	Pumps	50–70

2.5.1.3 Estimate of Power Demand from Equipment List

Load data extracted from the equipment list prepared by the process and mechanical engineers is used for sizing the secondary distribution equipment, including transformers and motor control centers.

2.5.2 Factors Used for Load Estimate and Power Demand

$$\text{Average Power} = \frac{\text{Energy Consumed during a Specified Period (kWh)}}{\text{Operating Period in hours}}$$

$$\text{Load Factor} = \frac{\text{Average Load for a Period}}{\text{Peak Load for the Same Period}}$$

$$\text{Demand Factor} = \frac{\text{Maximum Demand}}{\text{Connection Load}}$$

$$\text{Coincidence Factor} = \frac{\text{Maximum Demand}}{\text{Sum of Individual Demand}} < 1.0$$

$$\text{Diversity Factor} = \frac{\text{Sum of Individual Max. Demand}}{\text{System Max. Demand}} = \frac{1}{\text{Coincidence Factor}} > 1.0$$

2.6 VOLTAGE CONSIDERATIONS

2.6.1 Voltages Used in North America

The most common (preferred) system and utilization voltages used in the United States and Canada for industrial power systems are given in table 2.4. The power supply frequency is 60 Hz. Reference [S1] provides the voltages and their ranges. For utilization equipment, the voltage variation is defined in the relevant standards.

TABLE 2.4
Preferred Voltage Level for Industrial Plants (Used in North America)

Voltage Class	System Voltage		Transformer Voltage- Primary or Secondary at No Load [a]	Generator Rated Voltage [b]	Motor Rated Voltage	Remarks
	Highest	Nominal				
Low	...	120/208Y	208 primary 208 Y / 120 sec.		208	single phase in Canada
		240	240		220	
		480	480	480	460	
		600	600	600	575	
Medium		2 400	2 400	240	2300	not a preferred voltage
	4 760	4 160	4 160	4160	4000	
	8 250	7 200	7 200	7200	4160 6900	synchronous not a preferred voltage
	15 000	13 800	13 800	13800	13200 13800	synchronous
		25 000	25000			
		27 600	27600			
	38 000	34 500	34500			
	48 300	44 000	44000			utility
	72 500	69 000	69000			distribution
High	145 000	120 000	120 000			transmission
	170000	161 000	161 000			
	245 000	230 000	230 000			
EHV	362 000	245 000	345 000			transmission
	550 000	500 000	500 000			

[a] Transformer rated voltage (primary/secondary) is at no load.
[b] Generator rated voltage is higher (15 to 24 kV) for large machines.

2.6.2 Voltages Used in Europe

Preferred system voltages used in Europe are given in ref. [S6]; power supply frequency is 50 Hz. The preferred system voltages are summarized in table 2.5.

2.6.3 Voltage and Frequency Used in Other Countries

Most of the countries have adopted IEC standards for preferred voltages for system and equipment. However, some countries use a combination of old and new standard voltages. Power system voltages used in countries other than North America and Europe are given in table 2.6. For a complete list and current status, refer to Siemens or other relevant publications.

System Planning

TABLE 2.5
Preferred Voltage in IEC Standard

Voltage Classification	Normal Voltage	Highest System Voltage
>100 V, <1000 V	230/400	...
	400/690	
>1 kV, <35 kV	3	...
	6	
	10	
	20	
	35	
>35 kV, <245 kV	69	72.5
	115	123
	138	145
	230	245
>245 kV		362
		420
		550
		800
		1050
		1200

Source: IEEE 446-1987, IEEE Recommended Practice for Emergency and Standby Power Systems.

2.6.4 Voltage Drop and Flicker

The performance of the utilization equipment (motors, lighting, etc.) is guaranteed when the voltage and frequency applied to its terminals is within the limits specified in the standards. Because the variation in power frequency is negligible during steady-state conditions, the voltage spread from no load to full load, transient voltage dips during switching operations (such as starting a large motor), and voltage flicker caused by cyclic loads such as reciprocating compressors need to be checked. Chapter 2 of ref. [1] has covered the effect of voltage spread on different utilization equipment.

2.6.4.1 Steady-State Voltage Drop

The steady-state voltage drop is caused by the variation in utility power supply and by voltage drop in the transformer and feeders connected to the utilization equipment. The spread from the utility power supply is generally remedied by utilizing main step-down transformers with load tap changers.

The electrical safety codes, NEC and CSA, have specified the allowable voltage drop at the maximum load current; these are 5% for feeders and 3% for branch circuits. The steady-state voltage drop can be calculated or estimated using one of the following methods.

- **Formula**

   Phase-to-neutral voltage drop is given by the formula:

$$\Delta V_{LN} = E_S + (I_R \times \cos\theta + I_X \times \sin\theta) + \sqrt{E_S^2 - (I_X \times \cos\theta - I_R \times \sin\theta)^2}$$

TABLE 2.6
System Voltage Used in Countries Other than Europe and North America

Country	Frequency, Hz	Transmission, kV	Distribution, kV	Utilization, V
Afghanistan	50	110	6, 15, 20, 33	220/380
Algeria	50	30, 60, 90, 150, 220, 400	5.5, 10, 15, 20, 33	220/380
Argentina	50	27, 33, 66, 132, 145, 170, 220, 500	6.6, 13.2, 13.8, 27	220/380
Australia	50	22, 33, 66, 88, 110, 132, 220, 275, 330, 500	3.6, 7.2, 11, 22, 33	250/440, 240/415
Bahamas	60	33, 66	2.4, 7.2, 11	115/200, 120/208
Bahrain	50, 60	66, 220	6.6, 11, 33	240/415
Bangladesh	50	33, 66, 132, 230	11, 33	230/400
Barbados	50	24.9	3.3, 11	110/190, 120/208
Brazil	60	24, 33, 44, 69, 88, 138, 230, 345, 440, 500, 750	2.4, 4.16, 6.6, 11.5, 13.2, 13.8	110/220, 220/440, 127/220, 220/380
Chile	50	23, 44, 66, 110, 154, 220, 500	3.3, 5, 6.0, 6.9, 12, 13.2	220/380
China	50	33, 60, 110, 220, 380, 420, 500	6, 10	127/220, 220/380
Colombia	60	34.5, 44, 66, 115, 230, 500	2.4, 4.16, 6.6, 7.2, 11.4, 13.8	110/220, 150/260, 480
Costa Rica	60	24.9, 34.5, 69	13.2	120/208, 120/240, 127/220, 254/440
Cuba	60	33, 66, 110, 220	2.4, 4.16, 6.0, 6.3, 7.2, 13.2, 22	120/240, 231/400, 277/480, 440
Dominican Republic	60	34.5, 69, (138)	4.16	110/220
Ecuador	60	24, 34.5, 46, 69, 115, 138, 230, 380	2.4, 4.16, 6.9, 13.8	120/208, 127/220
Egypt	50	33, 66, 132, 220, 500	3, 6, 10, 11, 33	400, 380/220
Ethiopia	50	45, 132, 220	15, 33	380/220
Federation of Arab Emirates	50	132, 220	6.6, 11, 33	220/380, 240/415
Ghana	50	33, 161	3.3, 6.6, 11	380/220
Haiti	50, 60	...	7.2, 12.5	110/220, 220/380
Honduras	60	34.5, 69, 138, 230	2.4, 4.16, 6.6, 13.8	110/220, 127/220
India	50	33, 66, 132, 220, 400	3.3, 6.6, 11	242/420
Indonesia	50	20, 69, 150, 500	6, 20	127/220, 220/380
Iran	50	66, 132, 230, 400	6, 11, 20, 33	220/380
Iraq	50	33, 66, 132, 400	(3.3), (6.6), 11, 33	220/380
Israël	50	33, 110, 161, 380	6.3, 12.6, 22	220/380
Ivory Coast	50	30, 90, 220	5.5, 15	380/220
Jamaica	50	66	2.2, 3.3, 11	110/220, 440
Japan	50, 60	33, 66, 77, 110, 138, 154, 187, 220, 275, 500	3.3, 6.6, 11, 13.8, 22, 33	100/200, 400, 110/220, 440
Jordan	50	33, 66, 132, 230, 400	6.6, 11, 20	220/380

TABLE 2.6
System Voltage Used in Countries Other than Europe and North America

Country	Frequency, Hz	Transmission, kV	Distribution, kV	Utilization, V
Kenya	50	66, 132, 275	3.3, 11, 33	415/240
Korea (north)	60	66, 110, 220, 400	3.3, 6, 10	220/380
Korea (south)	60	66, 154, 345	3.3, 6.6, 11, 13.2, 23, 33	100/200, 220/380, 440
Kuwait	50	33, 132, 300	3.3, 6.6, 11	240/415
Lebanon	50	33, 66, 150, 230	5.5, 6, 11, 20	110/190, 220/380
Liberia	60	24.5, 69	12.5	120/240, 120/208
Libya	50	33, 66, 150, 220, 400	6, 10	380/220
Malaysia	50	33, 132, 275	6.6, 11, 22	230/400
Mexico	60	23, 34.5, 69, 85, 115, 230, 400	2.4, 4.16, 6.6, 13.8	120/208, 120/240, 127/220
Mongolia	50	110, 220	6, 10	220/380
Morocco	50	60, 72.5, 150, 220	5.5, 6, 20, 22	220/380
New Zealand	50	22, 33, 50, 66, 110, 220	3.3, 6.6, 11, 22, 33	230/400
Nigeria	50	66, 132, 330	3.3, 6.6, 11, 33	415/240
Pakistan	50	33, 66, 132, 230, 500	6.6, 11	230/400
Panama	60	34.5, 115, 230	2.4, 4.16, 13.8	110/220, 120/240
Philippines	60	34.5, 69, 115, 138, 230, 500	2.4, 3.3, 4.16, 4.8, 6.24, 13.8, 23, 34.5	110/220, 440
Qatar	50	66, 132	6.6, 11, 33	240/415
Saudi Arabia	60	33, 69, 115, 132, 380	4.16, 11, 13.8	220/380, 240/415, 127/220
Singapore	50	66, 230	6.6, 11, 22	230/400
South Africa Republic	50	88, 132, 220, 275, 400	3.3, 6.6, 11, 22, 33, 44, 66	550, 500, 380/220
Sri Lanka	50	33, 66, 132	3.3, 11	230/400
Sudan	50	66, 110, 220	6.6, 11, 33	415/240
Syria	50	66, 230, 400	6, 6.6, 11, 20	115/220, 220/380
Taiwan	60	69, 161, 345	3.3, 6.6, 11, 22, 33	110/220, 220/380
Thailand	50	69, 115, 132, 230, 500	3.3, 11, 22, 33	220/380
Trinidad	60	33, 66, 132	2.3, 4, 12, 13.8	110/220, 115/230, 230/400
Tunisia	50	30, 90, 150	10, 15, 22, 30	380/220
Turkey	50	66, 154, 380	6.3, 10, 20, 34.5	220/380, 110/190
Venezuela	60	34.5, 69, 115, 138, 230, 400	2.4, 4.16, 4.8, 6.6, 8.3, 12.4, 13.8, 24	120/,208, 120/240, 240/480
Vietnam	50	66, 132, 230	6.6, 13.2, 15, 35	220/380

where

E_S = sending end voltage, phase to neutral
$\cos \theta$ = power factor
R = sum of the resistance component in the circuit, ohms
X = sum of the reactance component in the circuit, ohms

A commonly used formula for percent voltage drop is

$$\Delta V = I \times (R \times \cos\theta + X \times \sin\theta)$$

or

$$= kVA \times \frac{(R \times \cos\theta + X \times \sin\theta)}{(10\ kV)^2}$$

- **Ampere-foot Method**

  This method can be used in computing the voltage drop in low-voltage feeders and branch circuits that have not been included in load-flow runs. In this method, the load current is multiplied by the one-way cable length to get the value in ampere-feet. The voltage drop can be read from the graphs in ref. [1] or estimated from the tables in ref. [11] against the cable size. The conductor size can be increased or the load can be fed from a different bus to reduce the voltage drop. Table 2.7, extracted from ref. [11], can be used for estimating voltage drop in a three-phase system with the load current at 0.8 power factor.

- **Load Flow**

  This is a system planning and study activity, and it is covered in chapter 3. The output files provide bus voltage summary for light or no load, full load, and contingency operations. The output files also highlight the bus voltage when it is in violation of the specified range. The transformer voltage spread or voltage drop from no load to full load is reduced by adjusting the transformer taps.

2.6.4.2 Voltage Flicker

The voltage changes of a transient nature, such as turning loads on and off, which last only a short duration, are generally referred to as voltage flicker. The rapid voltage fluctuations affect the light output from incandescent lamps, which can irritate the human eye. Flicker-limit curves adopted by IEEE and used by many power utilities in North America are shown in fig. 2.9 [S1]. These curves are the composite of the flicker on incandescent lamps studied by General Electric Co., Kansas City Power and Light Company, T&D Committee, Detroit Edison Company, and West Pennsylvania Power Company.

Flicker is divided into four groups based on their frequency of occurrence:

- *Cyclic flicker*: is that resulting from periodic voltage fluctuations, the range of frequency of fluctuation is 10 per second to 2 per second. Reciprocating compressors and pumps, arc furnaces, and automatic spot welders fall into this category.

System Planning

TABLE 2.7
Voltage Drop for Copper Conductor in Nonmagnetic Conduit (Power Factor 80%, Three Phase, Three or Four Wire, 60 Cycles)

Wire Size AWG/KCmil	1000	750	500	350	300	250	4/0	2/0	1/0	1	2	4	6	8	10	12
Amp.-Feet									Voltage Drop							
500	26.85	29.44	36.37	44.17	48.50	54.56	60.62	86.60	104.79	127.30	155.01	235.55	358.52	...	...	...
300	16.11	17.67	21.82	26.50	29.10	32.73	36.37	51.96	62.87	76.38	92.66	141.16	213.90	327.35	...	...
100	5.37	5.89	7.27	8.83	9.70	10.91	12.12	17.32	20.96	25.46	31.00	47.11	71.70	109.12	171.47	271.06
90	4.76	5.37	6.58	7.88	8.75	9.79	10.91	15.59	18.88	23.04	27.89	42.52	64.60	97.86	154.35	244.21
70	3.72	4.16	5.11	6.15	6.75	7.62	8.49	12.12	14.72	17.84	21.65	32.99	50.23	76.38	120.37	189.65
50	2.68	2.94	3.64	4.42	4.85	5.46	6.06	8.66	10.48	12.73	15.50	23.56	35.85	54.47	85.73	135.96
40	2.17	2.25	2.94	3.55	3.90	4.33	4.85	6.93	8.40	10.22	12.38	18.79	28.49	43.65	68.59	108.25
30	1.56	1.73	2.17	2.68	2.94	3.29	3.64	5.20	6.32	7.62	9.27	14.12	21.39	32.73	51.44	81.32
20	1.04	1.21	1.47	1.73	1.91	2.17	2.42	3.46	4.16	5.11	6.24	9.44	14.29	21.82	34.29	54.21
10	0.52	0.61	0.69	0.87	0.95	1.13	1.21	1.73	2.08	2.51	3.12	4.68	7.19	10.91	17.15	27.11
9	0.52	0.52	0.69	0.78	0.87	0.95	1.13	1.56	1.91	2.34	2.77	4.24	6.50	9.79	15.41	24.42
8	0.43	0.52	0.61	0.69	0.78	0.87	0.95	1.39	1.65	2.08	2.51	3.81	5.72	8.75	13.68	21.65
7	0.35	0.43	0.52	0.61	0.69	0.78	0.87	1.21	1.47	1.82	2.17	3.29	5.02	7.62	12.04	18.97
6	0.35	0.35	0.43	0.52	0.61	0.69	0.69	1.04	1.30	1.56	1.91	2.86	4.24	6.58	10.31	16.28
5	0.26	0.26	0.35	0.43	0.52	0.52	0.61	0.87	1.04	1.30	1.56	2.34	3.55	5.46	8.57	13.60
4	0.26	0.26	0.26	0.35	0.43	0.43	0.52	0.69	0.87	1.04	1.21	1.91	2.86	4.33	6.84	10.83
3	0.17	0.17	0.26	0.26	0.26	0.35	0.35	0.52	0.61	0.78	0.95	1.39	2.17	3.29	5.11	8.14
2	0.09	0.09	0.17	0.17	0.17	0.26	0.26	0.35	0.43	0.52	0.61	0.95	1.47	2.17	3.38	5.46
1	0.09	0.09	0.09	0.09	0.09	0.09	0.09	0.17	0.17	0.26	0.35	0.43	0.69	1.13	1.65	2.68
900	0.09	0.09	0.09	0.09	0.09	0.09	0.09	0.17	0.17	0.26	0.26	0.43	0.69	0.95	1.56	2.42
800	0.09	0.09	0.09	0.09	0.09	0.09	0.09	0.17	0.17	0.17	0.26	0.35	0.61	0.87	1.39	2.17
700	...	0.09	0.09	0.09	0.09	0.09	0.09	0.09	0.17	0.17	0.26	0.35	0.52	0.78	1.21	1.91
600	...	...	0.09	0.09	0.09	0.09	0.09	0.09	0.17	0.17	0.26	0.26	0.43	0.69	1.04	1.65
500	...	...	...	0.09	0.09	0.09	0.09	0.09	0.09	0.17	0.17	0.26	0.35	0.52	0.87	1.39

Note: For single-phase circuit, multiply the values by 1.155.
Source: Kaiser Aluminum, "Tables for the Determination of Voltage Drop in Aluminum and Copper Conductors," Kaiser Aluminum publication, 1996.

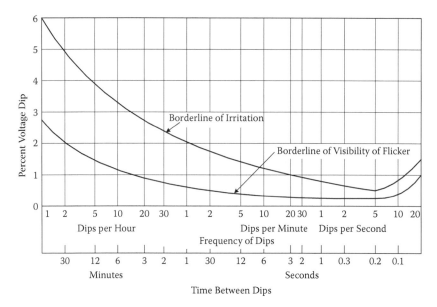

FIGURE 2.9 Range of observable and objectionable voltage flicker vs. time (From IEEE 141-1993, IEEE Recommended Practice for Electric Power Distribution for Industrial Plants (IEEE Red book), chap. 2. With permission.)

- *Cyclic low frequency*: The frequency of fluctuation ranges from 2 per second to 12 per minute. Arc welders, drop hammers, saws, and manual spot welders fall into this category.
- *Noncyclic frequent*: The frequency of fluctuation ranges from 12 per minute to 1 per minute. Hoists, cranes, single elevators fall into this category.
- *Noncyclic infrequent*: The frequency of fluctuation ranges from 1 per minute to 3 per hour. Motor starting falls into this category.

The utilization of equipment that may cause voltage flicker in industrial power plants is summarized here:

- *Starting of large motors*: Starting of large synchronous or induction motors draws four to seven times their rated current at a very low factor (0.10 to 0.3). The highly inductive current causes a voltage dip that may exceed the flicker limit. This can be remedied by utilizing an assisted starting method. The method of calculation for voltage drop due to motor starting and starting methods is covered in chapter 9.
- *Arc furnace*: An electric arc furnace is a problem load for the network, especially during the period when the scrap metal is converted into molten steel for refining. The meltdown period lasts about 70 to 90 minutes, followed by the refining period. The refining period lasts between 0.5 and 2.0 hours, depending on the quality of the finished steel. The characteristics of the meltdown period are:

System Planning

For an optimum arc, the active power must be at 0.707 power factor. The circuit constants for maximum arc power are:

Optimum current, $I_0 = 1.45 \times I$, where I is the rated rms current
Short-circuit current, $I_{SC} = 1.58 \times I_0 = 2.3 \times I$
Power factor for maximum arc = 0.707
Power factor during short circuits = 0.25

Frequently recurring short circuits occur at the electrodes; two phases are generally short-circuited, and one phase is with no current.

The frequent short circuits during the meltdown period cause voltage fluctuations in the power system at a frequency of about four to seven times per second.

The rapid change in current during meltdown period causes voltage flicker. The furnace is connected to a dedicated bus, and a static VAR compensator is utilized to eliminate the flicker (fig. 2.10). The voltage flicker for an arc furnace is defined as a short-circuit voltage depression (SCDV) and can be estimated using the formula:

$$SCDV = \frac{2 \times MW_F}{MVA_{SC}}$$

where
MW_F = the furnace rating in MW
MVA_{SC} = the three-phase fault level at the point of common coupling
The recommended short-circuit voltage depression is
SCDV < 1.9% for systems where lamps are at 230 V
SCDV < 2.5% for systems where lamps are at 120 V

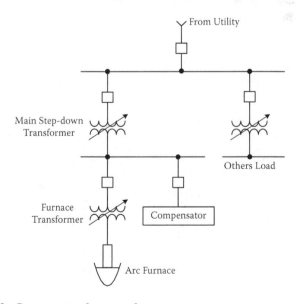

FIGURE 2.10 Power system for an arc furnace

- *Reciprocating compressors or pumps*: Reciprocating compressors and pumps operate at a very low speed (300 to 450 rpm), and the torque requirement varies in each revolution. This motor load fluctuates in each revolution, thus causing current pulsations. The standards allow a variation of up to 66% of the maximum motor full-load current. The motor size must be selected based on the system short-circuit capacity, or a lower limit of 25% or 40% on current pulsation must be specified.
- *Electric shovels*: Mining operations using electric shovels cause problems for the power distribution system. A typical duty cycle lasts about 25 to 30 s. This includes digging, hoisting, swinging back to the truck, dumping in the truck, and swinging back to the pit. The power drawn from the system is about 200% of the equipment rating during digging and hoisting periods, and there is some regeneration during lowering and stopping. Power system design and equipment rating must take the duty cycle into consideration, and voltage flicker must be checked to ensure that the limits are not exceeded.
- *Other loads*: These include electric welders, electric shovels and strippers, heavy rolling, etc. The system must be studied to check whether the voltage fluctuations exceed the flicker limit, and if they do, corrective measures must be taken accordingly.

2.6.4.3 Voltage Drop Due to Motor Starting

This subject is covered in chapter 9.

2.6.4.4 Voltage Drop/Rise Due to Switching

Switching on or off a large block of load causes voltage change. The approximate value can be estimated by:

voltage change $\cong$ load in MVA/fault level in MVA

Switching a capacitor bank causes voltage change, which can be estimated by

voltage change $\cong$ capacitor bank rating in MVA/system fault level in MVA

2.6.5 VOLTAGE SAG AND THRESHOLD VOLTAGE

A voltage sag is defined as a decrease in rms voltage of 0.1 to 0.9 P.U. at the power frequency for a duration of 0.5 cycles (8.3 ms) to 1.0 min. The threshold voltage is the critical voltage at which the equipment stops working; for example, a starter coil may drop out at 0.6 per-unit (P.U.) voltage, and a variable-frequency drive may trip at 0.8 to 0.85 P.U. voltage. Disturbances such as faults on utility transmission or distribution systems cause momentary voltage dips or voltage sags. The utilization equipment in the plant may trip or stop functioning if the voltage sag is below the equipment threshold voltage. Sag response curves have been published in ref. [S15]. The following measures can help to survive such voltage dips or provide ride-through capability:

System Planning   25

- High breakdown torque for induction motors, $T \cong V^2$
- Higher pull-out torque for synchronous motors and field-forcing capability
- Lower threshold voltage for control system, drives, monitoring, etc.
- Uninterruptible power supply (UPS) for critical control and monitoring

REFERENCES

Standards

S1. IEEE 141-1993, IEEE Recommended Practice for Electric Power Distribution for Industrial Plants, IEEE Red Book, chap. 2, 1993.
S2. IEEE 142-1991, IEEE Recommended Practice for Grounding of Industrial and Commercial Power Systems, IEEE Green Book, 1991.
S3. IEEE 241-1990, IEEE Recommended Practice for Electric Power Systems in Commercial Buildings, IEEE Gray Book, 1990.
S4. IEEE 242-2001, IEEE Recommended Practice for Protection and Coordination of Industrial and Commercial Power Systems, IEEE Buff Book, 2001.
S5. IEEE 399-1997, IEEE Recommended Practice for Industrial and Commercial Power System Analysis, IEEE Brown Book, 1997.
S6. IEEE 446-1987, IEEE Recommended Practice for Emergency and Standby Power Systems, 1987.
S7. IEEE 487-1992, IEEE Recommended Practice for the Protection of Wire Line Communications Facilities Serving Electric Power Stations, 1992.
S8. IEEE 493-1997, IEEE Recommended Practice for Design of Reliable Industrial and Commercial Power Systems, IEEE Gold Book, 1997.
S9. IEEE 519-1992, IEEE Recommended Practices and Requirements for Harmonic Control in Electrical Power Systems, 1992.
S10. IEEE 602-1996, IEEE Recommended Practice for Electric Systems in Health Care Facilities, IEEE White Book, 1996.
S11. IEEE 739-1995, IEEE Recommended Practice for Energy Conservation and Cost-Effective Planning in Industrial Facilities, IEEE Bronze Book, 1995.
S12. IEEE 1100-1995, IEEE Recommended Practice for Powering and Grounding Sensitive Electronic Equipment, IEEE Emerald Book, 1995.
S13. IEEE 1159-1995, IEEE Recommended Practice for Monitoring Electric Power Quality, 1995.
S14. IEEE 1250-1995, IEEE Guide for Service to Equipment Sensitive to Momentary Voltage Disturbances, 1995.
S15. IEEE 1346-1998, IEEE Recommended Practice for Evaluating Electric Power System Compatibility with Electronic Process Equipment, 1998.
S16. ANSI C84.1-1995, Electrical Power Systems and Equipment: Voltage Rating (60 Hz), 1995.
S17. ANSI/NFPA 70-2002, National Electrical Code (NEC), 2002.
S18. CSA C22.1-2002, Canadian Electrical Code (CEC), 2002.
S19. IEEE C2-1999, National Electrical Safety Code (NESC), 1999.
S20. NFPA-780-1995, Lightning Protection Code, 1995.
S21. NEMA MG1-2003, Motors and Generators, 2003.
S22. IEC 60038-2002-07, IEC Standard Voltages, 2002.
S23. IEC 1000-3-3, Limitation of Voltage Fluctuations and Flicker in Low Voltage System for Equipment with Rated Current <16 A, IEC, 1994.
S24. IEC 1000-3-5, Limitation of Voltage Fluctuations and Flicker in Low Voltage System for Equipment with Rated Current >16 A, IEC, 1994.

S25. IEC 1000-3-7, Limitation of Voltage Fluctuations and Flicker for Equipment Connected to Medium and High Voltage Power Supply Systems.

S26. IEC Publication 868, Flicker Meter: Functional and Design Specification, 1986.

BIBLIOGRAPHY

1. General Electric Co., *Industrial Power Systems Data Book*, sections 0.210, 0.211, 0.62, Schenectady, NY: General Electric.
2. D. Beeman, *Industrial Power Systems Handbook*, New York: McGraw-Hill, 1955.
3. M. K. Walker, "Electric Utility Flicker Limits," *IEEE Trans. IAS* 15 (6): 644–655 (1979).
4. V. Wagner, T. Grebe, R. Kretschmann, L. Morgan, and Al Price, "Power System Compatibility with Industrial Process Equipment," *IEEE Ind. Applic. Mag.* Jan./Feb. (1996): 11–15.
5. Van E. Wagner et al., "Power Quality and Factory Automation," *IEEE Trans. IAS* 26 (4): 620–626 (1990).
6. Edward L. Owen, "Power Disturbance and Quality: Light Flicker Requirements," *IEEE Ind. Applic. Mag.* Jan./Feb. (1996): 20–27.
7. S. R. Mendis, M. T. Bishop, and J. F. Witle, "Investigation of Voltage Flicker in Electric Arc Furnace Power Systems," *IEEE Ind. Applic. Mag.* Jan./Feb. (1996): 28–34.
8. B. Bhargava, "Arc Furnace Measurements and Control," *IEEE Trans. Power Delivery* 8: 400–409 (1993).
9. M. A. Neslin and W. G. Wright, "Factors Affecting Distribution Equipment Size for Multiple Shovels," publication GEZ-2756, Schenectady, NY: General Electric.
10. W. H. Schweded, "High Voltage Pit Power Distribution System," publication GEZ-4574, Schenectady, NY: General Electric.
11. Kaiser Aluminum, "Tables for the Determination of Voltage Drop in Aluminum and Copper Conductors," Kaiser Aluminum publication.

3 Power System Studies

3.1 INTRODUCTION AND OVERVIEW

Power system studies are required to evaluate the performance of the proposed or existing power system. The objective is to ensure that the system is safe, reliable, easy to operate, easy to maintain, and is reasonable in cost. Power system analysis software such as load flow, short circuit, harmonic analysis, protective device coordination, and stability is available for use on a personal computer at a reasonable cost. Studies such as switching transients and insulation coordination are carried out using a TNA (transient network analyzer) or an EMTP (electromagnetic transient program). Considerable experience has been gained with EMTP and with practical or tested models, and most transient studies are carried out using this software. The following points must be taken into consideration for the selection of a software package for common studies:

- The software uses raw or nameplate data for input files and converts them into the selected base values. A package that uses a common database for more than one study such as load flow, short circuit, harmonic analysis, stability, etc., will save considerable time and minimize error.
- In the case of a short-circuit study for equipment rating based on ANSI/IEEE standards, the program adjusts the impedance multiplier of the rotating machines. The output files provide X/R ratios, adjusted fault current for local and remote contribution that can be directly compared with the equipment rating.

Load flow, short circuit, and protection-device coordination studies must be carried out during the development stage of the single-line diagram for new plants or for expansions/modifications of an existing power system. In addition to using a computer program, it is highly recommended that one carry out two or three short-circuit and protective-device coordination studies using hand calculations. This will serve as a check to the computer results and provide a better understanding of the subject. IEEE 399-1997, "IEEE Recommended Practice for Industrial and Commercial Power Systems Analysis" [S1], has provided an excellent coverage of the system studies that are carried out for industrial plants. The references provided at the end of each chapter can be used for a further understanding of the subject.

3.2 USEFUL FORMULAE

The following equations are used for converting raw data to per-unit values:

$$\text{Per-unit volt} = \frac{\text{Actual Volt}}{\text{Base Volt}} \quad (3.1)$$

$$\text{Per-unit ampere} = \frac{\text{Actual Ampere}}{\text{Base Ampere}} \qquad (3.2)$$

$$\text{Per-unit ohm} = \frac{\text{Actual Ohm}}{\text{Base Ohm}} \qquad (3.3)$$

$$\text{Base ampere } (I_b) = \text{Base MVA} \times \frac{10^3}{\sqrt{3} \times \text{kV}} \qquad (3.4)$$

$$\text{Base ohms } (Z_b) = \frac{\text{Base kV}^2}{\text{Base MVA}} \qquad (3.5)$$

$$\text{Per-unit ohm } (Z_{PU}) = \frac{\text{Actual Ohm} \times \text{Base MVA}}{\text{Base kV}^2} \qquad (3.6)$$

$$\text{Per-unit } Z_2 = \frac{\text{Per-unit } Z_1 \, (\text{Base kV}_1)^2}{(\text{Base kV}_2)^2} \times \frac{\text{Base MVA}_2}{\text{Base MVA}_1} \qquad (3.7)$$

Utility system fault levels are generally given in MVA, and the X/R ratio (or R + JX) in ohms or per unit on a 100 MVA base. Convert fault MVA to ohms using equation (3.5), and then convert to per-unit on the chosen base using equation (3.6).

3.3 LOAD FLOW

3.3.1 Overview

Load-flow studies for industrial power systems are based on static network models. Utilities use real-time on-line models in automatic supervisory control and data acquisition (SCADA) for the optimization of generation, VAR control, losses, and tie-line control.

A load-flow study is carried out to determine the steady-state bus voltages, active and reactive power flows, transformer tap settings, component or circuit loading, generator exciter regulator voltage set points, system performance under contingency or emergency operations, and system losses. Load flow can also be used to determine voltage profile at the time of starting a large motor. The starting motor is modeled as a constant-impedance shunt with the X/R ratio based on a locked rotor or starting power factor. The load-flow case is run with the starting motor disconnected, and the voltage at the relevant buses is recorded. The starting-motor locked-rotor impedance is connected as a shunt, and the new case is run. The difference in voltage at any bus is the voltage drop at the instant of starting the motor.

Power System Studies

Two algorithms, Gauss-Siedel and Newton-Raphson, are used to solve the load-flow equations. Both are options in commercially available programs. The Gauss-Siedel method gives a simple and stable solution and works well up to 100 buses. The solution iterates one bus at a time, corrects that bus voltage to the specified value, and continues until an error is detected. The solution may not converge for the following reasons:

- Error in the input data
- System is too weak to carry the load
- Insufficient VAR in the system to support the voltage

In the Newton-Raphson method, the n quadratic equations are first linearized by forming a Jacobian matrix. The present value of the bus voltage is then calculated, and then n linear equations are solved in steps. The number of iterations is small, between five and ten.

3.3.2 Load-Flow Runs

System and equipment data are common for load-flow and short-circuit studies, with the exception of the tolerance given in the standards. Apply positive tolerance for load flow and negative tolerance for short circuit. Suggested guidelines to help avoid error are:

- Enter the data with care, especially with units. This is the most common cause of error.
- Start with a small system, for example a 10-bus network, and expand the system as the solution is found.
- Do not use a very small impedance for ties and feeders.
- Add a dummy capacitor or a synchronous condenser for voltage support if the solution does not converge.

3.4 SHORT CIRCUITS

Short-circuit studies are required to check or select the equipment rating and determine the relay settings. Hand calculations with motor contributions can be used for small systems (up to six buses). However, for a complete plant electrical system, the use of engineering software becomes a necessity.

3.4.1 Contribution from Different Sources

A typical utility power supply arrangement at a large industrial complex, with fault contribution from different sources, is shown in fig. 3.1. A three-phase fault on the plant medium-voltage main bus will result in short-circuit contribution from the utility generators and the plant rotating equipment (generators, synchronous and induction motors). The current contribution from different sources is represented in (a), (b), (c), and (d), and the total current in (e) and (f):

(a): From utility sources
(b): From the plant generator
(c): From synchronous motors
(d): From induction motors
(e): Symmetrical short-circuit contribution from all sources
(f): Asymmetrical short-circuit contribution from all sources, including DC component

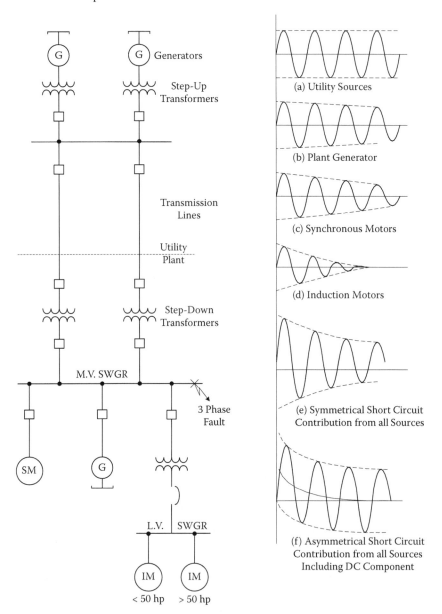

FIGURE 3.1 Fault contribution from different sources

Power System Studies

The first-cycle current from different sources is based on the saturated subtransient reactance of synchronous motors and generators and the locked-rotor impedance of the induction motors. The first-cycle current is seen at the time of the close-and-latch (CL/L) period of all circuit breakers and the interrupting period of fuses and low-voltage circuit breakers. Moreover, all switching and protective devices in the circuit see the first-cycle current, and thus these must be designed to withstand mechanical stress caused by the peak current.

Medium- and high-voltage circuit breakers clear the fault current in two, three, or five cycles and, depending on the type and voltage class, their contact parting times are 1.5, 2, and 3 cycles, respectively. There is decay in the short-circuit current at the time of interruption. In the ANSI/IEEE method, the first-cycle impedance is increased using the established multiplying factors to account for the decay in the current.

3.4.2 STEPS REQUIRED FOR A SHORT-CIRCUIT STUDY

Prepare the system one-line diagram, showing the equipment rating and relevant data:

- Collect the required system and equipment data. Typical data for various components are provided in tables 3.1 to 3.12. These values, or applicable data from other sources, can be used if nameplate data from the manufacturer is not available.
- For computer runs, prepare input data files as required by the software. The output files will provide fault current based on E/X or E/Z, X/R ratios, and multiplying factors to account for AC and DC decrement. The adjusted fault current for the specified duty can be compared against the equipment rating. Allow a 10% to 15% margin.
- For hand calculations:
  a. Convert the data to per-unit values on a common base (10 MVA for an industrial system and 100 MVA for utility systems).
  b. Adjust the rotating machine impedance (motors and generators) using the "rotating machine impedance multipliers" given in table 3.1.
  c. Prepare separate X and separate R diagrams. Calculate equivalent X and equivalent R at each fault location. Determine the X/R ratio.
  d. Compute the fault current: three-phase $I_f = \dfrac{E}{X}$, and line-to-ground
  $$I_{fg} = \dfrac{3 \times E}{(X_1 + X_2 + X_0)}.$$
  e. Adjust the fault current to comply with the relevant standards for comparison with the equipment rating.

3.4.3 MEDIUM- AND HIGH-VOLTAGE CIRCUIT BREAKERS

Applicable standards are [S2, S3] for circuit breakers (CB) rated on a symmetrical current basis, and [S4] for circuit breakers rated on a total, or asymmetrical, current basis.

TABLE 3.1
Rotating Machine Impedance Multipliers

Description	First-Cycle Duty (1)	1.5–4-Cycle Duty (2)	30-Cycle Duty (3)
Generators			
- Turbo, hydro with amortisseur windings, condensers	1.0 X''_d	1.0 X''_d	1.0 X''_d
- Hydro without amortisseur windings	0.75 X''_d	0.75 X''_d	0.75 X''_d
All synchronous motors	1.0 X''_d	1.5 X''_d	neglect
Induction motors			
- Above 1000 hp at 1800 rpm or less	1.0 X''_d	1.5 X''_d	neglect
- Above 250 hp at 3600 rpm	1.0 X''_d	1.5 X''_d	neglect
- All others, 50 hp and above	1.2 X''_d	3.0 X''_d	neglect
- Less than 50 hp	1.67 X''_d	neglect	neglect

Source: IEEE 141-1993, Electric Power Distribution for Industrial Plants, 1993.

3.4.3.1 Contact Parting or Interrupting Duty

- Use rotating machine impedance multipliers for 1.5–4-cycle duty. Contact parting times are: 1.5 cycles to 2.0 cycles for breakers, 2.0 cycles to 3.0 cycles for SF_6 or 5.0 cycles for vacuum circuit breakers, 3.0 cycles to 5.0 cycles for SF_6 or air circuit breakers, 4 cycles to 8 cycles for oil circuit breakers. Test results have shown that the opening time of a vacuum CB is less than that for an SF_6 or air CB of the same class.

- Compute the fault current: three phase $I_f = \dfrac{E}{X}$, and line-to-ground,

$$I_{fg} = \frac{3 \times E}{(X_1 + X_2 + X_0)}.$$

- Adjust the fault current for a higher X/R ratio and prefault voltage for comparison with the equipment rating. Multiplying factors for higher X/R ratios are given in figs. 3.2, 3.3, and 3.4.
- If the calculated fault current, I_f is <80% and I_{fg} is <70% of the circuit-breaker interrupting rating, no further derating is required. The interrupting capability (IC) of circuit breakers for single line-to-ground faults is 15% greater than the IC for three-phase faults.
- If the calculated X/R < 15%, no further adjustment is required for comparison with the circuit breaker rating.
- For circuit breakers rated on a symmetrical current basis, determine the multiplying factors using the following figures from [S2]:

Power System Studies

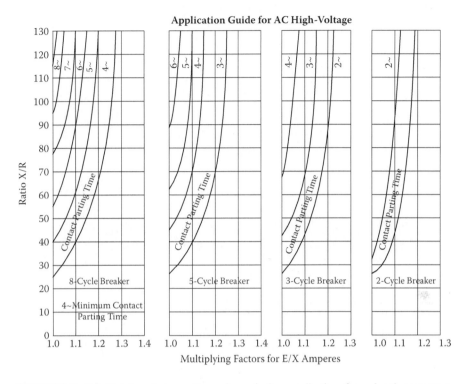

FIGURE 3.2 Multiplying factors: three-phase fault contribution from local generators through no more than one transformation

a. Figure 3.2 for three-phase fault contributions and fig. 3.3 for line-to-ground fault contributions from local generators
b. Figure 3.4 for three-phase and line-to-ground fault contributions from remote sources

Note: Because motor impedances have already been adjusted, consider their contribution as from a remote source.

3.4.3.2 Close-and-Latch Duty

- Use rotating machine multipliers for first-cycle duty.

- Compute the fault current: three-phase $I_f = \dfrac{E}{X}$, and line-to-ground

$$I_{fg} = \dfrac{3 \times E}{(X_1 + X_2 + X_0)}.$$

- Adjust the fault current for a higher X/R ratio (≥25) and prefault voltage. Multiplying factors for higher X/R ratios are given in fig. 3.5.
- Compare the adjusted fault current against the equipment rating.

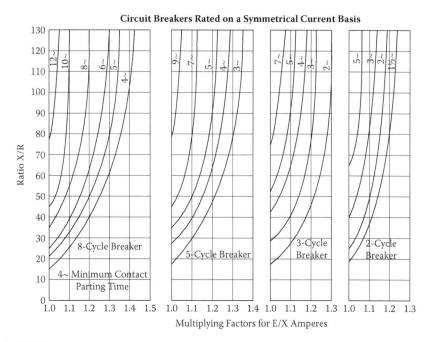

FIGURE 3.3 Multiplying factors: line-to-ground fault contribution from local generators through no more than one transformation

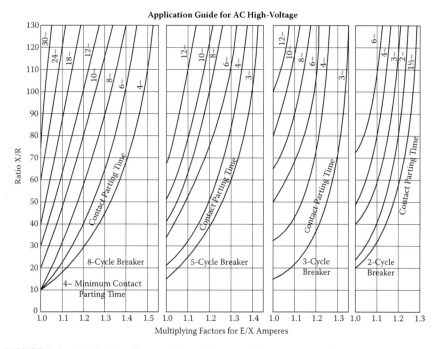

FIGURE 3.4 Multiplying factors: three phase and line-to-ground fault contribution from remote generators fed predominantly through two or more transformations

Power System Studies

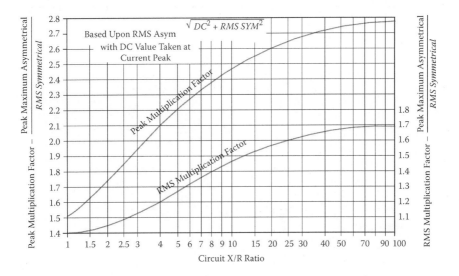

FIGURE 3.5 Multiplying factors: relation of X/R ratio

3.4.4 Interrupting Duty of Low-Voltage Circuit Breakers (<1000 V)

3.4.4.1 Metal-Enclosed Circuit Breakers [S6, S7]

Use rotating machine multipliers for first-cycle duty:

- Compute the fault current: three phase, $I_f = \dfrac{E}{X}$ or $= \dfrac{E}{Z}$ and $\dfrac{X}{R}$ ratio.

- If the calculated $\dfrac{X}{R} \leq 6.6$, the first-cycle rms current can be directly compared with the three-phase short-circuit current rating given in the standards or manufacturers' tables (without a multiplying factor to account for the DC component).
- If the short-circuit X/R ratio at the fault location has not been calculated, a ratio of 20 should be assumed, and the calculated value of rms symmetrical current should be multiplied by the appropriate multiplying factor (MF) from table 3.2 (MF = 1.14 for nonfused circuit breakers and 1.26 for fused circuit breakers).
- Adjust the fault current for a higher X/R ratio from table 3.2 and prefault voltage.
- Compare the adjusted current against the equipment rating.

Note: The short-circuit rating given in the standards and manufacturers' data sheets is based on a short-circuit power factor of 15% or an X/R ratio of 6.6. When the short-circuit power factor is less than 15% (X/R > 6.6), the computed fault current must be increased by a factor given in table 3.2.

TABLE 3.2
Short-Circuit Current Multiplying Factors for Low-Voltage Metal-Enclosed Circuit Breakers

System Short-Circuit PF (%)	System X/R Ratio	Multiplying Factor	
		Nonfused CB	Fused CB
20	4.9	1.00	1.00
15	6.6	1.00	1.07
12	8.27	1.04	1.11
10	9.95	1.07	1.15
8.5	11.72	1.09	1.18
7	14.25	1.11	1.21
5	20.0	1.14	1.26

Source: ANSI/IEEE C37.13-1990, IEEE Standard for Low-Voltage AC Power Circuit Breakers Used in Enclosures, table 3, 1990.

3.4.4.2 Molded-Case Circuit Breakers

Low-voltage molded-case circuit breakers are assumed to have a contact parting time of 0.5 cycles. The first-cycle symmetrical rms current for molded-case and insulated-case circuit breakers can be directly compared with the interrupting ratings, without a multiplying factor for the DC component, if the short-circuit X/R is equal to or less than the appropriate test circuits as follows:

Rated Interrupting Current (symmetrical rms current)	Power Factor	X/R
10,000 and less	0.45–0.5	1.7–2.0
10,001 to 20,000	0.25–0.3	3.2–3.9
Above 20,000	0.15–0.2	4.9–6.6

If the short-circuit X/R ratio exceeds the above values, the appropriate multiplying factor from table 3.3 shall be used to determine the required capability of the circuit breaker.

3.4.5 Interrupting Duty of Medium- and High-Voltage Fuses [S5]

- Use rotating machine impedance multipliers for first-cycle duty.
- Compute fault current and X/R ratio.
- Adjust (increase) fault current for higher X/R ratio and prefault voltage.
- Compare the adjusted fault current against the equipment rating.

3.4.6 For Time-Delayed Relaying Devices

- Use rotating machine impedance multipliers for 6–30-cycle duty.
- Compute the fault current for applications in relay settings.

3.4.7 Rotating Machine Impedance Multipliers [S9]

For short-circuit studies to comply with ANSI/IEEE standards, the rotating machine impedance must be adjusted in accordance with table 3.1.

TABLE 3.3
Short-Circuit Current Multiplying Factor for MCCB and ICCB

Power Factor (%)	X/R Ratio	Interrupting Rating (A)		
		10,000 or Less	10,000 to 20,000	Over 20,000
		Multiplying Factor		
5	19.97	1.59	1.35	1.22
10	9.95	1.49	1.26	1.13
12	8.27	1.45	1.23	1.10
15	6.59	1.39	1.18	1.06
20	4.90	1.31	1.11	1.00
22	4.43	1.28	1.09	1.00
25	3.87	1.24	1.05	1.00
27	3.57	1.22	1.03	1.00
30	3.18	1.18	1.00	1.00
40	2.29	1.08	1.00	1.00
50	1.98	1.04	1.00	1.00

Source: IEEE 1015-1997, IEEE Recommended Practice for Applying Low-Voltage Circuit Breakers Used in Industrial and Commercial Power Systems, IEEE Blue Book, 1997.

- For interrupting duty of fuses and low-voltage (LV) circuit breakers and for closing and latching (momentary) duty of medium-voltage (MV) and high-voltage (HV) circuit breakers
- For interrupting duty of MV and HV circuit breakers
- For time-delayed relay-setting applications

3.4.8 SHORT-CIRCUIT CURRENT MULTIPLYING FACTORS FOR HIGHER X/R RATIO

The short-circuit rating of the power switching or interrupting equipment (circuit breaker, fuse) is based on the test carried out at a short-circuit power factor or X/R ratio. If the short-circuit study shows that the system X/R ratio at the point of application of the switching device is higher than the value given in the standards, then the computed fault current shall be increased using the relevant multiplying factor for comparison against the equipment rating. Multiplying factors for circuit breakers and fuses are given in the following figures and tables:

- Figure 3.2: Multiplying factors: three-phase fault contribution from local generators through no more than one transformation
- Figure 3.3: Multiplying factors: line-to-ground fault contribution from local generators through no more than one transformation
- Figure 3.4: Multiplying factors: three phase and line-to-ground fault contribution from remote generators through two or more transformations
- Figure 3.5: Multiplying factors: relation of X/R ratio

- Table 3.2: Short-circuit current multiplying factor for low-voltage metal-enclosed circuit breakers
- Table 3.3: Short-circuit current multiplying factor for MCCB and ICCB

3.4.9 X/R Ratio of System Component

The resistance of a system component using the X/R ratio can be obtained or calculated from the following:

- Power transformers: refer to fig. 3.6
- Three-phase induction motors: refer to fig. 3.7
- Synchronous motors and small generators: refer to fig. 3.8
- Reactors: $X/R \cong 80–120$, typical value 80
- Large generators: 80–120, typical value 80
- Power cable: refer to tables 3.9, 3.10, 3.11, 3.12
- Open-wire distribution line: refer to table 3.8

FIGURE 3.6 60 Hz X/R ratio power transformer

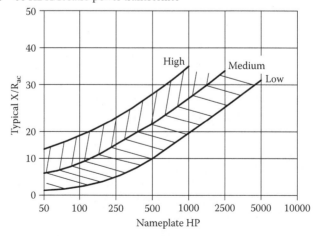

FIGURE 3.7 60 Hz X/R ratio of induction motors

Power System Studies

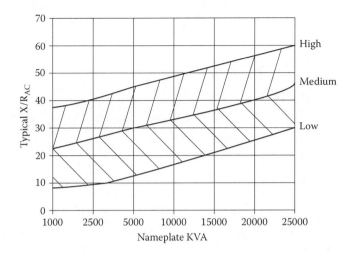

FIGURE 3.8 60 Hz ratio of synchronous motors and small generators

3.4.10 System and Equipment Data

Data from the following tables, graphs, and curves can be used when actual data is not available:

- Table 3.4: Reactance of rotating machines
- Table 3.5: Impedance of oil-immersed power transformers
- Table 3.6: Impedance of dry-type power transformers
- Table 3.7: Conductor spacing for overhead lines
- Table 3.8: Electrical characteristics of open-wire distribution lines
- Figure 3.9: External reactance of conductor beyond 1-ft spacing
- Table 3.9: AC resistance of copper and aluminum conductors
- Table 3.10: 60 Hz inductive reactance of 600/1000 V power cable
- Table 3.11: 60 Hz inductive reactance of 5 kV power cable
- Table 3.12: 60 Hz inductive reactance of 15 kV and 28 kV power cables

TABLE 3.4
Typical Reactance Values for Induction and Synchronous Machines (per unit of machine kVA rating)

Description	X_d''	X_d'
Turbine generator		
2 poles	0.09	0.15
4 poles	0.15	0.23
Salient-pole generators with damper windings		
12 poles or less	0.16	0.33
14 poles or more	0.21	0.33
Synchronous motors		
6 poles	0.15	0.23
8–14 poles	0.20	0.30
16 poles and more	0.28	0.40
Synchronous condensers	0.24	0.37
Synchronous condensers		
600 V direct current	0.20	
250 V direct current	0.33	
Individual large induction motors, usually above 600 V	0.17	
Small motors, usually 600 V and below	0.17	

Source: IEEE 141-1993, Electric Power Distributions for Industrial Plants, table 4 A-1, 1993.

TABLE 3.5
Percent Impedance Voltage of Oil-Immersed Power Transformers at Self-Cooled ONAN Rating

High-Voltage BIL (kV)	Without Load Tap Changing		With Load Tap Changing (2400 V and above)
	Low Voltage (480 V)	Low Voltage (2400 V and above)	Low Voltage (2400 V and above)
60–110	5.75 [a]	5.5 [a]	...
150	6.75	6.5	7.0
200	7.25	7.0	7.5
250	7.75	7.5	8.0
350	...	8.0	8.5
450	...	8.5	9.0
550	...	9.0	9.5
650	...	9.5	10.0
750	...	10.0	10.5

[a] For transformers greater than 5000 kVA self-cooled, these values will be the same as those shown for 150 kV HV BIL.

Source: From IEEE C57.12.10-1987, American National Standard for Transformers, table 10, 1987.

TABLE 3.6
Percent Impedance for Dry-Type Transformers

KVA	15 kV Class (95 kV BIL and below) NEMA	15 kV Class (95 kV BIL and below) Suggested	25 kV Class (125 kV BIL) Suggested	34.5 kV Class (150 kV BIL) Suggested
300	4.50	4.50	...	...
301–500	5.75	5.50	6.00	6.50
501–1500	5.75	6.00	6.00	6.50
2000	5.75	6.00	6.00	6.50
2500	5.75	6.00	6.50	6.50
3000	...	6.50	6.50	7.00
3750	...	6.50	7.00	7.00
5000	...	6.50	7.00	7.00
7500–10000	...	7.00	7.00	7.50

Note: The standard tolerance is ±7.5% for two-winding transformers and ±10% for three-winding transformers of the design impedance.

TABLE 3.7
Representative Conductor Spacing for Overhead Lines

Nominal System Voltage (V)	Equivalent Deltas Spacing (in.)
120	12
240	12
480	18
600	18
2 400	30
4160	30
6 900	36
13 800	42
23 000	48
34 500	54
69 000	96
111 500	204
161 000	238
230 000	300

Note: When the cross section indicates that conductors are arranged at points of a triangle with spacing A, B, and C between pairs of conductors, the following formula can be used:

$$\text{equivalent delta spacing} = \sqrt{A \times B \times C}$$

When the conductors are located in one place and the outside conductors are spaced at distance A from the middle conductors, the equivalent is 1.26 times the distance A:

$$\text{equivalent delta spacing} = \sqrt{A \times A \times 2A} = 1.26A$$

Sources: IEEE 141-1993, table 4A-2, 1993; and IEEE 399-1997, table 11-11, 1997.

TABLE 3.8
Electrical Characteristics of Open-Wire Distribution Lines

Conductor		Ampacity (current-carrying capacity)		Resistance at 75C° (ohm/1000 ft)	Reactance X_1 at 1-ft spacing (ohm/1000 ft)
Type	Size (AWG or KCmil)	Still Air	Wind (2 ft/s)		
HD Copper	6	70	110	0.493	0.121
	4	110	161	0.314	0.114
	2	145	210	0.200	0.109
	1	170	245	0.157	0.106
	1/0	200	285	0.125	0.104
	2/0	240	335	0.099	0.101
	4/0	330	450	0.063	0.095
	250	375	510	0.054	0.092
	300	425	575	0.044	0.090
	350	475	635	0.038	0.088
ACSR	4 (6/1)	75	120	0.509	0.126
	2 (6/1)	110	165	0.320	0.125
	1/0 (6/1)	150	225	0.201	0.124
	2/0 (6/1)	175	260	0.160	0.122
	4/0 (6/1)	245	355	0.101	0.110
	266.8 (26/7)	290	410	0.079	0.088
	336.4 (26/7)	340	480	0.063	0.086
	397.5 (26/7)	380	535	0.054	0.084
	477.0 (26/7)	430	605	0.044	0.082
	556.5 (26/7)	480	670	0.038	0.080
	795 (26/7)	620	850	0.026	0.076

X_1 = 60 Hz reactance of conductor at 1-ft spacing (ohm/1000 ft);
X_2 = 60 Hz external reactance of conductor beyond 1-ft spacing (ohm/1000 ft);
total 60 Hz reactance = $X_1 + X_2$ ohm/1000 ft (obtain X_2 from fig. 3.9);
$X_2 = 0.0528 \log_{10} D - 0.057$, where D = equivalent delta spacing (in.).

Source: General Electric Co., *Distribution Data Book*, publication GET-1008L, Schenectady, NY: General Electric.

TABLE 3.9
AC Resistance of Copper and Aluminum Conductors (ohm/1000 ft)

Conductor Size (AWG or KCmil)	Single-Stranded Conductor				Three-Conductor Cable [a]	
	Copper		Aluminum (ACM)		Factor	
	75°C	90°C	75°C	90°C	CU	AL
12	1.97	2.067	3.254	3.411	1.00	...
10	1.24	1.300	2.402	2.141	1.00	1.00
8	0.78	0.817	1.284	1.346	1.00	1.00
6	0.49	0.514	0.809	0.848	1.00	1.00
4	0.31	0.323	0.508	0.533	1.00	1.00
2	0.194	0.204	0.32	0.266	1.01	1.00
1	0.154	0.161	0.253	0.211	1.01	1.00
1/0	0.122	0.128	0.201	0.167	1.02	1.00
2/0	0.097	0.101	0.160	0.105	1.03	1.00
4/0	0.061	0.064	0.100	0.089	1.05	1.01
250	0.052	0.054	0.085	0.074	1.06	1.02
300	0.043	0.045	0.071	0.064	1.06	1.02
350	0.037	0.039	0.061	0.045	1.07	1.03
500	0.026	0.028	0.043	0.037	1.11	1.05
750	0.018	0.019	0.029	0.030	1.16	1.10
1000	0.014	0.014	0.022	0.021	...	1.16

[a] Multiply the single-conductor values by these factors to determine the AC resistor of three-conductor cables.

Source: Philips Cables Catalogue, table 2, 1 Nov. 1990, p. 5-10.

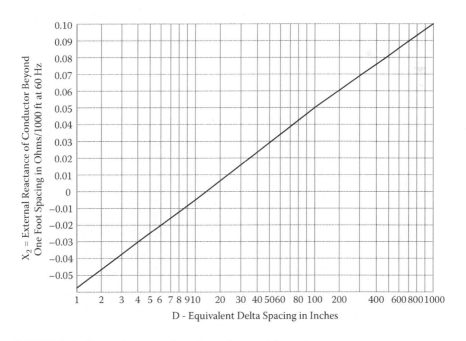

FIGURE 3.9 External reactor of conductor beyond 1-ft spacing

TABLE 3.10
60 Hz Inductive Reactance of 600/1000 V Power Cable (ohm/km)

Conductor Size (AWG or KCmil)	Voltage (V)	Three Single Cables One-Cable Diameter Spacing RW 90	Three Single Cables One-Cable Diameter Spacing TECK 90	Three Single Cables In Alum. Conduit RW 90	Three Conductors Aluminum Armor CU
12	600	...	...	0.1395	0.1155
10	600	...	...	0.1315	0.1090
8	600	...	...	0.1455	0.1145
6	1000	...	...	0.1370	0.1175
4	1000	...	...	0.1290	0.1105
2	1000	...	...	0.1280	0.1060
1	1000	...	...	0.1230	0.1020
1/0	1000	0.1695	0.2190	0.1200	0.0995
2/0	1000	0.1660	0.2140	0.1165	0.0945
4/0	1000	0.1630	0.2080	0.111	0.0950
250	1000	0.1620	0.2085	0.1105	0.0930
300	1000	0.1600	0.2045	0.1085	0.0915
350	1000	0.1565	0.2010	0.1070	0.0890
500	1000	0.1550	0.1955	0.1035	0.0890
750	1000	0.1535	0.1915	0.1000	0.0900
1000	1000	0.1520	0.1875	0.0985	0.0880

Source: Philips Cables Catalogue, 1 Nov. 1990, table 3, p. 5-10.

TABLE 3.11
60 Hz Inductive Reactance of 5 kV Power Cable (ohm/km)

Conductor Size (AWG or KCmil)	Single Conductor One-Cable-Diameter Spacing		Three Conductors [a] 5 kV Unshielded	Three Conductors [a] 5 kV Shielded
4	...	...	0.1290	0.1400
2	...	...	0.1205	0.1305
1	...	...	0.1145	0.1220
1/0	0.2240	0.2330	0.1100	0.1185
2/0	0.2225	0.2270	0.1065	0.1150
4/0	0.2130	0.2170	0.1010	0.1075
250	0.2110	0.2150	0.1000	0.1060
300	0.2075	0.2110	0.0980	0.1038
350	0.2040	0.2075	0.0960	0.1020
500	0.1980	0.2015	0.0925	0.0980
750	0.1935	0.1975	0.0935	0.0990
1000	0.1895	0.1920	0.0910	0.0955

[a] For steel armor, multiply table values by 1.25.

Source: Philips Cables Catalogue, 1 Nov. 1990, table 4, p. 5-10.

Power System Studies

TABLE 3.12
60 Hz Inductive Reactance of 15 kV and 28 kV Power Cables (ohm/km)

Conductor Size (AWG or KCmil)	Three Conductors [a]			
	15 kV (100%)	15 kV (133%)	28 kV (100%)	28 kV (133%)
1	0.1425	0.1500	0.1605	0.1725
1/0	0.1365	0.1435	0.1540	0.1665
2/0	0.1315	0.1385	0.1485	0.1605
4/0	0.1225	0.1285	0.1385	0.1490
250	0.1205	0.1275	0.1360	0.1450
300	0.1170	0.1235	0.1320	0.1410
350	0.1155	0.1205	0.1290	0.1370
500	0.1105	0.1150	0.1220	...
750	0.1080	0.1120	...	...
1000	0.1045	...	...	...

[a] For steel armor, multiply the values by 1.25.
Source: Philips Cables Catalogue, 1 Nov. 1990, table 5, p. 5-11.

3.4.11 EXAMPLE: SHORT-CIRCUIT CALCULATION FOR EQUIPMENT RATING

A small system consisting of medium- and low-voltage equipment is shown in fig. 3.10. A short-circuit study based on IEEE standards and equipment-rating selection is made using hand calculations. This example will help the reader understand the methodology used in the IEEE standards.

3.4.11.1 Hand Calculations

3.4.11.1.1 Calculate Per-Unit (P.U.) Reactance and Resistance
Base MVA = MVA_b = 10
Base kV = 120, 13.8, 0.6
Base current, I_b:

$$I_{b(13.8)} = \frac{10}{\sqrt{3}*13.8} = 0.418 \text{ kA}$$

$$I_{b(0.6)} = \frac{10}{\sqrt{3}*0.6} = 9.623 \text{ kA}$$

1. Utility system:

$$P.U.\ X = \frac{10}{2000} = 0.0050 \quad P.U.\ R = \frac{PUX}{15} = 0.0003$$

2. 30 MVA transformer, T_1 (use negative tolerance of −7.5 %):

$$P.U.\ X_{T1} = \frac{0.085}{30}*10*\underline{0.925} = 0.0262 \quad P.U.\ R_{T1} = \frac{0.0262}{25} = 0.001$$

3. Generator, *G*:

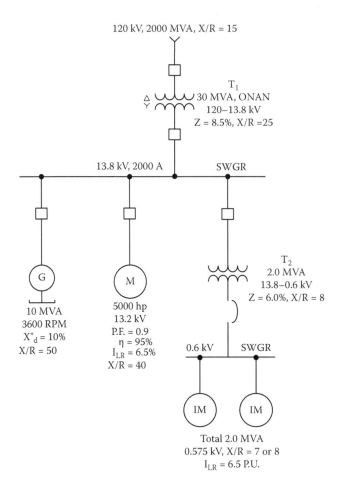

FIGURE 3.10 Short-circuit study: one-line diagram

$$P.U.\ X_G = \frac{0.10}{10} * 10 = 0.1 \quad P.U.\ R_G = \frac{0.10}{50} = 0.002$$

4. 5000-hp induction motor, M:

$$MVA = \frac{5000 * 0.746}{0.9 * 0.95} * 10^{-3} = 4.363$$

X at motor base $= \dfrac{1}{6.5} = 0.154$

(a) For first-cycle duty at bus, kV:

$$P.U.\ X_M = \frac{0.154}{4.363} * 10 * \left(\frac{13.2}{13.8}\right)^2 = 0.323 \quad P.U.\ R_M = \frac{0.323}{40} = 0.0081$$

Power System Studies

(b) For interrupting duty at bus, kV:

$$P.U. \ X_M = 1.5 * 0.323 = 0.485 \ P.U. \ R_M = 1.5 * 0.0081 = 0.0121$$

5. 600 V bus (575 V motors):
Assumptions:
  Total connected motor hp = transformer base kVA = 2000
  1 hp ≈ 1 kVA
  Motors rated <50 hp = 30% = 600 kVA
  Motors rated ≥50 hp = 70% = 1400 kVA

$$P.U. \text{ reactance at motor base} = \frac{1}{6.5} = 0.154$$

(a) For interrupting duty of LVCB and CL/L duty of MVCB,

$$P.U. \ X_{M1} = \frac{0.154}{0.6} * 10 * \left(\frac{575}{600}\right)^2 * 1.67 = 3.937, \quad P.U. \ R_{M1} = \frac{3.937}{7} = 0.562$$

$$P.U. \ X_{M2} = \frac{0.154}{1.4} * 10 * \left(\frac{575}{600}\right)^2 * 1.2 = 1.212, \quad P.U. \ R_{M2} = \frac{1.212}{8} = 0.151$$

(b) For interrupting duty of MVCB at bus kV,
Neglect M_1

$$P.U. \ X_{M2} = \frac{0.154}{1.4} * 10 * \left(\frac{575}{600}\right)^2 * 3.0 = 3.03, \quad P.U. \ R_{M2} = \frac{3.03}{8} = 0.379$$

6. 2.0-MVA transformer, T_2 (use negative tolerance):

$$P.U. \ X_{T2} = \frac{0.06}{2.0} * 10 * 0.925 = 0.278, \quad P.U. \ R_{T2} = \frac{0.278}{8} = 0.0350$$

3.4.11.1.2 Fault Calculations

1. 13.8 kV switchgear
   (a) Interrupting duty of circuit breakers (see fig 3.11)
       Calculate equivalent Xe at F:

$$\frac{1}{P.U. \ Xe} = \frac{1}{(0.005 + 0.0262)} + \frac{1}{0.1} + \frac{1}{0.485} + \frac{1}{(0.278 + 3.03)} = 44.415$$

$$P.U. \ Xe = 0.0225$$

Calculate fault current:
  Total fault current at 13.8 kV bus (I_F):

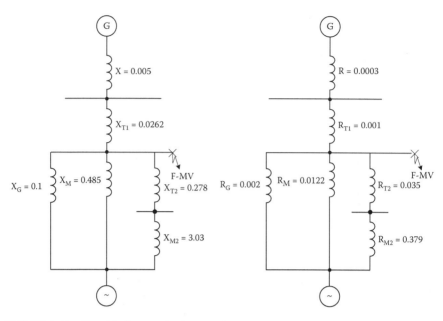

FIGURE 3.11 X and R diagram for interrupting duty

$$I_F = \frac{1}{0.0225} * 0.418 = 18.58 \text{ kA rms symmetrical current}$$

Fault contribution from utility (remote):

$$I_{F(R)} = \frac{1}{(0.005 + 0.0262)} * 0.418 = 13.4 \text{ kA}$$

Fault contribution from local:

$$I_{F(L)} = 18.58 - 13.4 = 5.18 \text{ kA rms symmetrical current}$$

Calculate equivalent R and X/R ratio:

$$\frac{1}{P.U.\text{Re}} = \frac{1}{(0.0003 + 0.001)} + \frac{1}{0.0121} + \frac{1}{0.002} + \frac{1}{(0.035 + 0.379)} = 1354.29$$

$$P.U. \text{ Re} = 0.0007$$

$$\frac{Xe}{\text{Re}} = \frac{0.0225}{0.0007} = 32.1$$

Determine multiplying factors for X/R:

Power System Studies

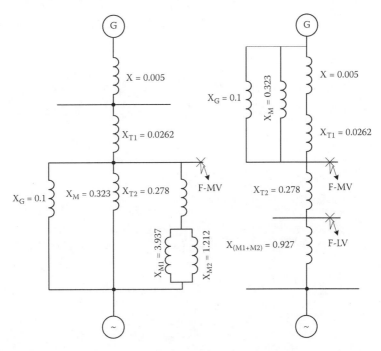

FIGURE 3.12 X diagram for CL/latch duty: MVCB and interrupting duty of LVCB

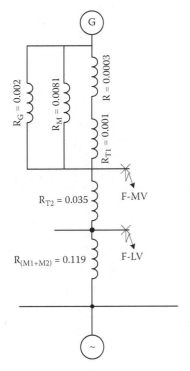

FIGURE 3.13 R diagram for CL/latch duty: MVCB and interrupting duty of LVCB

- From fig. 3.2 for local (generator) contribution (contact parting at three cycles), $MF_L = 1.03$
- From fig. 3.3 for remote contribution (contact parting at three cycles), $MF_R = 1.13$
- Fault current for interrupting duty (CB required capability)

$$I_{F(Int.)} = 4.18 * 1.03 + (13.4+1.22) * 1.13 = 20.83 \text{ kA rms symmetrical current}$$

- CB required capability if X/R is not known

$$I_{F(Int.)} = 18.58 \times 1.25 = 23.225 \text{ kA rms symmetrical current}$$

(b) Close/latch duty (first-cycle current)
Calculate P.U. Xe for M_1 and M_2:

$$Xe_{(M1+M2)} = \frac{1.212 * 3.937}{(1.212 + 3.937)} = 0.927$$

Calculate equivalent P.U. Xe:

$$\frac{1}{Xe} = \frac{1}{(0.005 + 0.0262)} + \frac{1}{0.1} + \frac{1}{0.323} + \frac{1}{(0.278 + 0.927)} = 45.97$$

$$\text{P.U. Xe} = 0.0217$$

Calculate fault current for circuit breaker CL/L duty:

$$I_{F(CL/L)} = \frac{1}{0.0217} * 0.418 = 19.263 \text{ kA} \quad I_{F(CL/L)} = 19.263 * 2.6 = 50.1 \text{ kA crest}$$

2. Interrupting duty of 600 V circuit breakers
Calculate P.U. Xe:
P.U. Xe of $(X_S + X_{T1})$, X_M, X_G:

$$\frac{1}{Xe_1} = \frac{1}{(0.005 + 0.0262)} + \frac{1}{0.323} + \frac{1}{0.1} = 45.147$$

$$Xe_1 = 0.0221, \quad Xe_{(M1+M2)} = 0.927$$

P.U. for combined Xe:

$$Xe = \frac{(0.0221 + 0.278) * 0.927}{(0.0221 + 0.278) + 0.927} = 0.227$$

Calculate fault current:
- Through-fault current (from MV bus)

$$IF(TH) \frac{1}{0.0221 + 0.278} \times 9.623 = 32.07 \text{ kA rms symm.}$$

- LV motor contribution

$$I_{F(M)} = \frac{1}{0.927} * 9.623 = 10.38 \ kA$$

- Total fault current

$$I_F = I_{F(TH)} + I_{F(M)} = 32.07 + 10.38 = 42.45 \ kA \ \text{rms symmetric current}$$

Calculate P.U. R and X/R ratio:

$$\text{Re}_{(M1+M2)} = \frac{0.151*0.562}{0.151+0.562} = 0.119$$

Re of $R_G + R_M + (R + R_{T1})$

$$\frac{1}{\text{Re}_1} = \frac{1}{0.002} + \frac{1}{0.0081} + \frac{1}{(0.0003+0.001)} = 1392.69$$

P.U. Re = 0.0007

$$\text{P.U. Re} = \frac{(0.0007+0.035)*0.119}{(0.0007+0.035)+0.119} = 0.0275$$

$$X/R = \frac{0.227}{0.0275} = 8.25$$

Determine multiplying factor [S6]: MF = 1.04
- CB required capability

(a) For X/R = 8.25, MF = 1.04 ⇒ I_F = 42.45 * 1.04 = 44.15kA rms symmetric current
(b) For X/R = not known, MF = 1.14 for X/R = 20 [S6]

$$I_F = 42.45 * 1.14 = 48.39 \ kA \ \text{rms symmetric current}$$

3.4.11.1.3 Equipment Selection
1. 13.8 kV switchgear
   Bus continuous rating:

$$I = \frac{(30+10)}{\sqrt{3}*13.8} = 1674 \ A$$

Select bus continuous current rating = 2000 A
Circuit breakers continuous current rating:
- Main (30 MVA transformer) CB

$$I = \frac{30}{\sqrt{3} * 13.8} = 1255 \ A$$

Select CB rating = 2000A
- CB required capability for interrupting duty

(a) If X/R is not known
    Interrupting current = 18.58*1.25*1.05 = 24.38 kA, rms symmetric current

(b) For X/R = 31.3
    - I_{int} = ((13.4+1.22)*1.13 + 4.18*1.03)*1.05 = 21.86 kA, rms symmetric current
    - CB required capability for close/latch duty:

    I_{CL} = 19.623*2.6*1.05 = 52.58 kA crest
    - Circuit breaker selection

(a) From table 1 in ANSI/IEEE C37.06-2000, K = 1.0
    - CB rated short-circuit and short time current = 31.5 kA rms symmetric current
    - Rated CL/L current = 82 kA crest

(b) From table 1 in ANSI/IEEE C37.06-1997, K > 1.0
    - Rated SC current at max (15 kV) = 28 kA rms symmetric current
    - Rated SC current at 13.8 kV = $\frac{28 * 15}{13.8}$ = 30.4 kA rms symm.
    - Rated CL/L current = 97 kA crest

Note: The above ratings provide enough margin for future growth.

2. 600 V switchgear, PFV = 1.0
   Required interrupting duty:
   (a) X/R is not known, I_{int} = 42.45*1.14 = 48.39 kA rms symmetric current
   (b) X/R = 8.25, MF = 1.04
   I_{int} = 42.45*1.04 = 44.148 kA rms symmetric current
   Circuit breaker selection:
   Interrupting rating (I_{int}) = 50 kA rms symmetric current

3.4.11.2 Computer Solution

Power system analysis software was used to carry out the short-circuit study of the system shown in fig. 3.10. The results of the study, and the comparison with the equipment ratings, are given in table 3.13. The computer printout for the relevant buses is also provided here.

TABLE 3.13
Study_SC_NF_01: Maximum Fault Levels and Equipment Ratings

Bus No.	Duty	Upstream Contribution, kA (S)	Downstream Contribution, kA (S)	Total Fault, kA (S)	X/R	MF	Required Capability	P S AS
13.8 kV switchgear	CL/L	13.41	5.71	19.12	29.38	2.6	49.71	P
13.8 kV switchgear	INT	13.41	5.1	18.51	29.42	1.0	18.51	S
0.6 kV switchgear	INT	32.1	10.4	42.5	9.66	1.06	45.05	S

Note: CL/L, closing/latching; INT, interrupting duty; MF, multiplying factor; P, peak; S, symmetrical current basis; AS, asymmetrical current basis.

3.5 PROTECTIVE DEVICE COORDINATION

3.5.1 Introduction

A preliminary protective-device coordination study must be carried out after the load-flow and short-circuit studies have been performed and before the single-line diagram or equipment specifications are issued. The study is performed to determine instrument transformer ratios and accuracy, characteristics and initial settings of the protection relays, the fuse ratings, and compliance with the electrical codes. The study enables the design engineer to develop a reliable and coordinated system.

The protective devices used in industrial systems are phase and ground overcurrent (50/51, 50/51G), differential (87), directional overcurrent (67), under/overvoltage (27/59), voltage restraint overcurrent (51 V) or impedance (21) for generators, under/overfrequency (81 U/O) relays, fuses, and static-trip units for low-voltage circuit breakers. With the exception of the differential protection, which is closed zone, the other protections are open zone and need to be time coordinated to ensure discrimination in that the faulted circuit trips first. The inverse-time overcurrent (51), static-trip devices and fuses also have operating times that are dependent on the fault current flowing. The plotting of such time–current coordination curves can be performed by superimposing transparent graph paper over the manufacturer's published curves and tracing them using French curves, or by using a computer program. Manually drawing time–current coordination curves for a large system is laborious and time consuming.

The application of computer software offers a better alternative, saves time, and makes it easier to make and record changes. To gain confidence and a better understanding of the subject, the designer must also learn how to perform the task manually. See chapter 11 for more details.

Selection of suitable relays also needs to be done carefully, since a large number of industrial-grade relays are on the market. The relays must be of utility grade, tested to IEEE or equivalent IEC switchgear standards for harmonics, electromagnetic interference (EMI), radio frequency interference (RFI), and fast transients for use on the primary system and any areas of a plant where high reliability and security are required. At the utility interface, the relays used will probably require the approval of that utility.

3.5.2 Data Required for a Study

The following documentation and data are required to carry out the study:

1. A single-line diagram showing the rating of instrument transformers and equipment rating, including power transformers and rotating machines, relays, fuse ratings, and feeder size.
2. Maximum and minimum values of short-circuit current that are expected to flow through each protective device whose performance is to be studied.
3. Maximum load current on all circuits, including starting current requirements of large motors.
4. Manufacturer's time–current characteristic curves (TCC) of the relays, fuses, static-trip units, and other protective devices. The relay TCC is a plot of current in multiples of the setting versus time, whereas a fuse TCC is a plot of absolute current versus time. The fuse curve is based on the test results in open air, and must be adjusted by a safety margin to compensate for the higher ambient temperature in an enclosure. The medium- and high-voltage fuse time–current characteristics are represented by two curves — minimum melting and total clearing. For low-voltage fuses, only an average time–current curve is provided.
5. Equipment capability (limit) curves, such as transformer through short-circuit withstand, motor hot and cold thermal limits during rotor and stator acceleration, and cable thermal damage or short-circuit withstand.
6. Current transformer (CT) performance curves to check the saturation during symmetrical and asymmetrical faults.
7. Decrement curves showing the rate of decay of the fault current supplied by the generators with the exciter operating at the ceiling voltage.
8. Protective-device setting requirements stipulated by the power utility company at the utility interface.
9. Relevant code and equipment standards.

3.5.3 Time Interval for Coordination of Overcurrent Elements

Guidelines for coordination between upstream and downstream protective devices in industrial plants are provided here. The software may make an allowance for the Delta-Wye and Wye-Delta transformer connections when coordinating primary and secondary devices for phase-to-phase and ground faults. Otherwise, this effect may need to be introduced manually.

3.5.3.1 Between Relays Controlling Circuit Breakers

Medium-voltage circuit breakers (five cycles): 0.3 s
High-voltage circuit breakers (two or three cycles): 0.25 s

3.5.3.2 Between Upstream and Downstream Fuses

Pre-arcing I²t of upstream fuse > total I²t of downstream fuse. I²t values are available from fuse manufacturers. In the majority of applications, a two-to-one current rating ratio (e.g., 200 A upstream and 100 A downstream) provides coordination, although a ratio at 3:1 is preferable.

3.5.3.3 Upstream Fuse and Downstream Circuit Breaker

For medium- or high-voltage fuses, use the minimum value of the melting curve of the upstream fuse. Allow a margin of safety to account for tolerance, preloading, and changes in ambient temperature. The circuit breaker or relay time–current curve shall not exceed 75% of the time on the fuse minimum melting time–current curve to prevent fuse aging.

For a low-voltage fuse, the total clearing time can be estimated by adding 15% up to 0.03 s and 10% above 0.03 s to the average time.

3.5.3.4 Static Trip Units of Low-Voltage Circuit Breakers

The time–current characteristic curves of static-trip units (LSIG) are represented by an operating band. These devices are coordinated properly when there is a clear space between their operating bands. This is also true for adjustable and nonadjustable low-voltage initiative circuit breakers (MCCBs), which often also have time–current trip characteristics.

3.5.4 COMPUTER PROGRAMS FOR OVERCURRENT COORDINATION

These programs come with a library of time–current curves of relays, fuses, transformers, and cable short-circuit withstand curves. The designer can select the device and plot the time–current curve that protects the equipment and coordinates with downstream devices. The user is advised to verify from the manufacturer's literature that the device is still available and the set points (tap, time dials, etc.) are valid for the selected model. The software also checks the maximum fault current levels of the downstream buses and cuts off the operating curve at that value to confirm coordination with upstream devices. A plot of a small system was generated through typical computer software, and this is shown in fig. 3.14.

3.6 ARC-FLASH HAZARD CALCULATIONS

3.6.1 INTRODUCTION

NFPA 70E, "Electrical Safety Requirements for Employee Work Place" [S13], has recognized the hazard of arc flashes and requires that employees be protected from the flash hazard at the workplace. The standard covers the need for flash protection boundaries and the use of personnel protective equipment (PPE), such as special suits and gloves, etc. If an employee works within the established flash-protection boundary, the standard requires that a flash hazard analysis be carried out to determine the level of protection required.

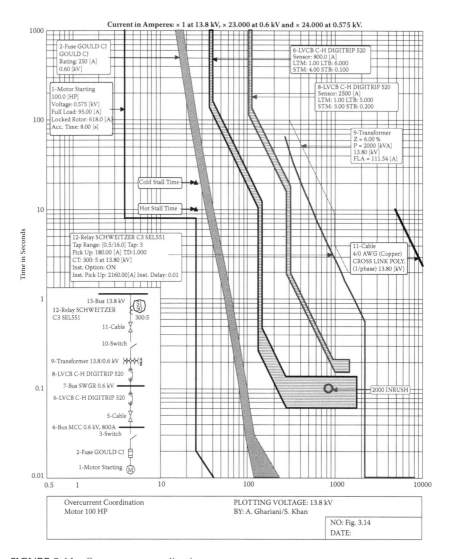

FIGURE 3.14 Overcurrent coordination

Arc-flash hazard analysis is performed using the results from the short-circuit and protective coordination studies. Perform the arc-flash hazard analysis, and calculate the incident energy and flash-protection boundary using the latest issue of ref. [S12] or any other guide adopted by the plant and approved by NFPA or OSHA (Occupational Safety and Health Administration). Computer software is available as a part of the standard [S12] and from other software developers, and can be used to perform the calculations at different buses on the system. However, practical data must be used to get meaningful results.

3.6.2 Steps Required for the Analysis

- Obtain the latest and updated one-line diagram, collect all necessary data, and identify the equipment that needs to be analyzed. The one-line diagram must include cable size and length for all circuits.
- Determine the system configuration for different plant operating modes that will provide maximum and minimum fault currents.
- Determine the rms symmetrical bolted fault currents, maximum and minimum, at each location or bus of concern. Include cable impedance and do not add any margins.
- Calculate the arc fault current. For system voltages less than 1000 V, the calculated arc fault current will be less than the bolted fault current due to arc impedance. For medium-voltage systems, calculate the current using the equations in ref. [S12] or other relevant guides.
- Determine the arc fault clearing times. This can be obtained from the time–current curve (TCC) plots prepared for the protective-device coordination study. For fuses, use the manufacturer's TCC and add a 15% margin to the average curve if the total clearing time is not available. Use 0.01 s if the fault current is above the total fault clearing line at 0.01 s. For circuit breakers, use 0.025 s for molded case; 0.05 s for low-voltage (<1000 V), metal-enclosed or insulated case; 0.08 s for medium-voltage five-cycle; and 0.05 s for two- or three-cycle HV breakers.
- Select the working distances by using values from the relevant standards and guides.
- Determine the incident energy for all buses using a software program such as the arc-flash calculator included in IEEE 1584-2002 [S12].
- Determine the flash-protection boundary for all equipment using the equations provided in the relevant standard and guides.

3.7 HARMONIC ANALYSIS

3.7.1 Introduction

A harmonic analysis is required when one of the following conditions is present or is anticipated in the future:

- The possibility of exceeding the harmonic limits at the point of common coupling with the power company. These limits are established by IEEE 519 [S11] or by the utility company responsible for the power supply.
- Application of power factor correction capacitors, either in the existing or the newly planned system.
- Malfunctioning of power, electronic equipment, or control systems.

3.7.2 Methods and Techniques

Harmonic analysis can be carried out using one of the following techniques:

- *Manual or hand calculations*: these are restricted to small networks, are tedious and time consuming, and are vulnerable to errors.
- *Transient network analyzer (TNA)*: an analog simulation with small-scale system components (generators, transformers, etc.) with responses close to those of the actual rating. This method is restricted to small networks and is costly and time consuming.
- *Field measurements*: these are used effectively to validate and calibrate the modeling of digital simulations.
- *Digital simulations*: these are load-flow runs using harmonic models for system components. This is the most convenient method for harmonic analysis, provided that the system component is modeled accurately and verified through field measurements.

3.7.3 Necessary Steps for the Study

The following steps shall be carried out for the system, including base case, contingencies or different modes of operation, future system expansions, and changes to the utility short-circuit levels:

1. Prepare or obtain the single-line diagram.
2. Collect system component data as required by the harmonic analysis software. Some of the programs use the same raw data as used for load-flow and short-circuit analysis. Include cable impedance (R + JX) at all voltage levels and for all feeders; this will provide damping. For locked-rotor impedance of motors, the X/R ratio shall be calculated from the starting power factor and not the X/R ratio used for the short-circuit study.
3. Obtain relevant data from the power utility company, including:
   - Minimum and maximum system impedance as a function of frequency for different system conditions (frequency scan)
   - Harmonic limits, including distortion factors and IT values, if they are different from the IEEE 519 standard [S11]
4. Prepare input data files as required by the software.
5. Perform the harmonic analysis for the base case and compute the impedances at the harmonic source buses and capacitor locations.
6. Compute the harmonic distortion and IT values at the point of common coupling with the power company.
7. Examine the results, then go back to steps 4 and 5.
8. Compare the loading (fundamental plus harmonics) of each capacitor bank permitted by the relevant standards.
9. Relocate the capacitor bank or change the rating if it is overloaded. Apply a detuning reactor to shift the resonance frequency, and go back to step 5.
10. Add harmonic filters if the harmonic distortion factors and IT values at the point of common coupling exceed the established limits.

3.8 POWER SYSTEM STABILITY

3.8.1 An Overview

Power system stability studies are required for an industrial power system under one or more of the following conditions:

- On-site generation exists or is planned.
- Synchronous motors are a major part of the load.
- Disturbance on the utility network causes plant shutdown.
- Validation or determination of relay settings is needed to avoid uncoordinated trips.

The information derived from the stability study includes the following:

- Rotor angle of generators and synchronous motors
- Active and reactive power flow in all machines, feeders and lines, and transformers.
- Slip of critical induction motors.
- Bus voltage throughout the system.
- System frequency and frequency decrement.

3.8.2 Transient Behavior of Synchronous Generators and Motors

The steady-state torque equation for a synchronous motor or generator is given by

$$T = \frac{\pi \times P^2}{8} \times \phi_{SR} \times F_R \times \sin \delta_R$$

where

- T = mechanical shaft torque
- P = number of poles
- ϕ_{SR} = air-gap flux (approximately proportional to the stator current)
- F_R = rotor field MMF (proportional to the field current)
- δ_R = mechanical torque angle between rotor and stator field

A plot of torque and rotor angle is shown in fig. 3.15. For a synchronous motor, the rotor field (F_R) lags the rotating stator field (ϕ_{SR}) by the angle δ_R. For a generator, F_R leads the stator field.

During a system disturbance or a fault, the power output of the generator is reduced. Because the power input to the prime mover does not change, the excess energy is added to the rotating inertia, which causes an increase in the angle between the internal voltage of the generator and the system voltage. After the fault is cleared, the generator power exceeds the power input to the prime mover as the angle is advanced. This condition brings the angle back to the initial value, and the rotor

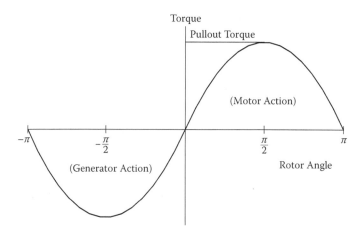

FIGURE 3.15 Torque vs. rotor angle relationship for synchronous machines in steady state

oscillates. In a stable system, the oscillations are damped and subside until the rotor angle is restored to a steady-state value.

3.8.3 Stability Study for an Ammonia Plant

An existing ammonia plant was experiencing more than the usual number of outages, initiated by system disturbances on the power utility subtransmission system. A decision was made to build a new ammonia plant in the same vicinity and design a reliable power system that would ride through the disturbances on the power utility system. The bus diagram for the new plant electrical system is shown in fig. 3.16 and comprises the following major equipment:

- A steam turbine driving an air compressor at one end and a synchronous generator at the other end
- Two ammonia compressors driven by synchronous motors and one common LCI drive for starting
- Medium- and low-voltage induction motors
- Power transformers, step-down and generator step-up
- Medium- and low-voltage switchgear and motor control centers
- Variable-frequency drives
- Uninterruptible power supply (UPS) for critical controls and LV starter coils

A transient-stability study was carried out from the solved load-flow cases. Three-phase and line-to-ground faults were created on the power utility substation bus and transmission lines feeding other customers. The fault-clearing time was tested from 6 cycles to 12 cycles. An islanding and load-shedding scheme was developed to island the new plant power system, tripping the large ammonia compressors upon detection of the fault on the utility network or existing plant power system. The results from the study were used to define the following, within practical limits:

Power System Studies

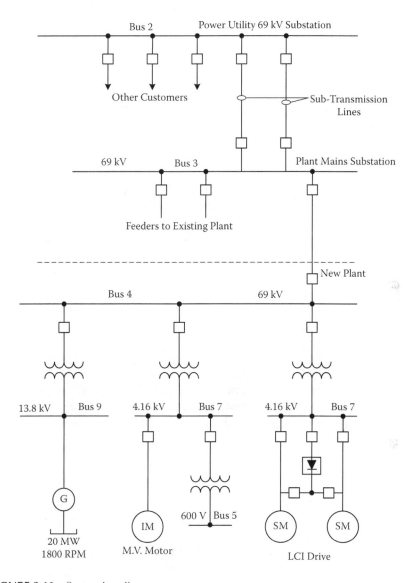

FIGURE 3.16 System bus diagram

- Generator reactance and inertia
- Exciter characteristics
- Critical induction motors with high starting and breakdown torques, and deep rotor-bar design
- Higher starting and pull-out torque for synchronous motors
- Low-threshold voltage for control component
- Low-threshold voltage and ride-through capability for low-voltage drives

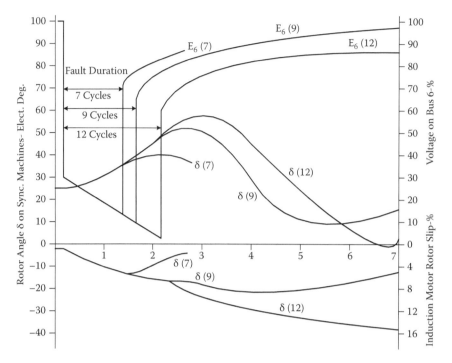

FIGURE 3.17 Composite plot of stability results

Figure 3.17 shows a composite plot made from one of the stability cases. The plot shows system behavior when a three-phase fault on the utility company line feeding other customers was cleared in 7, 9, and 12 cycles. The following are represented:

- Generator bus voltage
- Generator rotor angle
- Induction motor slip

The plant was built with the selected equipment and a properly designed distribution system. During the subsequent years of operation, the plant power system islanded successfully in the majority of cases, thus minimizing plant shutdowns.

3.9 OTHER COMMON STUDIES AND CALCULATIONS

Other studies and calculations performed during the engineering stage are listed in the following subsections.

3.9.1 SWITCHING TRANSIENTS

Switching-transient studies are carried out for industrial power systems when one or more of the following is present or planned:

- Switching of capacitor banks

Power System Studies

- Switching of harmonic filters
- Switching of autotransformers to assist motor starting
- Switching of medium-voltage circuit breaker with large inductive reactance behind

Two methods, transient network analyzer (TNA) or electromagnetic transient program (EMTP), are used for the study. Because more practical models are available in EMTP, this technique is widely used for the study.

3.9.2 Cable Ampacity Calculations

The majority of power cable installations are in cable trays, and only a small number of feeders are installed in duct banks or directly buried. Ampacity tables and adjustment factors for the common installation methods are provided in IEEE 835 [S14], NFPA 70 [S13], the Canadian Electrical Code, and manufacturers' catalogs.

The ampacity calculations for cables in duct banks or directly buried are more complex. Additional adjustment factors provided in chapter 11 can be used when the installation conditions are different from those provided in [S13, S14].

Commercially available computer programs can be used to calculate the feeder current-carrying capacity or anticipated temperature. The majority of the programs are based on the assumptions made in [10]. The application of these programs becomes necessary when one of the following situations arises:

- The duct bank has different-sized ducts.
- Soil resistivity, ambient temperature, and other parameters are different from those provided in standard tables.
- The cable shield is grounded at both ends.

The ampacity tables based on the planned installation methods and site conditions shall be prepared at an early stage of the project. This will provide a uniform basis for cable selection and avoid mistakes.

3.9.3 DC Power Supply

A power supply of 125 V DC is required for switchgear and circuit breaker control and protection relays. The 24 V DC required for some control systems can be derived through AC-to-DC converters. The DC power supply consists of a lead-calcium battery, battery chargers, and a DC distribution system. The sizing of battery and chargers, equipment selection, and system design is covered in chapter 8.

3.9.4 Motor Starting

Motor-starting studies are carried out to evaluate the following:

- Voltage drop (profile) caused by high inrush current during motor start
- Motor performance during acceleration

This subject is covered in chapter 9.

3.9.5 Ground Grid

A ground-grid study is carried out for high- and medium-voltage substations to check if the design will limit the step-potential, touch-potential, and ground-potential rise to a safe value, as outlined in IEEE 80 [S15]. This subject is covered in chapter 13.

REFERENCES

Standards

S1. IEEE 399-1997, IEEE Recommended Practice for Industrial and Commercial Power Systems Analysis, 1997.
S2. IEEE C37.010-1999, Application Guide for AC High Voltage Circuit Breakers Rated on a Symmetrical Current Basis, 1999.
S3. IEEE C37.06-2000, AC High Voltage Circuit Breakers Rated on a Symmetrical Current Basis, 2000.
S4. IEEE C37.5-1979, IEEE Guide for Calculation of Fault Currents for Application of High Voltage Circuit Breakers Rated on a Total Current Basis, 1979.
S5. IEEE C37.41-2000, IEEE Standard Design Tests for High Voltage Fuses, Fuse Disconnecting Switches and Accessories, 2000.
S6. ANSI/IEEE C37.13-1990, IEEE Standard for Low-Voltage AC Power Circuit Breakers Used in Enclosures, 1990.
S7. ANSI/IEEE C37.16-1997, Preferred Ratings, Related Requirements and Application Recommendations for Low-Voltage for Power Circuit Breakers and Power Circuit Protectors, 1997.
S8. IEEE C37.20.7-2001, IEEE Guide for Testing Medium Voltage Metal-Enclosed Switchgear for Internal Arcing Faults, 2001.
S9. IEEE 141-1993, IEEE Recommended Practice for Electric Power Distribution for Industrial Plants, IEEE Red Book, 1993.
S10. IEEE 242-2001, IEEE Recommended Practice for Protection and Coordination of Industrial and Commercial Power Systems, IEEE Buff Book, 2001.
S11. IEEE 519-1992, IEEE Recommended Practices and Requirements for Harmonic Control in Electrical Power Systems, 1992.
S12. IEEE 1584-2002, IEEE Guide for Performing the Arc Flash Hazard Calculations, 2002.
S13. NFPA 70E-2000, Electrical Safety Requirements for Employee Work Place, 2000.
S14. IEEE 835-1994, Standard Power Cable Ampacity Tables, 1994.
S15. IEEE 80-2000, IEEE Guide for Safety in AC Substation Grounding, 2000.
S16. IEEE 1015-1997, IEEE Recommended Practice for Applying Low-Voltage Circuit Breakers Used for Industrial and Commercial Power Systems. IEEE Red Book, 1997.
S17. ASTM F-1505-01, Standard for Performance Specification of Flame Resistance Textile Materials for Wearing Apparel for Use by Electrical Workers Exposed to Momentary Electric Arc and Related Thermal Hazards.

BIBLIOGRAPHY

1. Richard H. McFadden, "Power System Analysis: What It Can Do for Industrial Plants," *IEEE Trans. IAS* 7 (2): 181–188 (1971).
2. G. L. Nuss and T. W. Haymes, "Application of New Method for Calculating Short Circuit Duties and Power Circuit Breaker Capabilities," *IEEE Trans. IAS* 7 (2): 290–392 (1971).
3. General Electric Co., *Distribution Data Book*, publication GET-1008L, Schenectady, NY: General Electric.

4. Larry Conrad, Kevin Little, and Cliff Craig, "Predicting and Preventing Problems Associated with Remote Fault-Clearing Voltage Dips," *IEEE Trans. IAS* 27 (1): 167–172 (1991).
5. Philip A. Nobile, "Power System Studies for Cogeneration: What's Really Needed," *IEEE Trans. IAS* 23 (5): 777–785 (1987).
6. Richard H. McFadden, "Stability Considerations for Industrial Power Systems," *IEEE Trans. IAS* 13 (2): 107–115 (1977).
7. Charles L. Bencel, "Maintaining Process Continuity during Voltage Dips," *IEEE Trans. IAS* 18 (4): 324–328 (1982).
8. R. Lee, "The Other Electrical Hazard: Electrical Arc Blast Burns," *IEEE Trans. IAS* 18 (3): 246–251 (1982).
9. R. Lee and Dunki Jacobs, Jr., "Pressure Developed by Arcs," *IEEE Trans. IAS.* 23 (3): 760–764 (1987).
10. J. H. Neher and M. H. McGrath, "The Calculation of Temperature Rise and Load Capability of Cable Systems," *AIEE Trans. Power Applic. Syst.* 76 (pt. 3): 762–772.
11. Edison Institute, *Underground Reference Book*, EEI publication 55-16, Edison Institute, New York, 1957.

4 System Neutral Grounding

4.1 INTRODUCTION AND OVERVIEW

Power system neutral grounding has been practiced since the beginning of power system development to hold the phase voltages stable with respect to ground. However, the problem of system grounding is often not given the required attention. The grounding of many systems has been based on past experience, or opinion, or an extension to the grounding methods in existing installations. For this reason, system neutral grounding practice is found to vary widely in industrial plants.

Problems encountered with cable systems, such as insulation breakdown resulting from the transient overvoltages caused by arcing ground faults in ungrounded systems, and the devastating arcing ground-fault damage to switchgear and motor control centers (MCCs) in solidly grounded systems resulted in the widespread application of high-resistance (HR) and low-resistance (LR) grounding to 480 and 600 V systems in the late 1960s and the general provision of ground alarms on ungrounded systems, as discussed in Nelson and Sen's IEEE paper [1].

This has led to the development of preferred methods of system neutral grounding for industrial power systems, which are summarized as follows:

- *6.9 kV–34.5 kV systems*: Low-resistance (LR) grounding, using a 200–400 A rated resistor, is generally used at this voltage level. Ground-fault protection is provided by using zero-sequence current transformers (CTs), and the relays are set to trip at about 5–10% of the maximum ground-fault current. Resistors rated 1000–2000 A are used for systems that utilize overhead lines with bare conductors. For such systems, ground-fault relays are residually connected and are set to trip the lines at about 10–20% of the maximum ground-fault current.

  High-resistance grounding (HR) using 5–15 A rated resistors is applied for generators. HR grounding is also applied for motors fed from captive transformers. The resistor current must exceed the total system charging current of the system to which the generator or motor is directly connected.

- *4.16 kV–2.4 kV systems*: LR grounding, using 100 to 400 A rated resistors is generally used, and the ground-fault protection is provided with zero-sequence CTs.

  HR grounding is used for systems where power interruption resulting from single line-to-ground faults is detrimental to the process. Ground fault on one phase will not require removal of the faulted circuit. The resistor current rating must exceed the system charging current, and the vector sum of system charging current plus resistor current shall not exceed 8.0 A. Above this value, the ground-fault current (if maintained) will escalate to phase-to-phase or three phase.

- *600 V or 480 V systems*: These networks are relatively small and form separate subsystems, so the system charging current is usually less than 5.0 A. HR grounding is generally used for such low-voltage systems. Resistors rated at 3–5 A are common, and the system can be maintained for a ground fault on one phase.
- *Grounding for mine power systems*: LR grounding for a medium-voltage system using a 25 A or 50 A continuously rated resistor is used to limit the touch voltage $I_G \times R_G$ to 100 V or less. The ground-fault relay current pickup must not exceed one-third of the resistor rating. HR grounding is used for low-voltage systems using a 5 A continuously rated resistor.

4.2 UNGROUNDED SYSTEM

Power systems are always grounded through the system capacitance, even if the system neutral(s) is ungrounded. The system behavior during normal operation and ground fault on phase "a" is shown in figs. 4.1 and 4.2, respectively. A typical 4.16 kV system fed from a 10 MVA transformer is used in the example.

4.2.1 Normal Operation

System capacitance/phase = C_0 = 1.7684 µf (microfarad)
60 Hz capacitive reactance/phase

$$X_{C0} = \frac{1.0}{\omega \times C_0} = \frac{10^6}{2 \times \pi \times 60 \times 1.7684} = 1500\,\Omega$$

It is reasonable to assume that 60 Hz X_{C0} is the same for each phase.

$$I_a = I_b = I_c = \frac{E_{LN}}{X_{C0}} = \frac{4160}{\sqrt{3} \times X_{C0}} = 1.6\,A$$

Figure 4.1 shows the capacitive currents and respective voltages leading the currents in each phase. The vector sum of the three currents $I_a + I_b + I_c = 0$, so there is no ground current flowing in the system.

4.2.2 Ground Fault on Phase "a"

When X_{C0} in phase "a" is shorted,

$$I_a = 0, \quad I_b = I_c = \frac{V_{ba}}{X_{C0}} = \frac{V_{ca}}{X_{C0}} = \sqrt{3} \times 1.6\,A = 2.77\,A$$

The ground-fault current is

$$I_G = Vector\ Sum\ I_b + I_c = \sqrt{3} \times 2.77\,A = 4.8\,A$$

The voltage between healthy phases V_b and V_c and the ground is the phase-to-phase voltage, as shown in fig. 4.2.

System Neutral Grounding

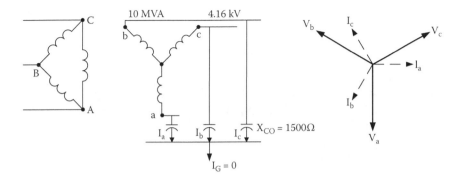

FIGURE 4.1 Normal operation

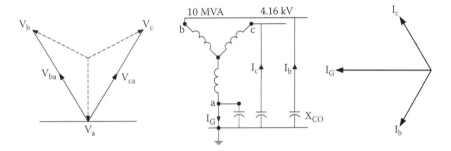

FIGURE 4.2 Ground fault on phase "a"

4.2.3 Derivation of Equivalent Circuit

Figure 4.3 shows the equivalent circuit of the 10 MVA transformer windings and the system capacitance.

Base resistance = $R_b = \dfrac{kV^2}{MVA} = \dfrac{(4.16)^2}{10} = 1.73\,\Omega$

Transformer impedance = $Z_T = 6\% = 0.06 \times 1.73 = 0.104\,\Omega$

Z_T is much smaller than X_{C0} and can be neglected

Equivalent system capacitance = $\dfrac{X_{C0}}{3}$

4.2.4 Overvoltage Due to Ground Fault through an Impedance

If an ungrounded system is grounded through a resistance or a capacitance, there will be no transient overvoltages. There will be, however, a high transient overvoltage if the fault is through an inductance, $X_L = \dfrac{X_{C0}}{3}$.

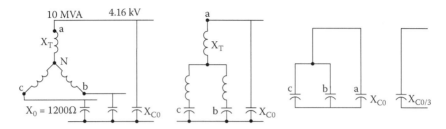

FIGURE 4.3 Derivation of equivalent circuit

Figure 4.4 shows the equivalent circuit referred to "a" phase (and the fault is through a resistance, capacitance, or an inductance) and a plot of R, X_C, or X_L over $X_{C0}/3$ versus E (times normal E_a).

4.2.5 Conclusion and Recommendation

The advantages and disadvantages of the ungrounded systems can be summarized as follows:

Advantage: An ungrounded system can ride through the ground fault on one phase.

Disadvantage: It can produce high transient line-to-ground overvoltages. The insulation system in rotating machines (motors and generators) is then severely stressed and can cause insulation breakdown.

The recommended solution to the above disadvantage is to add a resistor in the neutral, which will reduce potential phase-to-ground overvoltages:

$$R \leq X_{C0} \text{ or } I_R \geq 3 \times I_{C0}$$

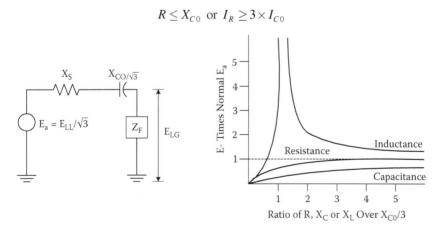

FIGURE 4.4 Overvoltage due to ground fault through an impedance

System Neutral Grounding

4.3 HIGH-RESISTANCE GROUNDED SYSTEM

For a high-resistance (HR) system, a resistor is added to the neutral of the 4.16-kV system, as illustrated in fig. 4.2. The system behavior during normal operation and during a ground fault on phase "a" are shown in figs. 4.5 and 4.6, respectively.

4.3.1 NORMAL OPERATION

Figure 4.5 shows the system conditions during normal operation. The conditions that exist without a ground fault are the same as for the ungrounded conditions, and the same values of capacitance are assumed, giving an X_{CO} of 1500 ohm, as calculated in section 4.2.1, and as in that case, the net ground current is given as:

Net ground current = $I_G = I_a + I_b + I_c = 0$

4.3.2 GROUND FAULT ON PHASE "a"

Figure 4.6 shows the systems conditions for a ground fault on phase "a," and the following conditions apply when X_{CO} = 1500 ohm, as shown in fig. 4.5:

- Maximum resistor rating $R = \dfrac{X_{CO}}{3} = 500\,\Omega$
- Capacitance current in phases "b" and "c" = $I_b = I_c = \dfrac{4160}{1500} = 2.77A$
- Total capacitance current = vector sum $I_b + I_c = 4.8A$
- Resistor current = $I_R = \dfrac{4160}{500 \times \sqrt{3}} = 4.8A$
- Ground-fault current $I_G = \sqrt{(4.8)^2 + (4.8)^2} = 6.8A$

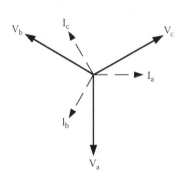
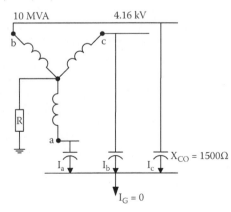

FIGURE 4.5 Normal operation (HR)

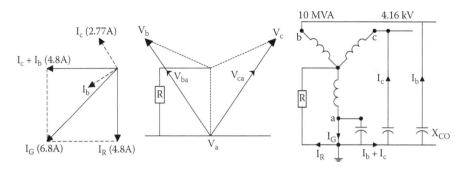

FIGURE 4.6 Ground fault on phase "a" (HR)

For medium-voltage systems, a directly connected resistor of high ohmic value with a low current rating is mechanically weak and is prone to damage. In such cases, it is recommended to use a single-phase distribution transformer in the system neutral connection and connect the low ohmic value and high-current-rated resistor to the transformer secondary. The secondary resistance is reflected into the neutral by the square of the single-phase transformer rating, hence providing the required high-resistance grounding.

4.3.3 Example of a High-Resistance Grounding Calculation

The objective is to design a HR grounding for the 4.16 kV system, as shown in fig. 4.7. The system is connected to a 10 MVA transformer and supplies power to three synchronous and five squirrel cage induction motors. The power supply to the motors is critical for the process.

The faulted circuit should not be removed quickly on a line-to-ground fault, and the tripping can be delayed to permit the operator to take corrective action. The following criteria must be met:

- The ground-fault current $I_R + 3I_{C0} \leq 8.0$ A
- The phase-to-ground insulation level for the motors, switching equipment, and cables needs to be rated for phase-to-phase voltage if the faulted circuit is maintained indefinitely, or 133% if the faulted circuit is removed within one hour.

Step 1: Estimate the system charging current

Data from section 4.9 can be used to estimate the system charging current if vendor data is not available. The details of the estimate are provided in table 4.1. Increasing the system charging current by 20 to 50% allows for contingencies and inaccuracies. The relevant formulae are:

$$X_{C0} = \frac{10^6}{\omega \times C_0} \text{ ohm/phase}$$

System Neutral Grounding

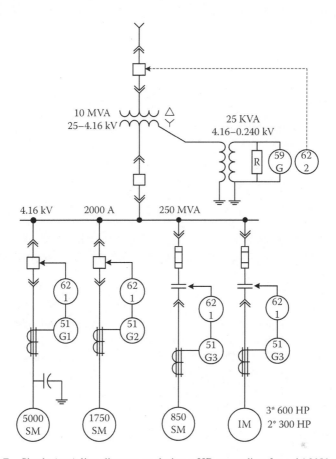

FIGURE 4.7 Single (one)-line diagram to design a HR grounding for a 4.16 kV system

TABLE 4.1
Estimate of System Charging Current

Equipment		C_0 µf/phase	$3I_{C0}$, A	Quantity, m	$3I_{C0}$, (Total) A
Surge capacitor		0.50	1.36	1	1.36
Synchronous motors	5500 hp	0.14	0.38	1	0.38
	1750 hp	0.035	0.10	1	0.10
	850 hp	0.02	0.054	1	0.054
Induction motors	600 hp	0.014	0.04	3	0.12
	300 hp	0.009	0.024	2	0.05
Cable	350 KCmil	0.185/m	0.502	2.3	1.15
	#2 AWG	0.1/m	0.27	3.41	0.92
Power transformer	10 MVA	0.004	0.01	1	0.01
Total system charging current					*4.15*

where

$\omega = 2 \times \pi \times f$, C_o = capacitance, µf/phase

$$3I_{CO} = \frac{\omega \times C_o \times kV \times \sqrt{3}}{1000} = 0.652 \times C_o \times kV \text{ at 60 Hz}$$

Increase the system charging current by 20% to account for contingencies and tolerances.

- Total system charging current: $1.2 \times 3I_{CO} = 1.2 \times 4.15 = 5.0 A$
- Resistor current referred to primary of the grounding transformer:

$$I_{RP} = 5.0 A$$

- Ground current: $I_G = \sqrt{5^2 + 5^2} = 7.07 A < 8.0 A$

Hence the system can continue to operate with one phase to ground

Step 2: Select grounding transformer and resistor rating

The transformer is generally rated for the system line-to-line voltage on the primary and 240 V or 120 V on the secondary. A higher voltage on the primary is selected to protect against possible transient overvoltage.
- Voltage rating = 4160 – 240 V

- Voltage ratio $N = \dfrac{4160}{240} = 17.33$

- Resistor rating referred to the primary = $R_P = \dfrac{4160 \text{ V} / \sqrt{3}}{5.0 A} = 480 \Omega$

- Resistor rating referred to the secondary = $R_s = \dfrac{R_P}{N^2} = \dfrac{480}{17.33^2} = 1.6 \Omega$
- Resistor current = $I_{RS} = I_{RP} \times N = 5.0 \times 17.33 = 86.65 A$
- Transformer kVA = 240 × 86.65 = 20.8
- Transformer rating: 25 kVA, 4160 – 240 V

4.3.4 HR Grounding for Delta-Connected Systems (System Neutral not Available)

In the majority of industrial power systems, the winding connection of medium- and low-voltage transformers is Delta primary and Wye secondary, with the neutral brought out. The grounding resistor or transformer/resistor combination is connected to the neutral of the secondary. However, for applications with Delta secondary winding, one of the following arrangements (fig. 4.8) can be used:

- Three-phase zig-zag grounding transformer with the resistor connected to the neutral. This is also a preferred arrangement for low-resistance grounding in Delta-connected systems or where the system neutral is not available.

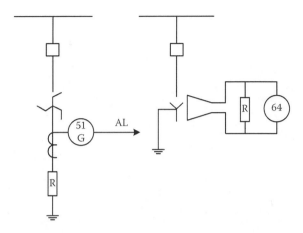

FIGURE 4.8 HR grounding for Delta connection system

- Three-phase or three single-phase transformer with Wye primary with solidly grounded neutral and Delta secondary. The grounding resistor is connected across the broken Delta secondary with a voltage relay in parallel.

4.3.5 HR Grounding for Low-Voltage (480 and 600 V) Systems

The system charging current for a 480 V or 600 V system is on the order of 0.1–1.0 A per 1000 kVA. This makes the HR grounding the best choice for systems with the majority of the load being three-phase. Typically a 3.0–5.0 A continuously rated resistor is connected to the neutral of the transformer secondary.

Microprocessor-based ground-fault (GF) detection schemes are available from several manufacturers as a package. These schemes include a pulsing system that gives an alarm for a single phase-to-ground fault and a trip signal for phase-to-phase or cross-country faults. A ground-fault element provided with the static trip unit provides a back-up for the phase-to-phase or cross-country faults. A typical scheme with HR grounding and GF detection system is shown in fig. 4.9.

4.3.6 Measurement of System Charging Current

Because the capacitance data on rotating machines and power cables is very approximate, it is highly recommended to measure the system charging current after the installation is complete. GE Publication GET 1354 [5] provides schemes for measuring the system charging currents in low- and medium-voltage systems. This approach ensures that the selected resistor current exceeds the total system charging current, thus avoiding the possibility of transient overvoltage on line-to-ground faults.

4.3.7 Conclusions

For a correctly installed and configured high-resistance grounded system, the following can be concluded:

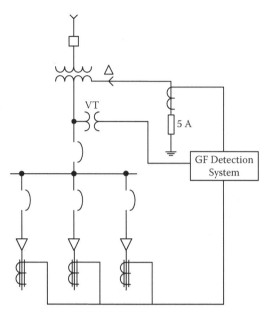

FIGURE 4.9 HR grounding for 480 and 600 V systems

- Ground fault on one phase does not require removal of the faulty circuit
- Suppresses transient overvoltage caused by line-to-ground faults
- Limited to systems with charging current not exceeding 5.5 A
- Not enough ground current for selective relaying

4.3.8 Recommendations

High-resistance grounding is recommended for the following systems, with the associated precautions listed:

- Low-voltage systems with predominantly three phase loads (Delta or ungrounded Wye)
- Medium-voltage systems up to 4.16 kV where power interruption will be detrimental to the process

Requirements:

- Requires 100% rated surge arresters
- Requires that the cable insulation must be 133% if the ground fault is cleared in less than 1 h and 173% or one voltage level higher if maintained indefinitely

4.4 LOW-RESISTANCE (LR) GROUNDED SYSTEM

4.4.1 LR Grounding for Wye-Connected System

In these systems, a resistor sized for and rated at 100–400 A is connected to the neutral of the transformer secondary winding. The resistor resistance value is reduced

System Neutral Grounding

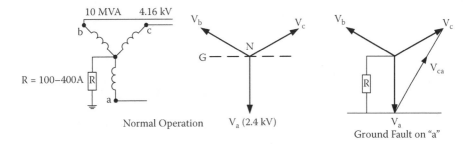

FIGURE 4.10 Low-resistance (LR) grounded system during normal operation

and the current rating is increased to 1000–2000 A for systems that utilize bare conductor overhead lines for distribution. The system behavior during normal operations and with a ground fault on phase "a" is shown in figs. 4.10 and 4.11, respectively. The same system, with an appropriate relaying scheme that will operate selectively to detect and trip the faulted circuit, is shown in fig. 4.12.

The maximum phase-to-ground voltage on the healthy phases (b and c) is equal to phase-to-phase voltage. The faulted circuit needs to be removed, and the power must be interrupted to the faulted circuit.

4.4.2 LR Grounding for Delta-Connected System

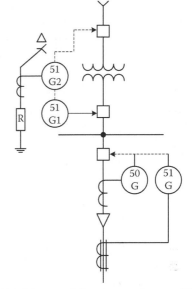

FIGURE 4.11 LR grounding (Wye system)

For applications where the transformer winding connection is Delta, the preferred arrangement is to use a three-phase zig-zag grounding transformer with a resistor connected to its neutral. The short time rating of the zig-zag grounding transformer is the same as the resistor rating; however, the continuous rating shall be at least 10% of the available ground-fault current.

4.4.3 Conclusions

A low-resistance grounded system works from the principle that the protective system is appropriately configured to clear the fault at a speed relative to the level of ground-fault current that is allowed to flow, to limit the damage at the point of fault. For low-resistance grounding the following can be concluded:

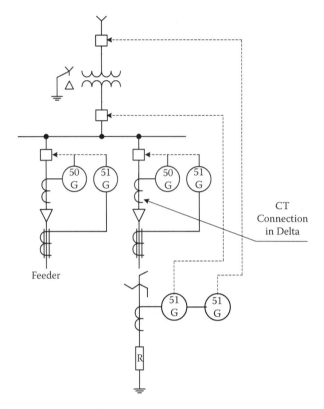

FIGURE 4.12 LR grounding (Delta system)

- The neutral can shift to the system phase-to-ground voltage, and the first ground fault must be detected to remove the fault circuit from the system.
- This grounding method suppresses the transient phase-to-ground overvoltages.
- Selective ground-fault relaying is possible.
- Burning damage at the point of fault is considerably reduced.
- Requires that the cable insulation be 133%, as the ground fault is cleared quickly.

4.4.4 Recommendations

Low-resistance grounding is recommended to be applied where a network is extensive and subsidiary switching stations exist, and hence where selective tripping is required to achieve the objective level of service reliability. It is also appropriate for overhead line feeders. To apply this system, the following are also needed:

- Requires 100%-rated surge arresters
- Requires zero-sequence CTs for ground-fault relaying when the ground-fault level is significantly lower than the phase fault level

System Neutral Grounding

4.5 SOLIDLY GROUNDED NEUTRAL SYSTEM

4.5.1 SOLIDLY GROUNDED NEUTRAL

The neutral is solidly connected to ground, and hence the neutral is always at ground potential. The system behavior during a bolted ground fault close to the source ($Z_G = 0$) and ground fault remote from the source is shown in fig. 4.13. As can be seen in the figure, for faults near the source, the healthy phase voltages during a ground fault are held to the rated phase-to-ground voltage. For faults more remote from the source, the healthy phase voltage will rise but will always be less than the phase-to-phase voltage.

4.5.2 POWER SYSTEM CONSIDERATIONS

In most power systems, the net zero-sequence (ground fault) impedances are lower than the phase or positive sequence impedances. This means that, for solidly grounded networks, it is likely that the fault current for ground faults will be higher than that for three-phase and phase-to-phase faults. It is therefore necessary to check the ground-fault current levels when performing switchgear rating studies in solidly grounded networks.

4.5.3 CONCLUSIONS

The solidly grounded system does not restrict the ground-fault current levels, and hence it needs to be applied together with fast protection for ground faults to limit the thermal damage in the network for such faults and to minimize damage at the point of fault. On the other hand, the fact that ground-fault current is not limited works in the favor of fuse-protected circuits because fuses are then equally effective in clearing both phase and ground faults quickly. The solidly grounded system therefore can be summarized as having the following characteristics:

- It eliminates transient phase-to-ground overvoltages during ground faults.
- The first ground fault, due to the unrestricted operation of the ground-fault protection, quickly removes the faulted circuit from operation.
- Can cause extensive damage from arcing ground faults.

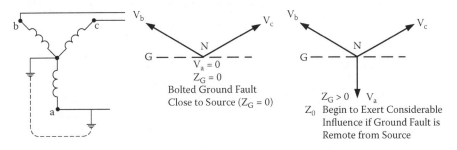

FIGURE 4.13 Ground fault on phase "a"

- The protection on low-voltage systems must function effectively on the self-extinction and restrict characteristics of 480 V and 600 V systems.
- The cable insulation can be 100% rated.

4.5.4 Recommendations

The use of solidly grounded neutrals is recommended for:

- Low-voltage systems (three-phase, four-wire) when the majority of the load is connected between phase and neutral
- HV, EHV, or UHV systems (above 34.5 kV) for better surge protection and for the application of lower BIL (basic insulation level) equipment, as the healthy phase voltages during faults are significantly reduced and lower-rated surge arrestors can be safely applied

4.6 GENERATOR NEUTRAL GROUNDING

The objective of generator neutral grounding is as follows to:

- Minimize the damage for internal ground faults
- Limit mechanical stresses in the generator for external ground faults
- Limit temporary and transient overvoltage on the generator insulation system
- Provide a means of system ground-fault detection
- Coordinate with the requirements of other equipment connected to the system

A selection of generator neutral grounding methods is shown in fig. 4.14, and these are discussed in the following sections:

4.6.1 Solidly Grounded Neutral

The generator neutral is solidly connected to the ground, as shown in fig. 4.14(a). This method is not recommended for industrial and utility generators, for the following reasons:

- Because the ground-fault current is high, and the damage is proportional to $I_g^2 \times t$, there is a possibility of extensive damage to internal stator winding ground faults.
- The generator is normally designed to withstand stresses associated with three-phase faults at the machine terminals. Because of the low zero-sequence impedance of generator windings, a solid phase-to-ground fault at the machine terminals will produce higher winding currents than those from a three-phase fault.
- There is a risk of abnormal third-harmonic current flow.
- There is a risk of mechanical damage from line-to-ground faults.

System Neutral Grounding

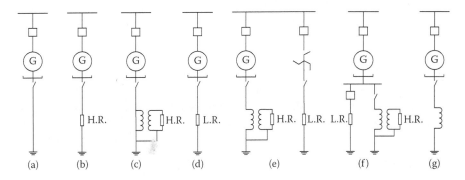

FIGURE 4.14 Generator neutral grounding methods

Solidly grounded neutrals may have to be applied for emergency generators that directly feed low-voltage systems, where phase-to-neutral loading is common, as noted in section 4.5.4.

4.6.2 High-Resistance Grounding

High-resistance (HR) grounding can be achieved as shown in fig. 4.14(b) or (c) either by connecting the resistor directly between the generator neutral and ground (b) or utilizing a distribution transformer/resistor combination (c). The neutral current is limited to 5–15 A, and the generator is shut down on the first line-to-ground fault. The main advantages are:

- Minimum damage from internal ground faults
- Limited transient overvoltage

The design criteria for HR grounding are that

$$R_n \leq X_{cg} \text{ or } \frac{X_{cg}}{R_n} \geq 1.0$$

where

R_n = effective neutral resistance
X_{cg} = capacitive reactance of the three phases

From the criteria above, it is evident that the successful application of HR grounding is dependent on limiting the capacitance, and inherently the extent, of the circuits directly connected to the generator.

4.6.2.1 Resistor Directly Connected to Generator Neutral

A directly connected resistor, as shown in fig. 4.14(b), is not recommended, as high resistance and low current make the resistor more fragile and prone to mechanical damage.

4.6.2.2 Distribution Transformer-Resistor Combination

In a transformer-resistor combination arrangement, as shown in fig. 4.14(c), a single-phase distribution-type transformer is used for the neutral grounding, and the resistor is connected to the secondary. This combination permits the application of a robust and higher-current-rated resistor.

The IEEE guide [S2] and Westinghouse [2] and GE [3] publications on generator grounding provide guidelines for sizing the transformer and resistor. The following example illustrates the steps required for necessary calculations and shows an example of equipment selection.

4.6.2.3 Example: Sizing of Generator Neutral Grounding Equipment

The application configuration is shown in fig. 4.15.

1. Design criteria
   - Make the resistance component of the ground-fault current greater than the capacitive current $I_R \geq 3I_{C0}$.
   - Make resistor power loss greater than the generator circuit three-phase capacitive VA.
2. System data
   - Generator: 70 MW, 0.85 PF, 60 Hz, 13.8 kV
   - System frequency: 60 Hz
   - System kV: 13.8 (line-to-line)

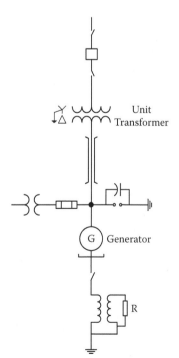

FIGURE 4.15 HR grounding of generator neutral

System Neutral Grounding

3. Capacitance data
   - Generator, $C_{gen} = 0.3900$ µf
   - Surge pack, $C_{sb} = 0.2500$ µf
   - Unit transformer, $C_{tr} = 0.0091$ µf
   - Isolated phase bus: main = 63.46×60 (L meters) $\times 10^{-6}$ $C_{mb} = 0.0038$ µf
   - Voltage transformer, $C_{pt} = 0.0002$ µf

   Total capacitance = 0.653 µf

4. Selection of resistor and transformer
   The first calculation is the total capacitive impedance:

$$X_{co} = \frac{1}{\omega(C_{gen} + C_{sb} + C_{tr} + C_{mb} + C_{pt}) \times 10^{-6}} = \frac{1}{2\pi \times 60 \times 0.653 \times 10^{-6}} = 4062 \text{ ohm}$$

Capacitive reactance (X_{cg}) seen at the neutral = 1/3 of the capacitive reactance of all phases:

$$X_{cg} = \frac{X_{co}}{3} = 1354 \text{ ohm}$$

Neutral resistor: $R_n \leq X_{cg}$
Grounding transformer voltage rating: 12 kV primary, 240 V secondary
Grounding transformer turns ratio: $N = \frac{12{,}000}{240} = 50$

Resistor, $R_{sec} = \frac{R_n}{N^2} = \frac{1354}{(50)^2} = 0.542 \, \Omega$.

Resistor current,

$$I_{sec} = \frac{V_{sec}}{R_{sec}} = \frac{V_{gen}(LL)}{\sqrt{3}} \times \frac{1}{N} \times \frac{1}{R_{sec}} = \frac{13{,}800}{\sqrt{3}} \times \frac{1}{50} \times \frac{1}{0.542} = 294 \, A$$

Increase the resistor current by 20% to account for contingencies and tolerances, say to 350 A·s. The resistor should be:

$$R_{sec} = \frac{0.542}{350} \times 294 = 0.455 \text{ ohm}$$

$$I_{sec} = 350, \; I_{prim} = \frac{350}{50} = 7.0 A > 3I_{CO} \, (5.88 A)$$

Power loss in the resistor (P_R)

$$P_R = I_{sec}^2 \times R_{sec} = (350)^2 \times 0.455 = 55.74 \text{ kW}$$

$P_R >$ charging kVA $[(5.88)^2 \times 1354 \times 10^{-3} = 46.81]$

5. Transformer rating

Transformer thermal rating (60 s) = $E_{sec} \times I_{sec} \times 10^{-3}$ kVA
$= 240 \times 350 = 84$ kVA
60 s overload factor = 4.7

Transformer continuous rating $= \dfrac{84}{4.7} = 17.87 kVA$
Select a 20 kVA transformer.

6. Summary

- System charging current $(3I_{C0}) = \left[\dfrac{13,800}{\sqrt{3}} \times \dfrac{1}{4062} \right] \times 3 = 5.88 A$
- Resistor current referred to primary, $I_{prim} = 7.0$ A > 5.88 A
- Power loss in the resistor $(P_R) = 55.74$ kW > charging kVA
- Ground-fault current $I_g = \sqrt{(7.0)^2 + (5.88)^2} = 9.14 A$
- Resistor rating 240 V, 350 A, 60 s
- Transformer 12,000-240/120 V, 20 kVA cont., 84 kVA, 1 min

Transformer insulation level (recommended)
- BIL = 95 kV, crest
- PFW = 36 kV, rms

4.6.3 Low-Resistance Grounding

The resistor is directly connected to the generator neutral, as shown in fig. 4.14(d). This method permits coordination with other equipment connected to the system and is generally used where

- A generator is connected directly to the plant load bus with outgoing feeders.
- Two or more generators are bused at generator voltage and are connected to the system through one step-up transformer.

The resistor rating is usually 100 A to 1.5 times generator full-load current with a short time rating of 10 s. For industrial applications, the upper limit is 400 A. Higher values are used in distribution systems utilizing bare overhead conductors for distribution feeders.

Although a LR grounding system with this arrangement has been used frequently for generators in industrial plants, there is a possibility of extensive damage from higher internal ground-fault current. For this reason, the arrangement shown in fig. 4.14(e) is applied.

System Neutral Grounding

4.6.4 Generator HR Grounded and System LR Grounded

This method, as shown in fig.4.14(e), is applied for direct bus-connected generators and is used to provide low-resistance grounding for the same purpose as above and offers some advantages. A zig-zag grounding transformer with a neutral resistor is connected to the bus. The ground current is not affected by the number of units connected to the bus or system interconnections.

In addition to the zig-zag transformer, each generator shall be provided with a high-resistance (HR) grounding. This arrangement will provide ground-fault protection during unit start-up and synchronizing. The combination of a zig-zag grounding transformer and HR grounding for the generator is highly recommended, and because it is effective once the generator circuit breaker has opened, it offers the following advantages:

- Damage from internal ground faults is reduced.
- The magnitude of the ground current is independent of the number of transformers or generators connected to the bus.

4.6.5 Hybrid High-Resistance Grounding

This method, as shown in fig. 4.14(f), is applied to a direct bus-connected generator in existing installations with LR grounding. In this method, HR grounding and LR grounding resistors are connected to the generator neutral. The LR interrupting device is opened as a part of the generator scheme for ground fault within the generator protection zone.

4.6.6 Low-Reactance Grounding

This method, as shown in fig. 4.14(g), is generally used where the generator is connected directly to a distribution system with a solidly grounded neutral. The inductive reactance is selected to give $X_0/X_f \geq 3$. Low-reactance grounding produces phase-to-ground fault currents ranging from 25 to 100% of the three-phase fault current. There is a possibility of significant winding iron core (in slot) damage for internal (stator) faults. This method is not recommended for industrial plants.

4.6.7 Resonant Grounding

This method is similar to HR grounding as shown in fig. 4.14(e), using a transformer-resistance combination, except that the resistor is replaced by a reactor. The reactor is selected so that the inductive reactance matches the three-phase capacitive reactance of the equipment. On line-to-ground fault, the system charging current is neutralized by an equal component of the inductive current. The system, also known as the Petersen coil system, is usually used for relatively small overhead-line networks but has the disadvantage that the reactor needs to be adjusted to "retune" the grounding any time there is a change in the system capacitance.

4.7 GROUNDING OF MINE POWER SYSTEM

4.7.1 An Overview

Electrical equipment in mines, such as electric shovels, draglines, and portable machinery, are not installed on a permanent foundation and hence do not have an attachment to a permanent ground grid. An operator touching or operating the equipment while standing on the ground during a line-to-ground fault will be subject to a touch voltage equal to $I_G \times R_G$. To safeguard against the shock hazard, the touch voltage should not exceed 100 V. The system grounding for mines shall be based on safety, established criteria, and comply with the applicable mining codes.

A safe touch voltage can be achieved by any combination of ground current and return ground resistance that gives a maximum voltage rise of 100 V. The ground resistance is variable, depends on the soil resistivity, and cannot be relied upon to achieve safe grounding conditions. A monitored metallic path, such as insulated ground wires integral with the phase conductors of the power cable, is used for continuity and to provide a dependable path for the return of the ground current.

Another important consideration for safety is to isolate the mine distribution system grounding from the high-voltage substation ground grid, where high ground potentials can appear during the ground faults in the substation. This usually requires a transformer dedicated to the mine loads adjacent to the mine. Because a resistance of less than 2.0 ohm is achievable from the insulated ground wires in the power cable, a 50 A neutral grounding resistor becomes a good choice for such a transformer.

4.7.2 Safety Considerations in HV Substation

A single-line diagram for a typical mine distribution system is shown in fig. 4.16. The following guidelines are recommended to prevent the transfer of potentials from HV-system ground faults to medium-voltage mine distribution systems:

- Isolate the HV ground grid/system from the mine distribution system neutral ground point by keeping them well separated; a minimum distance of 75 ft is recommended.
- Insulate the neutral grounding resistor using line-to-line rated insulators.
- Use insulated ground conductor from the load side of the resistor to the remote ground and other points in the distribution system.
- Limit frame-to-ground voltage to a maximum of 100 V on line-to-ground fault, considering ground wires connected between the portable equipment frame to earth.

4.7.3 Factors Involved in Shock Hazard

Factors involved in shock hazard upon touching the portable machine are shown in fig. 4.17. This diagram explains how different elements come into play. Because R_4 and R_5 are high, and R_3 is variable, the resistance of the metallic or ground wires shall be limited to 4.0 ohm for a 25 A neutral resistor or 2.0 ohm for a 50 A neutral resistor.

System Neutral Grounding

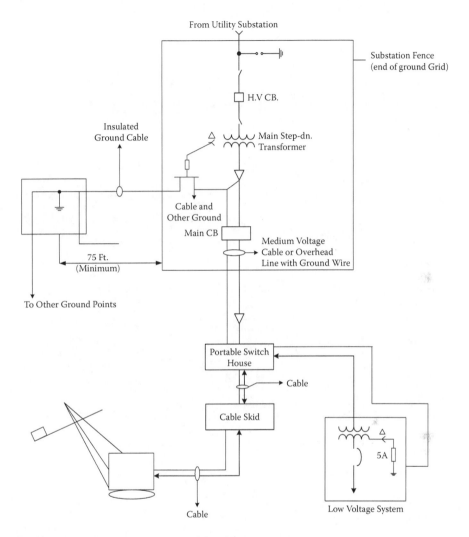

FIGURE 4.16 Typical mine distribution system

4.7.4 Neutral Grounding Resistor

The commonly used ratings for neutral ground resistors for mining power supplies are either 25 A or 50 A, a 50 A rating being a better choice. The other requirements are:

- Ensure that the resistor current exceeds the system charging current by a margin of 20% or higher.
- Use continuously rated, corrosion resistant, and insulated for line-to-line voltage.

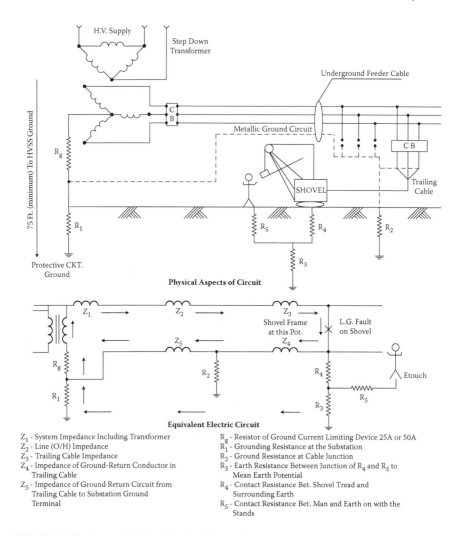

FIGURE 4.17 Factors involved in shock hazards

4.7.5 Ground-Fault Relaying and Ground-Conductor Monitoring

The ground protection relay pickup, including CT magnetizing current, shall not exceed one-third of the resistor current. The allowable pickup current is 8.0 A for a 25 A resistor and 16.0 A for a 50 A resistor. Continuous monitoring of the ground conductor is required. The power circuit shall be disconnected upon detection of a break in ground continuity.

4.7.6 Low-Voltage (600 V and 480 V) Systems

High-resistance grounding is a preferred method for low-voltage systems used for portable equipment. The power circuit must be disconnected upon detection of a line-to-ground fault. The step-down transformer is usually Delta primary-Wye secondary

System Neutral Grounding

with a 5.0 A continuously rated resistor. A zig-zag grounding transformer with a 5.0 A continuously rated resistor is used for a Wye primary-Delta secondary transformer.

4.8 NEUTRAL GROUNDING EQUIPMENT

A brief description using a typical rating of neutral grounding equipment is provided here to assist in the preparation of technical specifications.

4.8.1 HR Grounding for Wye-Connected MV Systems

For a typical MV system, as shown in fig. 4.7, the system neutral shall consist of equipment mechanically and electrically interconnected to form a complete assembly, and rated as per the system design, typically including, but not limited to, the following:

- *Grounding transformer*: Single-phase, dry-type, self-cooled (AN), epoxy-encapsulated with copper windings. Rated 25 kVA, 4.16 kV (60 kV BIL) primary and 240/120 V (10 kV BIL) secondary, Class 220 insulation system with a temperature rise not to exceed 115°C. Connect the HV winding to the transformer neutral and the LV winding to the resistor.
- *Grounding resistor*: Made of corrosion-resistant or stainless steel or cast iron, insulated for line-to-line voltage, completely protected by a metal screen cage. The rating shall be 1.6 Ω, 87 A continuous, temperature rise not to exceed 385°C.
- *Neutral cubicle*: Floor mounted, made of steel frame with bolted aluminum panels, stainless steel hardware hinged access door with lock, 25 mm × 6 mm copper ground bus bolted to the frame and ground, built to provide adequate ventilation for the heat dissipated from the equipment.

4.8.2 HR Grounding for Wye-Connected LV Systems

For a typical LV system, as shown in fig. 4.9, the HR grounded detection system shall be of the modular multicircuit type, to give an alarm when one phase of the system is faulted to ground, and provide a selective trip if a second ground fault occurs on a different phase of another feeder. The detection system shall have the following typical ratings and be complete with, but not limited to, the following:

- A neutral grounding resistor made of stainless steel or corrosion-resistant steel; 5 A continuous rating with 3 A, 4 A, and 5 A taps; temperature rise not to exceed 375°C above an ambient of 30°C; insulation system suitable for line-to-line voltage. Install the resistor in a screened metal enclosure. To minimize the risk of damage to the insulated conductor from the transformer neutral to the resistor, locate the resistor inside the enclosure of the dry-type transformer or on the top of the liquid-filled transformer.
- Zero-sequence sensors for each feeder and transformer neutral or resistor.
- Indication for faulted phase, faulted feeder, and blown fuse.
- Programmable second-fault priority.

4.8.3 LR GROUNDING FOR WYE-CONNECTED MV SYSTEMS

For a typical low-resistance grounded MV system, as shown in fig. 4.11, rate the neutral grounding resistor for the system voltage, typically a 4.16 kV system. The resistor rated typically at 200 A, 10 s, temperature rise not to exceed 760°C above an ambient of 30°C, made of corrosion-resistant or stainless steel, insulated for line-to-line voltage. Provide a free-standing ventilated enclosure made of steel frame with bolted aluminum panels, stainless steel hardware, 25 × 6-mm copper ground bus bolted to the frame and ground, built to provide adequate ventilation for the heat dissipated.

4.8.4 HR GROUNDING FOR WYE-CONNECTED GENERATOR NEUTRAL

For a typical 13.8 kV high-resistance grounded generator, as shown in fig. 4.14(b), the generator-neutral cubicle shall consist of appropriately rated equipment mechanically and electrically interconnected to form a complete assembly, including but not limited to the following:

- *Hook stick, with a 1.5–1.8-m-long hook stick, or a manually operated disconnect switch*: Rated 13.8 kV rms, 95 kV BIL, 36 kV PFW. The compartment containing the disconnect switch shall be equipped with a hinged door and a bolted (or lockable) screen over the disconnect to prevent accidental contact.
- *Grounding transformer*: Single-phase, dry-type, self-cooled, epoxy-encapsulated transformer with copper windings. Rated 25 kVA continuous, 84 kVA, 1 min, 12 kV (95 kV BIL) primary and 240/120 V (10 kV BIL) secondary, Class 220 insulation system with a temperature rise not to exceed 115°C. Connect the HV winding to the load side of the disconnect switch and the LV side to the resistor.
- *Grounding resistor*: Made of corrosion-resistant or stainless steel, insulated for line-to-line voltage, completely protected by a metal screen cage. The current rating shall be 350 A for 1 min, temperature rise not to exceed 760°C for a steel and 510°C for a cast-iron grid.
- *Neutral cubicle*: Floor mounted, made of steel frame with bolted aluminum panels, stainless steel hardware, hinged access door with lock, 50 mm × 6 mm copper ground bus bolted to the frame and ground, built to provide adequate ventilation for the heat dissipated from the equipment.

4.8.5 GROUNDING TRANSFORMER

Two types of grounding transformers, Wye-Delta and zig-zag, are used for this application. These transformers provide a low-impedance path for the zero-sequence current to flow from the point of the fault to the ground and return through its neutral to the faulted phase. The impedance of both types of transformers to normal three-phase current is high, so that under normal operating conditions, only a small amount of magnetizing current flows in the transformer winding.

4.8.6 Zig-Zag Grounding Transformer

Figure 4.18 shows a zig-zag grounding transformer connected to an ungrounded three-phase system. The two windings marked as "a_1" and "a_2" are on the same core leg; likewise, "b_1" and "b_2" are on another leg, and "c_1" and "c_2" are on the third core leg. All windings have the same number of turns, but each pair of windings on a leg is wound in opposite directions. The ground-fault current distribution for the fault on phase "A" is shown.

The short time rating of the zig-zag grounding transformer is equal to line-to-line voltage multiplied by the rated neutral current. The transformer is designed to carry its rated current for a short time only, such as 10 s or 1 min. The continuous current shall be at least 10 or 21% of the short-time 10 s or 60 s current rating, respectively.

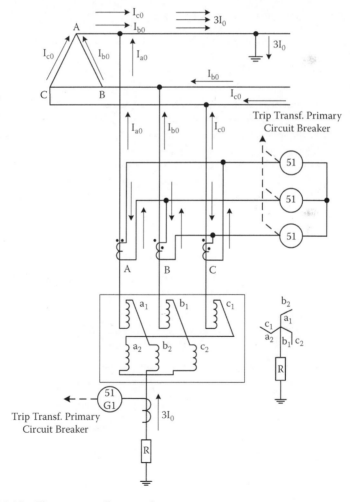

FIGURE 4.18 Zig-zag grounding transformer

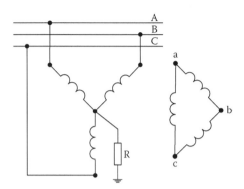

FIGURE 4.19 Wye-Delta grounding transformer for MV system

4.8.7 WYE-DELTA GROUNDING TRANSFORMER

Figure 4.19 shows a Wye-Delta grounding transformer connected to an ungrounded three-phase medium-voltage system. The transformer may or may not be used to serve other loads. The pattern of ground-current distribution is similar to the zig-zag grounding transformer. The short time current rating of the transformer should be checked with the manufacturer.

There is no direct relationship to determine the short time rating shown on the nameplate. This scheme is more economical on 480 V and 600 V systems and can provide enough resistive current to prevent transient overvoltage. A typical scheme for a LV system with three single-phase voltage transformers (VTs) in Wye–open-corner-Delta configuration is shown in fig. 4.20. An example for sizing the equipment for a LV system is provided.

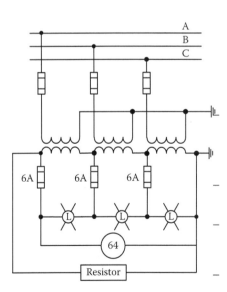

Example: LV system (see fig. 4.20)
A 600 V VT with an accuracy of 0.6
B has a thermal rating of 350 VA
Maximum current =

$$\frac{350 VA}{120 V} = 2.91\ A$$

- Limit the load (resistor) current to 2.0 A, resistor ohms = 100

- Maximum voltage across open-corner Delta = $120 \times \sqrt{3} = 208$ V

- Resistor current = $\frac{208}{100} = 2.08 A$, which is less than 2.91 A

FIGURE 4.20 Wye-Delta grounding transformer for LV system

System Neutral Grounding

- Resistor watts = $\dfrac{(208)^2}{100} = 432$
- Resistor rating = 208 V, 432 W

4.9 SYSTEM CAPACITANCE DATA

Typical values of system-capacitance data extracted from the Westinghouse design guide [2] and the GE Data Book [6], among other sources, are provided here for estimating the system charging current, where C_0 is the zero-sequence capacitance in µf (microfarad) per phase and kV is the line-to-line voltage:

- Capacitive reactance, $X_{C0} = \dfrac{10^6}{\omega \times C_0}$ ohm per phase, $\omega = 2 \times \pi \times f$

- Charging current in Amps at 60 Hz,

$$3I_{C0} = \dfrac{\omega \times C_0 \times kV \times \sqrt{3}}{1000} = 0.652 \times C_0 \times kV$$

4.9.1 SURGE CAPACITORS FOR ROTATING MACHINES

The surge capacitor ratings in microfarad (µf/phase) commonly used for motors and generators, and the calculated values of X_{C0} and $3I_{C0}$ at 60 Hz frequency, are given in table 4.2.

4.9.2 OVERHEAD TRANSMISSION/DISTRIBUTION LINES

For three-phase overhead lines, the sequence capacitance is negligible for the short length of lines used in industrial power distribution systems. The other typical values as per the GE Data Book [6] are:

TABLE 4.2
Surge Capacitors Rotating for Motors and Generators

Rated kV	C_0, µf/phase	X_{C0}, ohm/phase	$3I_{C0}$, A
0.48	1.0	2653	0.313
0.600	1.0	2653	0.391
2.4	0.5	5305	0.782
4.16	0.5	5305	1.356
6.9	0.5	5305	2.249
13.8	0.25	10610	2.249
13.8	0.5	5305	4.498
24.0	0.125	21221	1.956

Source: General Electric Co., *G.E. Industrial Power Systems Data Book*, Schenectady, NY: General Electric, 1964.

- $C_0 \cong 0.01$ µf /mile
- $C_1 = C_2 \cong 1.6\ C_0$ µf/mile

The corresponding 60 Hz susceptances are:

- $b_0 \cong 3.77$ µmho / mile
- $b_1 = b_2 \cong 6.03$ µmho/mile

- Line charging current $Ic_1 \cong 1.14 \times \left(\dfrac{L \times kV}{1000 \times \sqrt{3}} \right)$

- Line three-phase charging kvar = $\sqrt{3} \times kV \times I_{C1} \cong 1.14 \times \left(\dfrac{L \times kV^2}{1000} \right)$

where
  L = line length/1000 ft
  kV = line-to-line operating kV

4.9.3 Power Cable

Typical power cable data from the Westinghouse guide [2] is as follows:

- Nonshielded cable in trays: 0.02–0.05 µf/1000 ft, typically 0.03 µf/1000 ft
- Nonshielded cable in conduit or Teck type: 0.02–0.06 µf/1000 ft, typically 0.04 µf/1000 ft
- Shielded cable, thermoplastic (PVC), or thermoset (XLPE, EPR, PE) type, based on a dielectric constant of 3.3, are given in table 4.3.

TABLE 4.3
Zero-Sequence Capacitance Data for Shielded Power Cable

Conductor Size (AWG or KCmil)	Capacitance, µf/1000 ft			
	5 kV	15 kV	23kV	34.5 kV
2	0.0993	0.0401		
1	0.1096	0.0456	0.0345	
1/0	0.1202	0.0524	0.0388	
2/0	0.1321	0.0607	0.0441	0.0319
4/0	0.1600	0.0715	0.0546	0.0404
250	0.1598	0.0776	0.0588	0.0460
350	0.1846	0.0844	0.0636	0.0496
500	0.215	0.0920	0.0685	0.0532
750	0.241	0.0981	0.0743	0.0573
1000	0.274	0.1118	0.0789	0.0617

4.9.4 Power Transformers

Typically, industrial system power transformers can be assumed as 0.004 µf/phase, also from the Westinghouse guide [2], but can often be considered as negligible.

4.9.5 Outdoor Apparatus Bushings

Zero-sequence capacitance (C_0) in units of µf/ft is given in table 4.4

TABLE 4.4
Zero-Sequence Capacitance Data for Transformer Bushings (µf/ft)

Voltage, kV	Range	Typical Value
15	160–220 pf	200 pf
23	190–450 pf	300 pf
34.5	150–620 pf	400 pf

Source: Westinghouse, "System Neutral Grounding and Ground Fault Protection," publication PRSC-4B-1979, Westinghouse, 1979.

4.9.6 Isolated Phase, Segregated, and Nonsegregated Bus

Zero-sequence capacitance (C_0) in units of µf/ft is given in table 4.5.

TABLE 4.5
Zero-Sequence Capacitance Data for Bus

	Capacitance, µf/ft			
	Isolated Phase		Segregated	Nonsegregated
Current Rating A	15 kV	23 kV	15 kV	5–15 kV
1200	14.3	12	10	10
2000	14.3	12	10	15
2500	14.3	12		
3000	14.3	12	10	22
3500	14.3	12		
4000	14.3	14	13	29
5000	19	16	15	
6000	19	16	17	

Source: Westinghouse, "System Neutral Grounding and Ground Fault Protection," publication PRSC-4B-1979, Westinghouse, 1979.

4.9.7 Synchronous Motors with Class B Insulation System

The zero-sequence capacitance (C_0) in µf/phase extracted from Appendix C of the GE Data Book [6], is given in table 4.6.

TABLE 4.6
Zero-Sequence Capacitance Data for Synchronous Motors

MVA	kV	C_0, µf/phase			
		1800 rpm	1200 rpm	900 rpm	600 rpm
0.5	4.0	0.07	0.0078	0.01	0.014
1.0	4.0	0.026	0.03	0.035	0.04
2.0	4.0	0.045	0.052	0.064	0.08
3.0	4.0	0.058	0.068	0.074	0.084
5.0	4.0	0.06	0.10	0.12	0.15
10.0	13.8	0.08	0.09	0.095	0.10
20.0	13.8	0.10	0.13	0.15	0.18
30.0	13.8	0.12	0.16	0.19	0.23

Note: The multiplier for voltage ratings other than 4.0 kV and 13.8 kV is 0.45 for 13.2 kV; 0.75 for 6.6 kV, and 1.3 for 2.3 kV.

Source: General Electric Co., *G.E. Industrial Power Systems Data Book*, Schenectady, NY: General Electric, 1964.

4.9.8 Air-Cooled Generators

Zero-sequence capacitance (C_0) in µf/phase is given in table 4.7.

TABLE 4.7
Zero-Sequence Capacitance Data for Air-Cooled Generators (µf/phase)

Rating, MVA	kV	Turbo		Salient Pole		
		3600 rpm	1800 rpm	1800 rpm	1200 rpm	600 rpm
10.0	13.8	0.06	0.1	0.07	0.09	0.10
20.0	13.8	0.10	0.12	0.09	0.12	0.18
30.0	13.8	0.13	0.15	0.12	0.15	0.23
40.0	13.8	0.18	0.23	0.14	0.17	0.26
50.0	13.8	0.2	0.28	0.16	0.2	0.3

4.9.9 Induction Motors

Zero-sequence capacitance (C_0) in µf/phase is given in table 4.8.

TABLE 4.8
Zero-Sequence Capacitance Data for Induction Motors (µf/phase)

Rating, kW	kV	1800 rpm	1200 rpm	900 rpm	600 rpm
225	4.0	0.009	0.012	0.014	0.016
300	4.0	0.01	0.013	0.015	0.017
350	4.0	0.012	0.014	0.016	0.018
400	4.0	0.014	0.016	0.017	0.021

Source: Westinghouse, "System Neutral Grounding and Ground Fault Protection," publication PRSC-4B-1979, Westinghouse, 1979.

REFERENCES

Standards
S1. IEEE 142-1991, IEEE Recommended Practice for Grounding of Industrial and Commercial Power Systems, IEEE Green Book, 1991.
S2. ANSI/IEEE C62.92.2-1989, IEEE Guide for the Application of Neutral Grounding in Electrical Utility Systems, Part II: Grounding of Synchronous Generator Systems, 1989.
S3. IEEE 32-1972, Requirement, Terminology and Test Procedure for Neutral Grounding Devices, 1972.
S4. ANSI C57.12.01-1998, General Requirements for Dry Type Power Transformers, 1998.
S5. ANSI/IEEE C57.12.91-2001, Test Code for Dry Type Power Transformer, 2001.

BIBLIOGRAPHY

1. J. P. Nelson and P. K. Sen, "High-Resistance Grounding of Low-Voltage Systems: A Standard for the Petroleum and Chemical Industry," *IEEE Trans. IAS* 35 (4): 941–948 (1999).
2. Westinghouse, "System Neutral Grounding and Ground Fault Protection," publication PRSC-4B-1979, Westinghouse, 1979.
3. General Electric Co., "Generator Neutral Grounding," publication GET-1941, Schenectady, NY: General Electric.
4. General Electric Co., "Neutral Grounding of Industrial Power Systems," publication GET-1118F, Schenectady, NY: General Electric.
5. General Electric Co., "High Resistance Grounding System with Method for Locating Faults," publication GET-1354A, Schenectady, NY: General Electric.
6. General Electric Co., *G.E. Industrial Power Systems Data Book*, Schenectady, NY: General Electric, 1964.
7. J. R. Dunki-Jacobs, "The Reality of High-Resistance Grounding," *IEEE Trans. IAS* 13 (5): (1977).
8. D. Stoetzel, W. C. Heinz, R. B. Bennett, and S. C. Cooke, "Power Distribution System for Open Pit Mines," publication GET-2481A, Schenectady, NY: General Electric.
9. A. C. Lordi, "Grounding A-C Mining Equipment." American Institute of Mining and Metallurgical Engineers, 1953.

5 Power Transformers and Reactors

5.1 GENERAL

5.1.1 Basic Construction

The main parts of a transformer are: (a) iron core, magnetic circuit, (b) high-voltage (HV) and low-voltage (LV), or multiple secondary, windings, and (c) a tank for liquid-immersed transformers and an enclosure for dry-type transformers. The basic core and coil configurations are: core type and shell type. In a core-type transformer, the core is surrounded by the winding coils, and in the shell type, the core surrounds the winding coils. The normal design is a three-phase, three-leg core type, while a five-leg core design enables a reduction in transformer height and a higher zero-sequence impedance. Shell-type transformers were developed for very-high-magnitude short-circuit applications such as generator step-up transformers. Figure 5.1 shows three-phase core- and shell-type transformers.

The core is built from thin and cold-rolled grain-oriented steel laminations with low magnetizing losses. Each core lamination is insulated on both sides. The core legs of circular cross section with multiple steps are connected by yokes to form a complete magnetic path. Core clamping arrangements include few or preferably no through bolts, and the core stacking reduces noise. In large transformers, the heat loss in the core is removed by cooling ducts.

Transformer windings are manufactured of copper or aluminum, although the majority of users specify copper. In liquid-filled transformers, the winding terminals are brought out through bushings for connection to bare conductors or insulated power cable. In medium-voltage transformers, rated up to 35 kV, the winding terminals are brought out on the side, and the bushings are enclosed in a terminal box for cable terminations.

5.1.2 Normal (Usual) Service Conditions

Transformers are designed for operation at rated kVA under the following service conditions:

1. Cooling (ambient) air temperature for liquid-immersed or dry-type transformers
   - Average temperature for any 24 h period is not to exceed 30°C.
   - Maximum temperature is not to exceed 40°C.
2. Cooling-water temperature for water-cooled transformers
   - Average water temperature for any 24 h period is not to exceed 25°C.
   - Maximum water temperature is not to exceed 30°C.

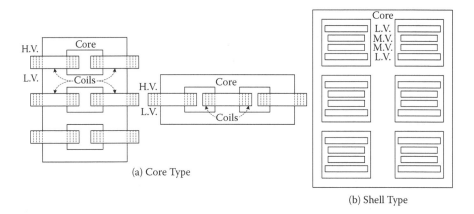

FIGURE 5.1 Core- and shell-type transformer configurations

3. Altitude (elevation)
   - Altitude is not to exceed 1000 m (3300 ft).
4. Load current
   - Load current is approximately sinusoidal.
   - Harmonic factor does not exceed 0.05 per unit.
5. Load power factor
   - Load power factor is 80% or higher.
6. Voltage and frequency
   - Secondary voltage and volts per hertz shall not exceed 110% at no load and 105% at rated load.
   - Frequency is at least 95% of rated value.

5.1.3 Unusual Service Conditions

When service conditions other than the usual (described in section 5.1.2) are present, correction factors need to be applied. The relevant IEEE loading guides are C57.12.91 [S8] for mineral-oil-immersed transformers and C57.96 [S9] for dry-type transformers. Special-duty transformers are discussed in sections 5.5 and 5.6 of this chapter.

5.1.3.1 Altitude Correction Factors

Because the air density decreases with the increase in altitude, the phase-to-phase and phase-to-ground spacings of live terminals need to be increased, and bushings need to have additional length or creep distance. The air spacing cannot be reduced for altitudes less than 1000 m. Dielectric correction factors for liquid-immersed and dry-type transformers are provided in table 5.1, which is extracted from table 1 of IEEE C57.12.00 [S1].

Dry-type transformers may be operated at altitudes above 1000 m, provided that the load is reduced and air clearances are increased. Table 5.2 provides the derating factor for each 100 m (330 ft) altitude above 1000 m and is extracted from table A2 of IEEE C57.96 [S9].

TABLE 5.1
Dielectric Correction Factors for Liquid-Immersed and Dry-Type Transformers for Altitudes Greater than 1000 m (3300 ft)

Altitude (m)	Altitude (ft)	Altitude Correction Factor for Dielectric Strength
1000 or less	3300 or less	1.0
1200	4000	0.98
1500	5000	0.95
1800	6000	0.92
2100	7000	0.89
2400	8000	0.86
2700	9000	0.83
3000	10 000	0.80
3600	12 000	0.75
4200	14 000	0.70
4500	15 000	0.67

Source: IEEE C57.12.00-2000, IEEE Standard General Requirements for Liquid Immersed Distribution, Power, and Regulating Transformers, 2000.

TABLE 5.2
Rated kVA Derating Factors for Each 100 m (330 ft) Altitude above 1000 m at 30°C Average Ambient Temperature

Type of Transformer	Type of Cooling	Derating Factor (%)
Dry-type, self cooled	AA or AN (IEC)	0.3
Dry-type, forced-air-cooled	AA/FA and AFA or AN/AF (IEC)	0.5

Source: IEEE C57.96-1999, IEEE Guide for Loading Dry-Type Distribution and Power Transformers, 1999

5.1.3.2 Unusual Cooling Air and Water Temperature

The transformer loading can be increased for lower ambient temperatures and decreased for higher ambient temperatures, without sacrificing the life expectancy of the transformer. IEEE loading guides are C57.12.91 [S8] for mineral-oil-immersed power transformers and C57.96 [S9] for dry-type power transformers. The approximate increase or decrease of transformer output (kVA rating) can be estimated from the values given in table 5.3, extracted from table 4 of IEEE C57.12.91 [S8] for mineral-oil-immersed transformers, and table 5.4, extracted from table 1 of IEEE C57.96 [S9] for dry-type transformers.

TABLE 5.3
Approximate Decrease or Increase in Loading of Mineral-Oil-Immersed Transformers for Ambient Temperature Different from Standard (Average = 30°C for air, 25°C for water)

	Percent of kVA Rating	
Type of Cooling	Decrease Load for Each °C Higher Temperature	Increase Load for Each °C Lower Temperature
Self-cooled: ONAN	1.5	1.0
Water-cooled: OFWF, ODWF	1.5	1.0
Forced-air-cooled: ONAN/ONAF/ONAF	1.0	0.75
Forced-oil-air-water-cooled: ONAN/ODAF, OFAF, OFWF	1.0	0.75

Source: IEEE C57.12.91-1995, IEEE Guide for Loading Mineral-Oil Immersed Transformers, 1995.

TABLE 5.4
Approximate Decrease or Increase in Loading of Dry-Type Transformers

Type of Unit	Max. Rated Hottest Spot Temp. (°C)	Hottest-Spot Temp. in a 30°C Ambient (°C)	Percent of Rated kVA/°C Increase for Average Ambient Less than 30°C, or Decrease for Ambient Temp. Greater than 30°C (%)
Ventilated self-cooled	150	140	0.57
	180	170	0.43
	220	210	0.35
Sealed self-cooled	150	140	0.65
	180	170	0.49
	220	210	0.4

Source: IEEE C57.96-1999, IEEE Guide for Loading Dry-Type Distribution and Power Transformers, 1999

5.1.4 Transformer Categories and Short-Circuit Withstand

IEEE Standards C57.12.00 [S1] and C57.12.01 [S2] have recognized four categories for mineral-oil-immersed transformers and three categories for dry-type transformers based on their size and configuration. These categories are defined in table 5.5.

Transformers are designed and constructed to withstand the thermal and mechanical stresses produced by the external short circuits to a level that is specified in IEEE C57.12.00 [S1] for mineral-oil-immersed transformers and C57.12.59 [S4]

TABLE 5.5
Transformer Categories

Transformer Type	Category	Single Phase (kVA)	Three Phase (kVA)
Mineral oil or dry	I	1–500	15–500
Mineral oil or dry	II	501–1667	501–5000
Mineral oil or dry	III	1668–10,000	5001–30,000
Mineral oil	IV	above 10,000	above 30,000

for dry-type transformers. The duration of short circuit and short-circuit withstand for different categories is covered in IEEE C57.109 [S13] for oil-immersed and IEEE C57.12.59 [S4] for dry type. These withstand characteristics, also known as damage curves, are discussed further and illustrated in chapter 11, section 11.3.5.

5.1.5 Short-Circuit Impedance or Impedance Voltage (Z_T)

The short-circuit impedance that is derived from the impedance voltage, as measured during the transformer test series, is the voltage measured across the HV terminals of the winding when the LV winding terminals are short circuited while carrying the rated current. This voltage is expressed as a percentage of the rated voltage of the HV winding, but it can also be given as a per-unit (P.U.) value. This value is used to calculate the short-circuit current for faults fed through the transformer. The impedance, reactance, and resistance are all measured during the same short-circuit test.

Standard impedance values based on the impulse level (BIL) are given in section 5.2.3 (table 5.6) for oil-immersed transformers and in section 5.4.5 (table 5.14) for dry-type transformers. The impedance of autotransformers can be much lower than for two- or three-winding transformers, depending on the winding configuration and the short-circuit duty of the Delta tertiary. The tolerances for the variation of impedance as given in the standards are:

- ±7.5% for two-winding transformers
- ±10% for three-winding transformers

5.1.6 Vector Group, Angular Displacement, and Terminal Markings

In North American standards (IEEE, CSA), the reference vector is shown at 8 o'clock as related to an analog clock face. For a two-winding Wye-Delta or Delta-Wye transformer, the low voltage lags the high voltage by 30° (i.e., at 9 o'clock), based on the standard counterclockwise phase rotation. Other vector groups, if required for paralleling, need to be specified specifically. Angular displacement and terminal markings for three-phase Delta-Wye and Wye-Delta transformers are shown in fig. 5.2.

The bushing arrangement for oil-immersed transformers built to IEEE and IEC standards is shown in fig. 5.3.

In IEC standards, any transformer vector group can be defined so that vectors can lead or lag and can be reversed by 180°. This is defined by a clock vector system, also based on the analog clock face hour display, where the reference vector is

TABLE 5.6
Impedance Voltage of Oil-Immersed Power Transformers, ONAN Rating (%)

High-Voltage BIL (kV)	Without Load Tap Changing		With Load Tap Changing
	Low Voltage, 480 V	Low Voltage, 2400 V and above	Low Voltage, 2400 V and above
60–110 (<5000 kVA)	5.75	5.5	...
150 (≥5000 kVA)	6.75	6.5	7.0
200	7.25	7.0	7.5
250	7.75	7.5	8.0
350	...	8.0	8.5
450	...	8.5	9.0
550	...	9.0	9.5
650	...	9.5	10.0
750	...	10.0	10.5

Source: ANSI C57.12.10-1997, American National Standard for Transformers — 230 kV and below 833/958 through 8333/10 417 kVA Single-Phase; and 3750/862 through 60 000/80 000/100 000 kVA without Load Tap Changing; and 3750/4687 through 60 000/80 000/100 000 kVA with Load Tap Changing: Safety Requirements, 1997.

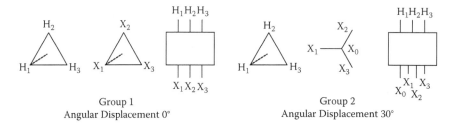

FIGURE 5.2 Angular displacement and terminal markings

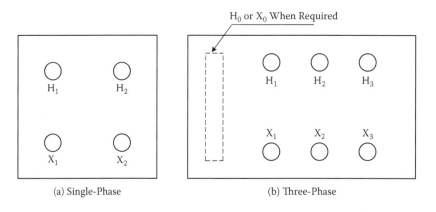

FIGURE 5.3 Bushing arrangement for oil-immersed transformers

Power Transformers and Reactors

at 12 o'clock on the clock face representation, and a 30° phase shift is represented by one hour. This number is prefixed by a letter, the first being in capitals to define the winding configuration of the primary winding, while the second, or more for multi-secondary windings, is in lower case to represent the winding configuration of the secondary winding(s). The most commonly used letters are: Y and y for Wye, D and d for Delta, and Z for zig-zag winding arrangements. For example, the two North American transformers Delta-Wye and Wye-Delta would be Dy 1 and Yd 1, respectively. A typical subset of the available IEC vector groups for transformers is shown in fig. 5.4.

Although the phase representation in fig 5.4 shows the phase arrangements as A, B, C, etc., the actual operating phases, usually designated R, S, T, can be connected to different phase connections on the transformer itself, so if the phase rise order is changed, and a similar rearrangement is made for the secondary connections, it is possible to connect a Yd 1 transformer to give an effective Yd 11 connection, and vice versa.

5.1.7 Taps or Tap Changer

Taps are provided on a transformer to be able to adjust the voltage ratio of the transformer. These taps are provided along the winding (those associated coils are known as the regulating winding) with connections to a tap-changing device that makes the physical change in the in-service tap. The tap-changing device can be "Off-Circuit" or "On Load" type and is usually placed on the primary winding to minimize the current to be switched. To maintain constant voltage on the secondary, these taps compensate for the variation in the primary voltage and voltage drop in the transformer under load. When the primary voltage is low, the tap changer reduces correspondingly the number of primary turns to maintain the secondary voltage constant.

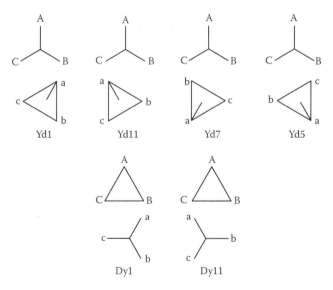

FIGURE 5.4 IEC vector group transformers

5.1.7.1 Off-Circuit Taps

These are used with liquid-immersed and dry-type transformers in industrial power systems that are not connected directly to the utility power supply. Four full-capacity taps (five positions) on the primary (HV) of step-down transformers are provided in four 2.5% steps, two above and two below normal. The tap-changer mechanism should change the taps on all three phases simultaneously and be operable from ground level, with the tap-changer position visible from ground, and shall have a provision for padlocking. Because it is an "Off-Circuit" tap changer, a nameplate specifying that "the transformer must be de-energized before the tap changer mechanism is operated" is required.

5.1.7.2 Load Tap Changer

Load or on-load tap changers (LTC) are usually used with oil-immersed transformers connected to the utility power supply at a voltage level exceeding 34.5 kV. Because the majority of power companies stipulate a voltage variation of ±10% in the power contract, the tap changer is provided with an equivalent range of voltage regulation of ±10% in 16 or 32 steps. A tap changer with 32 steps is preferred, as it provides a 5/8% voltage change in each step. It is usual to have the center tap at the nominal ratio, but in some special cases, where the system voltage has a fixed bias, an off-nominal center tap can be specified.

Two basic methods are in use for changing taps under load: the resistance method and the reactance switching method. In these two methods, the taps can be changed from one tap to another by using either a switched resistor or reactor to bridge the taps that are being transitioned temporarily, during the period while the taps are being switched, so that the load current is not interrupted during the switching process. These bridging features are usually short-time rated, so the switching devices used must be both fast and reliable. Three types of switch are typically used:

1. An arcing switch and a tap selector
2. An arcing tap switch that combines the arcing switch and tap selector
3. A vacuum interrupter

In a resistance-type LTC, switching methods 1 and 2 are used. In a reactance-type LTC, all three methods are used. The performance and test requirements for load tap changers installed in power transformers are detailed in IEEE C57.131 [18]. Further equipment details associated with the tap changer are provided in section 5.2.7.

5.1.8 Parallel Operation

Transformers can operate in parallel without any problems of loading balance, provided that the following conditions are met:

- Phase relations on the HV and LV sides must be the same.
- Voltage ratios must be the same, including the tap positions.
- Impedance of each transformer on its own base must be the same.

In other cases, some current circulation can be expected. Transformers in parallel with equal impedance on their own base permit proportional load sharing. If not equal, the load is divided in inverse proportion to the magnitude of the impedance. For estimating the load division (P_1 and P_2) between two transformers T_1 and T_2, convert the impedance to a common MVA base, say 10:

- $P_1 + P_2 = P$ (total)
- $P_1/P_2 = Z_{T1}/Z_{T2}$ (on 10 MVA$_b$)

Circulating current will flow between the two HV and LV windings, and load division will not be in proportion to its kVA rating if there is a slight difference in any one of the above. A difference of up to 10% in no-load circulating current or in load division is acceptable. A rough estimate can be made by assuming that (a) the impedance is approximately equal to the leakage reactance and (b) transformer load loss is neglected. The no-load circulating current caused by different voltage ratios can be estimated as follows:

- Circulating current = (difference in induced voltage)/(sum of % impedances).
- If the transformers are connected at different taps, then the circulating current = (difference between regulating taps)/($Z_1 + Z_2$).

IEEE C57.105 [S12] provides a more complete example to illustrate the calculations.

5.2 OIL-FILLED (IMMERSED) TRANSFORMERS

5.2.1 Classification of Mineral-Oil-Immersed Transformers

Transformers are classified as power or distribution class. Power transformers are further grouped into Class I and Class II. Class I power transformers include transformers with an HV winding voltage rating up to 69 kV, while Class II transformers include transformers with voltage ratings above 69 kV.

Distribution-class transformers are used in utility distribution systems at voltage ratings up to 69 kV and power ratings up to 500 kVA. The impedance and no-load loss of distribution-class transformers are kept low. For industrial power systems, the transformer duty is such that it is more appropriate for most transformers to be rated as power transformers.

5.2.2 Voltage, Power Rating, and Temperature Rise

Small transformers for the more common voltages used in industrial and utility distribution systems are usually manufactured to a small number of standardized designs and power ratings. This policy allows manufacturers to offer fast delivery and replacement and to avoid the necessity of spending detailed design effort for each of the smaller designs. Design effort is then limited to design enhancements as materials are improved and for specialized applications.

The international standards have recommended the use of the ISO number series, which increases ratings in uneven steps to get the most usage of different sizes. For sizes less than 10 MVA, this range is 100, 125, 160, 200, 250, 315, 400, 500, 630, 800, and 1000 kVA and then repeated for tenths and ten times these values. The equivalent IEEE numbers from C57.12.00 [S1] are 112.5, 150, 225, 300, 500, 750, and 1000 kVA, but for the one-tenth and larger sizes, the number series are different. In both cases, these ratings are "preferred" rather than "standardized," so large distribution and industrial organizations may use a different set of sizes.

The larger and specialized transformers are often designed on an individual or batch basis, as these tend not to have many repeat configurations. Although the larger transformers can be made to any rating, with the exception of specific applications to match generator or rectifier ratings, it is customary to use ratings rounded to the nearest 10s or 100s.

Most mineral-oil-immersed transformers sized for above 100 MVA are rated for a 65°C rise [S16], whereas units smaller than 100 MVA can be rated at 55°C or 65°C rise above ambient temperature [S32].

Transformers with additional cooling features, as described in section 5.2.4, can have multiple stages of cooling to achieve an economic design, especially where the higher ratings are not required on a continuous basis. On this basis, transformers with one additional stage of cooling can usually enhance their base rating by about one-third, whereas those with two stages of cooling have two one-third enhancements.

It should be noted when considering impedance values, as discussed in section 5.2.3, that the impedance of transformers manufactured to North American standards expresses the impedance in per unit or percentage on the base, naturally cooled rating, whereas international practice is to express that on the fully rated, all-cooling-on basis.

5.2.3 Impedance

The preferred standard impedance voltages of three-phase oil-immersed power transformers, as extracted from table 10 of ANSI C57.12.10 [S3], is provided in table 5.6.

5.2.4 Cooling Designation

IEEE, CSA, and IEC standards have harmonized and use the same cooling designations. The cooling method is designated by four letters in combination to define the cooling medium and the method of circulation:

- The first letter is used for internal cooling medium in contact with the winding, **O** for mineral oil or synthetic liquid with fire point less than 300°C, and **K** (or **L** for IEC) for insulating liquid with fire point more than 300°C. These include silicone or high-molecular carbon products. **G** is also included for gas-filled designs.
- The second letter is the circulation mechanism for the internal cooling medium, **N** for natural convection flow through cooling equipment and winding, **F** for forced circulation through cooling equipment, and **D** for

forced circulation through cooling equipment (directed from the cooling equipment to the winding).
- The third letter is for the external cooling medium, **A** for air, and **W** for water.
- The fourth letter is for the circulation mechanism for the external cooling medium, **N** for natural convection, **F** for forced circulation (fan for air cooling, pump for water cooling).

The present and previous IEEE cooling designations are given in table 5.7, which is based on table 2 of IEEE C57.12.00 [S1].

TABLE 5.7
Transformer Cooling Designations

Present Designation	Previous Designation (Obsolete)	Description
ONAN	OA	Oil-immersed, natural circulation, self-cooled
ONAF	FA	Oil-immersed, natural circulation, forced-air cooled
ONAN/ONAF/ONAF	OA/FA/FA	Oil-immersed, self-cooled, plus two stages of fan cooling
ONAN/ONAF/OFAF	OA/FA/FOA	Oil-immersed, self-cooled, plus one stage of fan and one stage with forced-air and forced-oil cooling
ONAN/ODAF	OA/FOA	Oil-immersed, self-cooled, plus one stage of directed fan cooling
ONAN/ODAF/ODAF	OA/FOA/FOA	Oil-immersed, self-cooled, plus two stages of directed fan cooling
OFAF	FOA	Oil-immersed, forced-oil, and forced-air cooled (pump and fans)[a]
OFWF	FOW	Oil-immersed, forced-oil, and forced-water cooled[a]
ODAF	FOA	Oil-immersed, directed forced-oil, and forced-air cooled[a]
ODWF	FOW	Oil-immersed, directed-forced oil, with forced-water cooling[a]

[a] These cooling systems must operate continuously, as such transformers have minimal or no natural cooling rating with the cooling systems out of service. It is therefore necessary to feed these cooling systems from a highly reliable power source or duplicated sources and, where deemed necessary, provide duplicate pumps, fans, and control systems.

Source: IEEE C57.12.00-2000, IEEE Standard General Requirements for Liquid Immersed Distribution, Power, and Regulating Transformers, 2000.

5.2.5 Oil or Liquid Preservation System

A preservation system is required for a liquid-immersed transformer to allow expansion and contraction of the liquid due to the changes in the temperature without exposing the insulating liquid to external contamination. The expansion space is known as "the oil preservation system." Four different designs: (a) sealed tank, (b) gas-oil seal, (c) conservator, and (d) conservator diaphragm are used. Preservation system designs are illustrated in fig. 5.5 and discussed in the following sections, noting that the term "oil" in those sections equally applies to any insulating liquid.

5.2.5.1 Sealed-Tank System

In a sealed-tank system, the transformer is sealed from the atmosphere. The gas or air space above the oil cycles through positive and negative pressures as the oil expands or contracts. When the air or gas space goes to a negative pressure or reduces in pressure due to a sudden change in temperature, the vapor pressure in the oil can cause the generation of gas bubbles. The gas detector or sudden-pressure relay is installed in the air or gas space. This system is used for small transformers, where expansion and the resultant pressure changes are limited.

5.2.5.2 Gas-Oil Seal System

A gas-oil seal system is similar to a sealed-tank system except for the addition of a regulated inert gas (usually nitrogen) supply. The gas supply prevents the

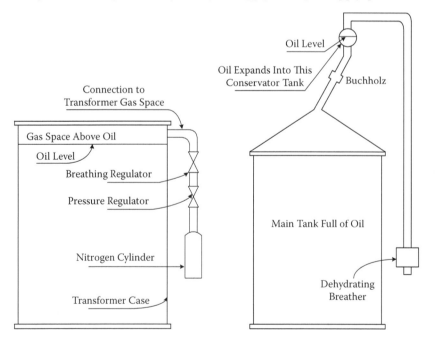

FIGURE 5.5 Oil-preservative system for power transformers

negative pressure and gas bubble formation. The pressure regulator and a relief valve maintain a pressure that will not exceed the allowable tank pressure. The gas-bottle gas-supply system must be monitored and maintained. There is also a version of the sealed-tank design where a space above the oil is permanently filled with an inert gas without a regulated gas supply. Such a design provides some of the advantages of the regulated system without the complexity of the external regulating system.

5.2.5.3 Conservator Design

The conservator is an oil-expansion tank mounted above the highest oil point on the transformer (see fig. 5.6). The transformer tank and the conservator are connected through a pipe, and a gas-detector relay is installed in the pipe. The transformer oil expands and contracts with the increase and decrease of the oil temperature.

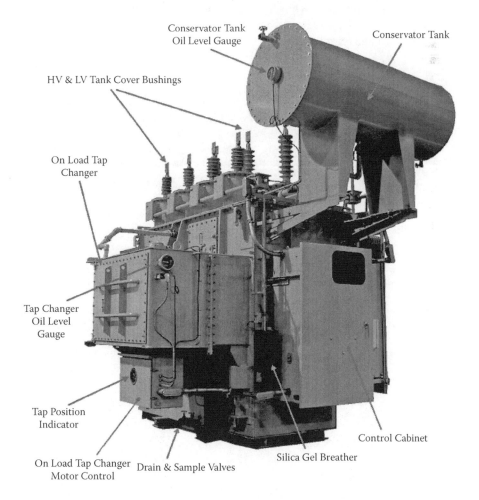

FIGURE 5.6 Conservator design 15/20 MVA 72kV–25kV

The conservator must therefore breathe to the surrounding air, and moisture in the air can be absorbed into the oil and insulation system. To prevent the moisture from entering the conservator, a silica gel breather is provided that will dry the air as the breathing takes place. As the silica gel absorbs the moisture, it will change from blue to pink in color and must be changed to retain its drying capability. Routine maintenance is therefore required, noting that monitorable electrical devices are now available for use where reliable auxiliary power is available.

5.2.5.4 Conservator Diaphragm

This is the same system conservator design, except that a rubber membrane is fitted inside the conservator to separate the air and oil. This can be done with an air bag or oil bag. The membrane is filled with air in the air-bag system and with oil in the oil-bag system. The oil and air do not come in contact with each other and, although the air side must still breathe to atmosphere, the oil has no direct contact with the air. Air drying is no longer critical, but it is common practice to install a silica gel or electric dryer in the breather system.

5.2.6 Winding Insulation Levels

The recommended winding insulation levels, basic impulse (BIL), and power frequency withstand (PFW) at the line terminals of oil-immersed transformers are provided in table 5.8. These values have been extracted from IEEE C57.12.00 [S1].

On effectively grounded systems, a reduced BIL, of one step below full basic insulation level, may be used for voltage ratings of 120 kV and above, provided that an insulation coordination study has been carried out. However, this approach is not recommended for industrial power systems.

TABLE 5.8
Winding Insulation Levels at the Line Terminals

No. System Voltage (kV, rms)	Max. Voltage (kV, rms)	BIL (kV, crest)	Low-Frequency (kV, rms)	Remarks
4.16	5.0	60	15	
13.8	15.0	95	26	
13.8	15.5	110	34	Direct connection to overhead line
25	27.5	150	50	
34.5	38	200	70	
44	48.3	250	95	
69	72.5	350	140	
120	145	650	275	
161	170	750	325	
230	245	900	395	

Source: IEEE C57.12.00-2000, IEEE Standard General Requirements for Liquid Immersed Distribution, Power, and Regulating Transformers, 2000.

Power Transformers and Reactors 113

5.2.7 Tap Changers

5.2.7.1 Tap-Changer Design

The basic principles of tap-changing design are dealt with in section 5.1.7, and the design concepts are also covered in that section. As on-load tap changers are almost exclusively applicable to the larger liquid-filled designs, their construction details are included in this section.

5.2.7.2 Basic Construction Features

The requirement of load-tap-changing equipment is covered in ANSI C57.12.10 [S3]. The tap-changer mechanism, including the diverter switch, shall be located in a separate compartment so that the oil is completely isolated from the transformer tank oil. Accessories provided with the diverter switch compartment are:

- Magnetic oil-level gauge with contacts for alarm
- Fault-pressure relay with contacts for trip and alarm
- Pressure-relief device
- Filling/filter press valve and drain press/filter press valve

When the tap-changer design requires an oil-expansion tank, it shall be piped to a separate compartment in the conservator. A separate tap-changer gas-detector relay shall be located in this pipe.

5.2.7.3 Controls and Indication

The tap-changing control equipment includes:

- Paralleling and control equipment. The preferred scheme for paralleling is the "circulating current method," incorporating an overcurrent relay or a device for lock-out and alarm.
- Automatic voltage-regulating relay with facility for setting any voltage from 110 V to 130 V (based on a VT secondary voltage of 120 V), adjustment of band width, and time delay from 0 to 120 seconds.
- Line-drop compensation equipment, including current transformers to provide voltage control at a point remote from the measuring point.
- Removable handle for manual hand operation, mechanical tap-position indicator with maximum and minimum resettable drag hands, mechanical operations counter, "Manual-Automatic" and "Remote-Local" selector switches, and an "Individual-Parallel" switch.
- Equipment for installation on a remote control panel, such as "Manual-Automatic" switch, "Raise-Lower" push button, and a dial-type position indicator or, alternatively, coil-coil-contact and transducer data interfaces for digital control systems.
- Weatherproof control cabinet, accessible from ground level, with devices such as a door-actuated light and thermostatically controlled heater.

Some designs of voltage and tap-change control logic are rated to be located in the transformer local control cubicle, while others are designed to be installed on a control panel located in a protected environment.

5.2.8 Transformer Bushings

Transformer bushings are covered in IEEE C57.19.00 [S21] and IEEE C57.19.100 [S22]. Transformer bushings are commonly of brown glazed porcelain, but other colors and synthetic types are now also available. Bushings rated 25 kV and above should be of the oil-filled type. Bushings rated 69 kV and above shall be of the condenser type with power factor test taps and, if oil-filled, provided with magnetic oil-level gauges legible from ground level.

The bushings shall have extra creepage distance and be of a higher insulation level than that of the winding, especially for medium-voltage transformers with weatherproof terminal boxes. The recommended insulation levels, which are extracted from table 1 of IEEE C57.19.01 [S29], are provided in table 5.9.

5.2.9 Bushing Current Transformers

Bushing-type current transformers (CT) are usually provided at the line and neutral ends of most large transformers. Each CT shall be removable without removing the transformer covers, but this usually requires the associated bushing to be removed. The secondary leads of the current transformers that are mounted in detachable wells shall be terminated in an oil-tight, weatherproof junction box on the side of the well and wired to the control cabinet. The current transformers have typically been the multiratio type to provide the ratio adjustments required with electromag-

TABLE 5.9
Electrical Insulation Characteristics of Transformer Bushings

Nom. System Voltage (kV, rms)	Max. System Voltage (kV, rms)	BIL (kV, crest)	PWF 1 m (Dry kV, rms)	PWF 10 s (Wet kV, rms)	Min. Creep Distance (mm)
4.16	5.0	75	27	24	
13.8	15.0	110	50	45	280
13.8	15.5	110	50	45	280
25	27.5	150	60	50	430
34.5	38	200	80	75	660
44	48.3	250	105	95	890
69	72.5	350	160	140	1220
120	145	650	310	275	2340
161	170	750	365	315	2900
230	245	900	425	350	3560

Source: IEEE C57.19.01-1991, Performance Characteristics and Dimensions for Outdoor Apparatus Bushings, 1991.

netic protection. However, modern digital relays offer significantly more flexibility in this respect and can correct for ratio mismatches within the relay, which allows single-ratio CTs to be used with a ratio based on 120% of the maximum transformer rating. Standard ratio and accuracy should be provided, as detailed in chapters 6 and 11.

5.2.10 Accessories

The accessories for mineral-oil transformers include:

- The conservator tank shown in fig. 5.7 (for conservator-type transformers) is of adequate capacity for the required ambient temperature range, mounted with a slope between the ends to facilitate draining.
- Dehydrating-type breather for conservator-type transformers containing a color indicator (usually silica gel), piped to the conservator and mounted so as to be accessible from ground level.
- Sudden-pressure relay for sealed-tank design transformers, generally when rated 7.5 MVA and larger. The relay is calibrated for mounting either in the oil or gas space above the oil, and is equipped with a microswitch for alarm and trip.
- Gas-detector relay (known as a Buchholz relay, shown in fig. 5.8) for conservator-type transformers, generally when rated 7.5 MVA and larger. The device is mounted in the pipe between the highest part of the transformer tank and the conservator. The relay is equipped with two sets of contacts, one for alarm upon gas accumulation and one for trip upon oil surge.
- Winding-temperature indicator, responsive to the combination of top oil temperature and winding current. The device is calibrated to follow the hottest spot temperature of the winding. The temperature indicator, for forced-cooled transformers, shall be equipped with adjustable contacts for starting the cooling fans and pumps, while all transformers should be equipped with at least two contacts for alarm and trip. These devices are fitted to the larger designs of transformer and, in

FIGURE 5.7 Conservator tank design

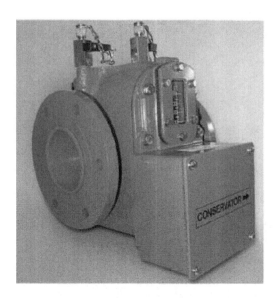

FIGURE 5.8 Buchholz relays for protection of oil-filled transformers

some cases, are applied to reflect, separately, the primary and secondary winding temperatures.
- Oil-temperature indicator with its sensing element located in the path of the hottest oil, and mounted adjacent to the winding temperature indicator. The temperature indicator is equipped with adjustable contacts that are normally used for alarm only, except for smaller transformers, which are not provided with winding-temperature devices, where the top oil temperature is used for tripping.
- Pressure-relief device with sealed contacts for alarm and, with some larger units, for trip.
- Valves with threaded plugs, including valves for oil drain, oil sampling, liquid treatment, and vacuum pulling.
- Metal ladder, which may be attached to the side of the transformer for units rated 20 MVA and larger, but which requires a lockable anti-use plate on the lower part to limit access.
- Lifting lugs for lifting the core and winding assembly from the main tank.
- Two pads for grounding, usually located on opposite sides of the tank.

Although temperature indicators have traditionally been mechanical devices, digital units with digital interfaces suitable for direct input into distributed control systems (DCS) are becoming available.

5.2.11 Cooling Equipment

Cooling equipment consists of radiators, fans for forced-air cooling, and pumps for forced oil- and water-cooled transformers. Radiators shall be of crevice-free

Power Transformers and Reactors 117

construction. Galvanized radiators should be provided for applications where the transformer will be installed in a corrosive environment. For transformers rated 7.5 MVA and larger, detachable radiators with shut-off valves, drain plugs, and vent plugs should be provided, together with lifting eyes to facilitate removal.

Fans and guards for forced-air cooling shall form an integral unit with the individual motor to form a totally enclosed construction. Horizontal mounting of fans at the top or bottom of the radiators should be avoided. Locate all cooler control equipment in the transformer control box.

5.2.12 Nameplate Information

Nameplates made of stainless steel are required with the information listed in table 10 of IEEE C57.12.00 [S1] plus the following:

- Primary system fault level for short circuit withstand
- Impedance at minimum and maximum taps of load-changing transformers

5.2.13 Core Grounding

The transformer core is grounded at one point through a connection to the tank. The core can move during transportation or unloading, which may develop a second ground path at some other location in the core. Current will then circulate between the two ground points and cause heating of the core, while a partial ground can cause local arcing and gas production. To facilitate the testing of core ground, the core ground connection shall be brought out to a bushing (1 kV minimum rating) and connected to the ground through a resistor (minimum rating 250 ohm, 25 watt continuous). The bushing and resistor should be mounted in an oil-tight box.

5.2.14 Transformer Gas Analysis: Dissolved Gases in Transformers

The transformer gas analysis is used to determine the probable condition of the transformer and, by tracking the results over time, is a method of predicting future problems. Gas analysis has therefore become a part of the preventive maintenance programs. For small- and medium-sized transformers, periodic oil samples are sent to laboratories for gas-in-oil analysis. The container used for the analysis is a syringe (not a jar), or a container provided by the laboratory, so that the air or exterior gases do not become a part of (contaminate) the sample.

On-line gas-in-oil monitors are recommended for large power transformers. These monitors extract gases from transformer oil and analyze a portion of the gases every 12 hours.

The detection of certain gases generated in a mineral-oil-filled transformer in service is frequently the first available indication of a possible malfunction that may lead to an eventual failure if not corrected. Arcing, corona discharge, low-energy sparking, severe overloading, and overheating in the insulation system can result in chemical decomposition of the insulating material. This decomposition can result in the formation of various combustible and noncombustible gases. Normal operation or loading may also result in the formation of some gases, but at a slower rate. IEEE C57.104 [S11] applies to mineral-oil-immersed transformers and addresses the following:

- The theory of gas generation in a transformer
- The interpretation of gas analysis
- Various diagnostic techniques, such as key gases and Dornenberg [8] and Rogers [9] ratios
- Instruments for detecting and determining the amount of combustible gases present

Some large power utilities have established norms for the safe values in parts per million (ppm) of dissolved gases. These benchmark or standard values are based on the oil volume and age of the transformer. The norm established by one such power utility is given in table 5.10.

5.3 NONFLAMMABLE LIQUID-FILLED TRANSFORMERS

5.3.1 Overview

These are similar to oil-immersed transformers but use special insulating fluids. The flash and fire point of the insulating liquid is higher than the oil, which permits an indoor installation without the need for fire vaults for voltage ratings up to 35 kV. Typical properties and thermal characteristics of liquids approved for use in such transformers are provided in table 5.11.

The different types of less inflammable liquid filled transformers are discussed in the following sections.

5.3.2 Askeral-Immersed Transformers

Askeral liquid, now known as PCB (polychlorinated biphenyl), was used for over 60 years. It has high dielectric strength and is fire resistant. However, the products

TABLE 5.10
Safe Values of Dissolved Gases in Power Transformers and Reactors (ppm)

Gas	Transformer Age (years)				
	0–3	3–6	6–12	12–15	>15
Hydrogen: H_2	110	150	250	500	500
Methane: CH_4	40	100	100	100	150
Ethane: C_2H_6	50	75	75	100	100
Ethylene: C_2H_4	50	125	150	150	150
Acetylene: C_2H_2	30	60	150	150	150
Carbon monoxide: CO	1,000	1,000	1,000	1,000	1,500
Carbon dioxide: CO_2	5,000	10,000	10,000	10,000	12,000

Note: Oil volume = 10,000 gallons. For a different volume of oil, the actual ppm = (given ppm × actual volume)/10,000.

TABLE 5.11
Typical Properties of Transformer Liquids

Property	Transformer Mineral Oil	Hydrocarbon, High Fire Point	Silicone(Dow Corning 561)
Dielectric strength at 25°C, kV	40	40	40
Dielectric strength	2.2	2.4	2.7
Viscosity:			
at 25°C	16	800	50
at 50°C	8	150	32
Specific gravity at 25°C	0.875	0.833	0.96
Flash point, °C	160	285	300
Fire point °C	170	312	350
Heat release rate, kW/M^2 (convective)	906	570	53

of decomposition have been found to be harmful, and its use has now been banned. Any transformers containing PCBs in North America should have been identified and replaced or decontaminated.

5.3.3 SILICONE-IMMERSED TRANSFORMERS

The silicone fluid (polidimethy siloxane) is a PCB replacement for less flammable transformers. Silicone-filled transformers have been manufactured since the early 1970s and are available in ratings up to 10 MVA, 34.5 kV, and 65°C rise. Silicone liquid is environmentally safe, reliable, and needs very little maintenance. It is specified in ASTM D4652-05 [S33].

5.3.4 OTHER AVAILABLE TRANSFORMER TYPES

Prototype transformers have been produced using SF_6 insulating gas, but these are not yet in common use. Another development that is gaining popularity is the use of vegetable oils for transformer insulation. The vegetable-oil-based insulating fluids have the advantage of being environmentally attractive and will be easier to dispose of and still provide a low fire risk.

5.4 DRY-TYPE TRANSFORMERS

In industrial plants, dry-type transformers are used for lighting, unit substations, and variable-frequency drives.

5.4.1 TYPE AND RATING

Dry-type transformers are available in ratings of up to 35 MVA and voltage ratings up to 44 kV at 250 kV BIL. Three different types of construction are available:

- Vacuum-pressure impregnation in polyester or silicone varnish. This is a less expensive option, needs a clean environment such as a pressurized electrical room, and is mechanically weak against through-fault current.
- Cast coil or cast resin. In this type of construction, the winding coils are cast in a resin that makes it mechanically short-circuit proof against through-secondary through-fault currents. The transformer can be installed in a harsh environment.
- VPI epoxy sealed or encapsulated. In this type of construction, the winding coils are sealed with epoxy and produce transformers that are mechanically strong against through-fault currents.

5.4.2 Insulation Class and Temperature Rise

Two insulation systems, Class 180 (F) and Class 220 (H), are available for dry-type transformers. VPI epoxy and polyester/silicone are available with a Class 220 insulation system, whereas cast coil is available with Class 185 insulation. Although standards allow a temperature rise of 115°C for Class 185 and 150°C for Class 220 insulation systems, an 80°C rise is recommended for both. A lower (80°C) rise reduces the load loss, extends the life of the transformer, and permits a 30% increase in power rating during emergency operations, such as a transformer outage in a secondary selective system. Insulation class and temperature rise recommendations for dry-type transformers are provided in table 5.12.

5.4.3 Primary System Fault Level for Short-Circuit Withstand

The current standards specify a primary system fault level of up to 1000 MVA for the design of dry-type transformers. However, to provide a better margin to prevent through-fault damage to the transformer, the following recommendations are made for the assumed system fault level at the various system voltages where such transformers are usually used:

- 4.16 kV system: 500 MVA
- 13.8 kV system: 1500 MVA
- 25–34.5 kV system: 2500 MVA
- 44 kV system: 3500 MVA

TABLE 5.12
Insulation Class and Temperature Rise for Dry-Type Transformers

Insulation Class	Ambient Temp. AV/Max. (°C)	Average Winding Temp. Rise (°C)	Winding Hottest Temp. Rise (°C)	Permitted Winding Temp. (°C)	Recommended Temp. Rise (°C)
180 (F)	30/40	115	140	180	80
185 (H)	30/40	115	145	185	80
220 (>H)	30/40	150	180	220	80

5.4.4 Insulation Level

The standard available options and recommended insulation levels (BIL) for dry-type transformers are provided in table 5.13.

5.4.5 Impedance

The standard and recommended percent impedance for dry-type transformers at the self-cooled rating is provided in table 5.14.

5.4.6 Tap Changer or Taps

Off-circuit taps are normally provided with four 2.5% full-capacity taps in the high-voltage winding, with two above and two below the rated voltage.

TABLE 5.13
Insulation Levels (BIL) for Dry-Type Transformers

Voltage Class (kV, rms)	Standard (kV, crest)	Options (kV, crest)	Recommended (kV, crest)
1.2	10	20, 30	20
5.0	30	45, 60	60
15	60	95, 110	95
25	110	125, 150	150
34.5	150	125, 200	150

TABLE 5.14
Percent Impedance for Dry-Type Transformers at Self-Cooled Rating

Power Rating (kVA)	15 kV Class 95 kV BIL and Below		25 kV Class 125 kV BIL	34.5 kV Class 150 kV BIL
	NEMA	Suggested	Suggested	Suggested
300	4.50	4.50		
301–500	5.75	5.50	6.00	6.50
501–1,500	5.75	6.00	6.00	6.50
2,000	5.75	6.00	6.00	6.50
2,500	5.75	6.00	6.50	6.50
3,000		6.50	6.50	7.00
3,750		6.50	7.00	7.00
5,000		6.50	7.00	7.00
7,500–10,000		7.00	7.00	7.50

Note: The standard tolerance is ±7.5% of the design impedance for two-winding and ±10% for three-winding dry-type transformers.

5.4.7 Accessories

The accessories provided with the transformer are:

- Provision for lifting, jacking, and skidding in both directions
- Three metal oxide surge arrestors on the HV winding
- Dial-type or digital-display-type temperature monitor with adjustable contacts for alarm and trip
- A ground bus run through the transformer enclosure and connected to non-current-carrying parts

5.4.8 Surge Arresters for Dry-Type Transformers

The insulation level of dry-type transformers used in industrial plants is the same as that of similar mineral-oil-immersed transformers. Even though dry-type transformers are less likely to be exposed to atmospherically induced surges, it is good engineering practice to install metal oxide surge arrestors on the primary. The recommended arrestors are station class, 100% rated, and should be installed by the manufacturer of the transformer.

5.5 TRANSFORMERS FOR NONLINEAR LOADS

The harmonics generated by nonlinear loads such as variable-frequency drives (VFD) will impose nonsinusoidal current on the power transformers that supply such loads, resulting in a substantial increase in losses and temperature rise. With the addition of harmonic currents, standard design transformers must be derated to limit the temperature rise to be within the insulation temperature-rise rating or the transformer needs to be replaced with a special "K"-rated transformer.

5.5.1 Standard Design

Transformers are designed to operate satisfactorily under the following conditions:

- Approximate sinusoidal and balanced voltage
- Maximum voltage at no load = 110%
- Maximum voltage at full load = 105%
- Load current with maximum harmonic distortion of 5.0%

5.5.2 Transformer Losses

Transformer load loss is given by the following equation:

$$\text{Load loss} = (I^2 \times R) + P_{ec} + P_{osl}$$

where
- I = rms load current
- R = DC resistance
- P_{ec} = winding eddy current loss
- P_{osl} = other stray losses

Power Transformers and Reactors   123

With the introduction of harmonic currents, the rms value of the load current will increase, and hence an increase in copper loss and winding eddy-current loss will occur. Higher-order (frequency) harmonic currents will also increase the AC resistance due to the skin effect in winding conductors, thus increasing the load loss and temperature rise.

Harmonic load currents are frequently accompanied by a DC component in the load current. The DC component of the load current will increase the magnetizing current and often increase the audible sound level substantially.

5.5.3 K-Rated Transformers

The "K" factor has been established by Underwriter Laboratories (UL) to define the ability of a transformer to serve varying degrees of nonlinear load current without exceeding the rated temperature rise. The "K" factor is based on the predicted losses as specified in ANSI/IEEE C57.110 [S14]. The K factor can be calculated from the equation:

$$K \text{ factor} = \Sigma\ I_h^2(PU) \times h^2$$

where I_h = rms current at harmonic order (h) per unit.

The features of K-rated transformers are:

- The core has a larger cross section to compensate for increased flux density.
- Delta primary winding utilizes a heavier conductor due to increased heating from circulating triplen harmonics.
- Secondary winding uses small parallel conductors to minimize the skin effect.
- An electrostatic shield is installed between the core and LV winding, and between the HV and LV windings.
- The transformer can operate at more than 10% system voltage without core saturation.
- A double-size neutral bar and lug pad are installed.

For standard transformers, K = 1 and 1.5 for single- and three-phase transformers, respectively. For nonlinear load applications, the standard K factors are 4, 9, 13, 20, 40, and 50. An example for calculating a nonlinear load K factor is given in table 5.15, based on table 4.5 of IEEE 1100 [S30]. The *J&P Transformer Book* [3] has also provided the following guidelines for estimating the K factor if the data on harmonic current is not available:

- K1.5 when the nonlinear load is about 15% of the transformer bank rating
- K4 when the nonlinear load is about 35% of transformer bank rating
- K13 when the nonlinear load is about 75% of transformer bank rating
- K20 when the nonlinear load is about 100% of transformer bank rating

TABLE 5.15
Example Calculation of a Nonlinear Load "K" Factor

Harmonic Number, h	Nonlinear Load Current, I^h (%)	I_h^2	i_h [a]	i_h^2	$i_h^2 \cdot h^2$
1	100	1.000	0.909	0.826	0.826
3	33	0.111	0.303	0.092	0.826
5	20	0.040	0.182	0.033	0.826
7	14	0.020	0.130	0.017	0.826
9	11	0.012	0.101	0.010	0.826
11	9	0.008	0.083	0.007	0.826
13	8	0.006	0.070	0.005	0.826
15	7	0.004	0.061	0.004	0.826
17	6	0.003	0.053	0.003	0.826
19	5	0.003	0.048	0.002	0.826
21	5	0.002	0.043	0.002	0.826
Total	...	1.211	...	1.000	9.083

K factor = 9.086

[a] $i_h = \dfrac{I_h}{(\sum I_h^2)^{1/2}}$

Source: IEEE 1100-1992, IEEE Recommended Practice for Powering and Grounding Electronic Equipment, 1992.

5.6 GENERATOR STEP-UP AND OTHER SPECIAL TRANSFORMERS

Traditionally, two-winding transformers used in industrial systems in North America have a Delta primary and a grounded Wye secondary to limit the effect of any ground fault in the feed on the upstream system's ground protections. For generator transformers, there are significant problems with having Delta (ungrounded) high-voltage transformers when they are used to connect synchronous generators to the system. The main problems occur if a line-to-ground fault on the upstream system causes the upstream breaker(s) to trip while the transformer remains energized from the generator. This can overstress the insulation of the connected equipment (surge arresters, VTs, etc.) and, if left so connected for an extended period, can lead to their failure.

It is possible to use open-Delta-type ground detection protection to trip the generator transformer. It is preferable for the generator transformer primary winding to be Wye connected, with the neutral solidly grounded when generation is connected. This connection will prevent overvoltages caused by line-to-ground faults. It also facilitates the detection of primary ground faults, which can then be isolated by tripping the generator (see fig. 5.9).

Power Transformers and Reactors

FIGURE 5.9 Generator step-up transformer

Although other configurations can be used, with suitable precautions, this is the preferred arrangement. The provision of a Delta secondary on the generator side also permits the application of HR grounding for the generator and a path to circulate the triplen harmonics that can be produced by the generator, and hence limit their injection into the general power system.

It is preferable to always use a dedicated generator transformer for each generator. Although it is possible to connect more than one generator and possibly some local load to the same transformer, there will be some limitations and higher damage possibilities than with a separate generator transformer.

Generator transformers are configured similarly to the other types of liquid-insulation transformers discussed earlier, but they have additional features to match the specific duty as a generator transformer. One major risk is the possibility of a generator being paralleled onto the system out of synchronism. Although this should be prevented by the synchronizing equipment, failures do occur, and major mechanical stresses are then reflected onto the transformer. Another main risk is that of the generator losing synchronism with the main system. In such a case, the generator will slip, and the pulsating currents produced will add to the mechanical stress in the transformer. There is also the risk that a generator will trip on full load and that the load rejection will cause an overfrequency and voltage transients, and that will also overstress the transformer. These special requirements are discussed further in the IEEE guide for generator transformers [S17], which notes that the following should be checked:

- The highest voltage on the high-voltage winding
- The transformer impedance requirements
- The optimal transformation ratio
- The reactive power transfer requirements
- The lowest voltage on the high-voltage winding
- The winding current levels required
- The overexcitation risk and temporary overvoltage withstand requirements

The enhanced mechanical stresses are one of the reasons why generator transformers often are constructed as shell type to more easily withstand such stresses. Similar special transformer designs include those designed for electric furnaces, where more frequent faults can occur than for normal power supply transformers and designs need to be short-circuit tested. Also, large rectifier transformers [S20] combine these risks with high harmonic levels and, sometimes, with the provision of filters on the rectifier transformer's tertiary winding. Aluminum smelters also require wide voltage ranges, and regulating transformers are often provided upstream of the associated rectifier transformers. These regulating transformers will often have off-line primary taps and wide-range secondary taps to cover pot-line start-up requirements.

In all of the above special cases, the transformer basic specifications need to be enhanced to provide the special performance and operating duties of that particular process and the purchase specifications adjusted accordingly. Although many such industries have developed good-practice designs, it is important to investigate any process to identify any special requirements that may be needed in the application of transformers to that specific situation.

5.7 INSTALLATION OF OIL-FILLED TRANSFORMERS AND REACTORS

5.7.1 GENERAL

Transformers and reactors that are configured to use liquids for insulation and cooling, which includes most large and high-voltage applications, will in most cases use the more flammable mineral oil. This entails significant fire, explosion, and environmental risks that need to be minimized. The use of fast protective systems can assist, but these transformers also require physical safeguards for the worst-case scenario of a fire or for the explosive ejection of oil occurring.

To minimize the risks entailed by the close proximity of such transformers to other transformers and other buildings, regulations have been prepared by the following:

- Electrical safety codes that include both the CEC and NEC codes and the various state, provincial, and municipal regulations and codes of practice
- Factory Mutual requirements and other codes that reflect the concerns of the groups that ensure electrical systems and the consequential damage caused by a serious event

5.7.2 INDOOR (INSIDE BUILDINGS)

These are usually relatively small transformers, but they need to be installed correctly to minimize the risks to adjacent systems equipment and operating staff as follows:

- An electrical equipment vault is required if the oil content exceeds 15 gallons.

- In Canada, such a vault must meet the National Building Code of Canada, and similar requirements will exist in other jurisdictions.
- The vault must meet a 2½-hour fire rating (minimum) to prevent the problem spreading outside the vault within that time.
- Automatic fire protection is required, and such fire protection will need to be designed specifically to handle oil-type fires efficiently.

5.7.3 Outdoor Installation

For outdoor installation, transformers must be located at the distance specified from buildings and between themselves to minimize the risk of a fire spreading to adjacent buildings or structures. For high-voltage large transformers, specially built fire walls are often used to separate transformers from adjacent structures or equipment, which allows closer spacing without increasing the risk of fire spread. The other factors that must be considered are:

- Prevent spilled oil from polluting streams
- Install fire-protection fill and an oil-retention system

To meet the above criteria, there are now very effective designs of oil-retention systems that will contain the oil that leaks from any transformer and prevent that oil from being discharged into the adjacent areas. It will also, by the use of pits around the transformer filled with crushed stone, or partly filled with stone over a grillage, ensure fire extinction of any such released oil as it flows through these stones. The oil is then directed into a storage reservoir. The oil retention systems are also designed to allow any water buildup, be that from rain or from fire-fighting operations, to be discharged without carrying over any significant quantity of leaked oil. The whole installation must also meet all local fire precaution codes, also taking note of the IEEE guide [S24] and the NFPA recommended practice [S25], as applicable.

5.8 REACTORS

Reactors for power systems in general use and for industrial use in particular exist as two distinct functional types — series reactors for fault-current limitation and shunt reactors, either for reactive compensation — and as two construction types: air-core air-insulated types and liquid-insulated iron-core types. Reactors, usually of the air-core type, are also used in combination with shunt capacitors to form harmonic filters or to detune reactive compensation capacitor banks.

A significant number of applications for reactors in the industrial environment now use the air-core design that is air insulated and hence requires air clearance and installation requirements similar to those of outdoor switchgear. For high-voltage and large-sized reactors and where space for the outdoor air-insulated design does not exist, the oil-filled design is still used.

Oil-filled designs, whether for shunt or series duty, present many of the design factors that are found in oil-filled transformers, and hence many of the standard requirements for such transformers also apply, and many of the standards quoted earlier in this chapter equally apply to these reactors.

The main differences between the requirements for reactors and transformers is that the shunt reactors do not have load-related characteristics, and hence their magnetizing or iron losses dominate, whereas the series type have no shunt losses, and the losses are all load- or copper-loss related. Also, as the function of a reactor is specifically related to its actual reactance value, this is of more significance than it is for a transformer. A typical series reactor manufacturer's test definition is provided in table 5.16.

TABLE 5.16
Typical Series-Reactor Factory Test Schedule

Current-Limiting Reactor Tests	Kind of Tests		
	R	O	D
1. Resistance	♦		
2. Impedance and load loss	♦		
3. Applied potential	♦		
4. Potential between turns	♦		
5. Temperature rise: rated I		♦	
6. Mechanical strength		♦	
7. Impulse		♦	

Note: R = routine (production), D = design, O = other.

5.9 INSPECTION AND TESTING

5.9.1 Factory Routine and Type Tests

To ensure good in-service reliability, transformers and reactors need to be tested before delivery. These tests fall into two categories: the design and type tests, which should be performed for every new design, and routine tests that should be performed on every unit manufactured. The majority of these tests are defined in the relevant standards, and an example of the tests required are detailed in ANSI/IEEE C57.12.00 [S1] and reproduced here in table 5.17.

Additional tests may be specified, particularly if special duties are envisaged. The type and design tests usually do not apply to the more common sizes of industrial transformers, but the purchaser should always check that such tests have been performed and that no significant deviations from the tested designs are involved.

The higher-voltage and special-service transformers tend to be designed specifically for each order, and hence both type and design checks are necessary, at least for the first unit produced.

TABLE 5.17
Routine, Design, and Other Tests for Liquid-Immersed Transformers

Tests	500 kVA and Smaller			501 kVA and Larger		
	Routine	Design	Other	Routine	Design	Other
Resistance measurements of all windings on the rated voltage tap and the tap extremes of the first unit made on one new design		♦		♦		
Ratio tests on the rated voltage connection and on all tap connections (for LTC units)	♦			♦		
Polarity and phase relation tests on the rated voltage connection	♦			♦		
No-load losses and excitation current at rated voltage and frequency on the rated voltage connection	♦			♦		
Impedance voltage and load loss at rated current and rated frequency on the rated voltage connection and at the tap extremes of the first unit of a new design (for LTC units)		♦		♦		
Zero-phase sequence impedance voltage						♦
Temperature rise at minimum and maximum ratings of the first unit on a new design; may be omitted if test of thermally duplicate or essentially duplicate unit is available		♦			♦	♦

-continued

TABLE 5.17 (continued)
Routine, Design, and Other Tests for Liquid-Immersed Transformers

Tests	500 kVA and Smaller			501 kVA and Larger		
	Routine	Design	Other	Routine	Design	Other
Dielectric tests						
Low frequency	♦			♦		
Lightning impulse		♦	♦		♦	♦
Form of wave impulse						♦
Switching impulse, phase-to-ground						♦
Radio influence voltage			♦			♦
Insulation power factor			♦			♦
Insulation resistance			♦			
Audible sound level		♦	♦		♦	♦
Short-circuit capability			♦			♦
Mechanical						
Lifting and moving devices		♦			♦	
Pressure		♦			♦	
Leak	♦			♦		
Telephone influence factor (TIF)			♦			

Source: IEEE C57.12.00-2000, IEEE Standard General Requirements for Liquid Immersed Distribution, Power, and Regulating Transformers, 2000.

5.9.2 Test of Transformer

1. Site and commissioning tests for power transformers include:
   - Making a general check of each transformer and its ancillary equipment
   - Checking the grounding connections, labeling, painting, cabling, valves, and pipe work
   - The following specific preservice commissioning tests
     1. Internal inspection
     2. Record vacuum/drying and oil tests during filling
     3. Insulation resistance of core and windings
     4. Dielectric strength of oil samples
     5. Ratio and no-load current at low voltage (e.g., 480–600 V) on all tappings
     6. Vector relation and polarity check
     7. Calibration check of all temperature instruments, including secondary current injection and inspection of all contact settings
     8. Operation checking of all gas/oil pressure-detection relays
     9. Setting check of all oil-level, oil-flow, and water-flow devices
     10. Test of all current transformers

11. Complete functional tests of cooling equipment and tap-change equipment, where included, including manual/automatic sequences, indications, alarms and interlocks, measurements of motor current, adoption of suitable motor protection settings, and proof-of-protection operation for stalled or single-phasing conditions
12. Insulation resistance of all secondary circuits
13. Testing and setting of auxiliaries and their controls (heaters, lighting, etc.)

2. Final checks before energizing:
   - Venting, position, and locking of valves; grounding of star point(s) and of tank; state of breathers and of pressure-relief devices; oil levels; absence of oil leakage; operation of kiosk heaters; tap-change counter readings; resetting of maximum temperature indicators; final proving of alarms and trips
3. Tests when energized:
   - Visual inspection and audible checks
   - On-load tap-changer operation throughout range (subject to not exceeding 1.1-PU volts on any windings)
   - Maintenance of 1.1-PU volts on untapped windings for 15 min (but not exceeding this value on tapped winding)
4. Tests on load:
   - Temperature instrument readings (temperature rise vs. load over a 24-h period)
   - Measurement of winding temperature (WTI) CT secondary currents
   - On-load tap-changer operation throughout range (subject to operational limits)
   - Check for oil and gas leaks; check that insulators are clean and free from external damage; check that loose items that are to be handed over to the owner, e.g., blanking plates, tools, and spares, are in order and are correctly stored or handed over

5.9.3 Routine Power Transformer Inspection

The routine service inspection should include the following tests. Unless specific problems have been noted then the appropriate installation tests may also need to be included:

- Inspect the control cabinet, indicators, control relays, contactors, and operating mechanism.
- Check oil level in the tanks, bushings, and tap-changer compartments.
- Check for loose terminals, oil leaks, and contaminated or damaged bushings.
- Inspect the inert gas system for leakage and proper pressure.
- Check, read, and record the operation-counter indicator reading associated with the load tap changer.

- Check the oil temperature, which should not exceed the sum of maximum winding temperature (as stated on the nameplate) plus the ambient temperature (not to exceed 40°C) plus 10°C. Oil temperature does not exceed 95°C and 105°C for 55°C and 65°C winding-temperature-rise units, respectively.

5.9.4 Internal Inspection of Transformer

Open-type transformers that have not been inspected for at least three years should, if time permits, be given an internal inspection consisting of the following:

- Check for evidence of corrosion of tank walls and other metal parts.
- Verify connections for secureness.
- Lower the oil level to expose the top of the core coil assembly.

REFERENCES

Standards
Liquid-Filled Transformers

S1. IEEE C57.12.00-2000, IEEE Standard General Requirements for Liquid Immersed Distribution, Power, and Regulating Transformers, 2000.
S2. IEEE C57.12.01-1998, IEEE Standard General Requirements for Dry-Type Distribution and Power Transformers Including Those with Solid Cast and/or Resin Encapsulated Windings, 1998.
S3. ANSI C57.12.10-1997, American National Standard for Transformers — 230 kV and below 833/958 through 8333/10 417 kVA Single-Phase; and 3750/862 through 60 000/80 000/100 000 kVA without Load Tap Changing; and 3750/4687 through 60 000/80 000/100 000 kVA with Load Tap Changing: Safety Requirements, 1997.
S4. IEEE C57.12.59-2001, IEEE Guide for Dry-Type Transformer Through-Fault Current Duration, 2001.
S5. IEEE C57.12.70-2000, IEEE Standard Terminal Markings and Connections for Distribution and Power Transformers, 2000.
S6. IEEE C57.12.80-2002, IEEE Standard Terminology for Power and Distribution Transformers, 2002.
S7. IEEE C57.12.90-1999, IEEE Standard Test Code for Liquid Immersed Distribution, Power and Regulating Transformers, 1999.
S8. IEEE C57.12.91-1995, IEEE Guide for Loading Mineral-Oil Immersed Transformers, 1995.
S9. IEEE C57.96-1999, IEEE Guide for Loading Dry-Type Distribution and Power Transformers, 1999.
S10. IEEE C57.98-1993, IEEE Guide for Impulse Tests, 1993.
S11. IEEE C57.104-1991, IEEE Guide for Interpretation of Gasses Generated in Oil Immersed Transformers, 1991.
S12. IEEE C57.105-1978 (R1992), IEEE Guide for Application of Transformer Connections in Three Phase Distribution Systems, 1992.
S13. IEEE C57.109-1993, IEEE Guide for Transformer Through Fault Current Duration, 1993.
S14. IEEE C57.110-1998, IEEE Recommended Practice for Establishing Transformer Capability When Supplying Non-Sinusoidal Load Current, 1998.
S15. IEEE C57.114-1990, Seismic Guide for Power Transformers and Reactors, 1990.

Power Transformers and Reactors 133

S16. IEEE C57.115-1991, IEEE Guide for Loading Mineral Oil Immersed Power Transformers Rated in Excess of 100 MVA (65°C Winding Rise), 1991.
S17. IEEE C57.116-1989, IEEE Guide for Transformers Directly Connected to Generators, 1989.
S18. IEEE C57.131-1995, IEEE Standard Requirements for Load Tap Changers, 1995.
S19. IEEE C57.120-1991, IEEE Standard Loss Evaluation Guide for Power Transformers and Reactors, 1991.
S20. IEEE C57.18.10-1998, IEEE Standard Practices and Requirements for Semiconductor Power Rectifier Transformers, 1998.
S21. IEEE C57.19.00-1991 (R1997), IEEE Standard General Requirements and Test Procedures for Outdoor Power Apparatus Bushings, 1997.
S22. IEEE C57.19.100-1995, IEEE Guide for the Application of Power Apparatus Bushings, 1995.
S23. IEEE C62.22-1997, IEEE Guide for the Application of Metal-Oxide Surge Arresters for Alternating Current Systems, 1997.
S24. IEEE 979-1994, IEEE Guide for Substation Fire Protection, 1994.
S25. NFPA 850-2000, Recommended Practice for Fire Protection for Electric Generating Plants and High Voltage Direct Current Converter Stations, 2000.
S26. IEC 60076-2:1993, Power Transformers, Part 2: Temperature Rise, 1993.
S27. CSA CAN/CSA-C88-M90, Power Transformers and Reactors.
S28. CSA C9-M1981, Dry Type Transformers.
S29. IEEE C57.19.01-1991, Performance Characteristics and Dimensions for Outdoor Apparatus Bushings, 1991.
S30. IEEE 1100-1992, IEEE Recommended Practice for Powering and Grounding Electronic Equipment, 1992.
S31. IEEE C57.12.91-2001, IEEE Standard Test Code for Dry Type Distribution and Power Transformers, 2001.
S32. IEEE C57.92-1981, IEEE Guide for Loading Mineral-Oil-Immersed Power Transformers up to and Including 100 MVA with 55°C or 65°C Average Temperature Rise, 1981.
S33. ASTM D4652-05, Standard Specification for Silicone Fluid Used for Electrical Insulation.

BIBLIOGRAPHY

1. Factory Mutual Insurance Co., Global Property Loss Prevention Data Sheets — Transformers, 5-4, Jan. 1997.
2. IEEE Tutorial, 84 EHO 209-7-PWR, Power Transformer Considerations of Current Interest to the Utility Engineer.
3. *J & P Transformer Book*, 10th ed., Newnes-Butterworths, 1973.
4. General Electric Co., "Automatic Parallel Operation of Load-Tap Changing Transformers," publication GET-1952B, Schenectady, NY: General Electric.
5. General Electric Co., "The Whys of Wyes: The Behavior of Transformer Y Connections," publication GET-3388B, Schenectady, NY: General Electric.
6. REA, "Evaluation of Large Power Transformer Losses," bulletin 65-2-1978, REA, 1978.
7. General Electric Co., "Transformer Connections," publication GET-2J, Schenectady, NY: General Electric, 1970.
8. E. Dornenburg and W. Strittmatter, "Monitoring Oil Cooled Transformers by Gas Analysis," *Brown Boveri Revue* 74 (5): 238 (1974).
9. R. R. Rogers, "IEEE and IEC Codes to Interpret Incipient Faults in Transformers Using Gas in Oil Analysis," *IEEE Trans. Electrical Insulation* 13 (5): 349 (1978).

6 Instrument Transformers

6.1 INTRODUCTION AND OVERVIEW

The selection of instrument transformers, current and voltage, must be carried out with care if the protection, instrumentation, and measuring systems are to function correctly. Current transformers (CT) for relaying should not saturate at maximum fault currents, and voltage transformers must not go into ferro-resonance during transient overvoltage resulting from line-to-ground faults. Most modern digital relays have been made more able to function correctly in the presence of CT saturation by using adaptive filtering to circumvent CT saturation. More details of the techniques used are provided in the Alstom (now Areva) *Network Protection and Application Guide* [1] and the Alstom relay technical guides [2].

The compact design of modern metal-clad switchgear has limited space for the number and size of current transformers that are required to perform satisfactorily in a high-fault-current environment. High rating CTs are physically large and, hence, optimizing CT design to meet the technical performance required while keeping the costs and size within limits needs a fine engineering balance.

6.2 CURRENT TRANSFORMERS

6.2.1 BUSHING TYPE

A bushing current transformer (CT) consists of secondary windings on an annular magnetic core. The core encircles the high-voltage (HV) bushings used on circuit breakers, power transformers, generators, and switchgear. The secondary turns of a bushing CT should be distributed to minimize leakage reactance. Physically, this is accomplished by distributing each section of the tapped secondary winding completely around the circumference of the core. In North American practice, CTs are tapped as per the IEEE and CSA standards [S1, S7], and are referred to as multiratio bushing current transformers, whereas international practice usually provides a 50% secondary tap only.

6.2.1.1 Bar Type

A bar-type CT is of similar construction as a bushing type. These current transformers have a single concentrically placed primary conductor, sometimes permanently built into the CT and provided with the necessary primary insulation.

6.2.1.2 Window Type

The window-type CT, which is used at medium- and low-voltage equipment, has a magnetic core with a center opening through which a power conductor passes to form the primary turns. The secondary is wound on the core, and in some styles the center assembly is encased in molded insulating material. The core may be annular in shape with a uniformly distributed secondary winding (similar to bushing CT), or rectangular in shape with the secondary winding either distributed or wound on only two legs.

6.2.1.3 Wound Type

A wound-type CT has a primary winding of one or more turns and a secondary winding on a common core, similar to power transformers. These are commonly used in medium-voltage (MV) and low-voltage (LV) starters.

6.2.1.4 Post Type

High-voltage CTs for external use need physical height to provide the required phase-to-ground insulation while providing the same insulation level between the primary and secondary windings. Depending on the supplier, these may consist of a single bar or multiple pass, primarily through a series of bar-type cores. In international practice, reconnectable primary tap arrangements are used to series- or parallel-connect two primary (or group of primary) passes through the CT cores; these are also used to provide a half tap ratio without reducing the knee point of the CT.

6.2.1.5 Auxiliary CT

Auxiliary CTs are sometimes used in the secondary circuits of other types of current transformers to change either, or both, the ratio and the phase angle of the secondary current. Such CTs are used in some electromagnetic designs of transformer differential protections to correct for differences in primary CT ratios to balance the scheme and avoid the need for multiratio CTs on transformer bushings. They can also provide the Wye-Delta connections required, allowing the main CT to be always connected in Wye and not connected to meet the Delta CT connection requirements of the protection of Wye-Delta transformers.

6.2.1.6 Linear Couplers (Air-Core Mutual Reactors)

A linear coupler consists of a toroidal secondary winding on a nonmagnetic annular core. Like the bushing CT, it is designed for mounting on a bushing, with the power conductor forming a single primary turn. The absence of iron eliminates core saturation problems. A linear relationship exists between primary current and secondary voltage.

6.2.1.7 Optical or Digital CT

Fully digital current and potential transformers have been under development since the early 1970s but only reached commercial service in the late 1990s. These CTs usually convert the current signal into a digital code that can be transmitted via fiber-optic links to the measuring device. Most designs used proprietary protocols to transmit this data, and hence development suffered from not having a universal standard to match the 1 A/5 A interface for the electromagnetic devices.

With the standardization in 2005 of the IEC 61850 protocols, which include a digital measurement data protocol, this restriction should be removed. Initial applications are tending to be in the extremely high voltage (EHV) area, where the advantages of high-accuracy linear measurement and the high cost of insulation can balance with the development costs and complexities, such as self-powering and fiber-cable protection. Such devices have maximum benefit when directly feeding relays, although there may, ironically, need to be some further development in the relay filtering algorithms to compensate for the high bandwidth and high transient accuracy of such devices.

The instrument optical current transformers (OCT) will bring a new level of accuracy to current instrument transformers. They operate on the principle that the magnetic field created as current flows through a conductor influences the polarization of light on a path encircling that conductor.

The sensor is based on an optic system that has a trusted reputation for accuracy and reliability in industrial applications. The sensor can be column mounted on an advanced polymeric insulating column, or bus mounted with a suspension insulator to bring the optical fiber to ground.

6.2.2 Equivalent Circuit

The equivalent circuit of a current transformer extracted from [S1, S4] is shown in fig. 6.1, where

V_S = secondary exciting voltage
V_B = CT terminal voltage across external burden
I_P = primary current, A
Z_E = exciting impedance, ohm
I_{ST} = total secondary current, A ($I_E + I_S$)
R_S = CT secondary winding resistance, ohm
I_S = secondary load current, A
X_L = leakage reactance (negligible in Class C CT), ohm
I_E = exciting current, A
N = CT turns ratio
Z_B = burden impedance, ohm (includes secondary devices and leads)

The secondary exciting voltage is given by:

$$V_S = I_S \times (R_S + Z_B)$$

The primary current is given by:

$$I_P = N \times (I_S + I_E)$$

The ratio correction factor is given by:

$$RCF = \frac{I_{ST}}{I_S}$$

The percent ratio error is given by:

$$\frac{I_E}{I_S} \times 100$$

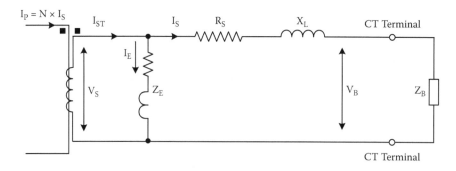

FIGURE 6.1 Equivalent circuit of a current transformer (From IEEE C57.13-1993, IEEE Standard Requirements for Instrument Transformers, 1993; and IEEE C57.13.3-1983 (R1990), IEEE Guide for the Grounding of Instrument Transformer Secondary Circuits and Cases, 1990. With permission.)

6.2.3 Excitation Characteristics

As in any magnetic device, the exciting impedance, Z_E, is nonlinear and is normally expressed as a CT excitation curve, which shows the relationship of secondary exciting voltage (V_S) to the exciting current (I_E). Figure 6.2, extracted from IEEE 57.13 [S1], shows typical excitation curves for a 1200:5 multiratio Class C current transformer. These curves were developed from test data.

The knee-point voltage of a nongapped CT is the point on the excitation curve where the tangent to the curve makes a 45° angle with the abscissa. The technical

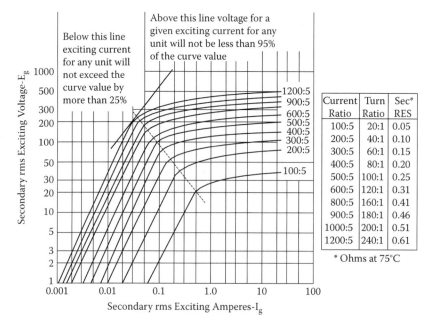

FIGURE 6.2 Typical exciter curve of a Class C or Class K multiratio CT (From IEEE C57.13-1993, IEEE Standard Requirements for Instrument Transformers, 1993. With permission.)

definition of this point is defined differently in the International Standards, but it is virtually the same for most practical purposes. The knee-point voltage of a gapped CT is where the tangent makes a 30° angle to the curve.

The multiratio-bushing-type CT built to IEEE standards [S1] has five taps, X_1, X_2, X_3, X_4, and X_5. Different current ratios can be obtained by connecting the appropriate device to the appropriate taps. Current ratings and taps for a 1200:5 multiratio CT are given in table 6.1. Only a single connection can be made, and the unused taps need to be left open and not shorted together or to ground.

6.2.4 Accuracy Class and Burden

The accuracy class of a North American CT is given by a letter and secondary voltage rating. The letter designation for the class is:

- *Class C*: Ratio error can be calculated, the secondary winding is distributed uniformly, and the leakage flux is negligible. The calculations assume that the exciting current and the burden are in phase. The knee-point voltage is 50–75% of the accuracy of the terminal voltage.
- *Class K*: This class of CT is the same as class C, but the knee-point voltage must be at least 70% of the secondary terminal voltage rating.
- *Class T*: Ratio error of T-class CT is determined by tests. A T-class CT has an appreciable core flux leakage that contributes to a high ratio error.
- *Class H, L*: This classification is not used any more and is applicable to old CTs manufactured before 1954. These current transformers have 2.5% or 10% ratio error, e.g., 2.5 H200 or 10 L200.

The secondary terminal voltage or accuracy voltage rating is the CT secondary voltage that the CT will deliver at 20 times the rated secondary current, when connected to a standard burden without exceeding a 10% error. The secondary voltage of a C100 CT is 20 * 5 A * 1.0 ohm (= 100 V). The percent error is calculated using the equation:

TABLE 6.1
1200:5-A Multiratio CT Current Rating and Secondary Taps

Current Rating (A)	Trans Ratio	Secondary Tap
100:5	20:1	$X_2 - X_3$
200:5	40:1	$X_1 - X_2$
300:5	60:1	$X_1 - X_3$
400:5	80:1	$X_4 - X_5$
500:5	100:1	$X_3 - X_4$
600:5	120:1	$X_2 - X_4$
800:5	160:1	$X_1 - X_4$
900:5	180:1	$X_3 - X_5$
1000:5	200:1	$X_2 - X_5$
1200:5	240:1	$X_1 - X_5$

$$\% \text{ Error} = (\frac{\text{Multiple of rated primary current}}{\text{Multiple of rated secondary current}} - 1) \times 100$$

The standard burden of relaying current transformers with 5 A secondary winding is given in table 6.2.

For a multiratio CT, the accuracy voltage is given for the full or 100% winding, and the voltage at other taps is directly proportional to the tap or percentage of the winding used. For a 1200/5 A C400 multiratio CT, the secondary terminal or accuracy voltage at 600 A tap (600/5 A) will be C200.

Because the leakage reactance of a T-class CT is not negligible, overcurrent-ratio error curves drawn on a rectangular paper plot are required to check the ratio or percent error. The plot is drawn in terms of primary versus secondary current from 1 to 22 times the rated current for all standard burdens up to the burden that causes a ratio error of 50%. A typical plot is shown in fig. 6.3.

TABLE 6.2
Standard Burden at 60 Hz for a Relaying CT with 5 A Secondary Winding

Burden Designation	Resistance, Ω	Inductance, mH	Impedance, Ω	Terminal Voltage C Rating, V	VA at 5 A	Power Factor
B – 1	0.5	2.3	1.0	C 100	25	0.5
B – 2	1.0	4.6	2.0	C 200	50	0.5
B – 4	2.0	9.2	4.0	C 400	100	0.5
B – 8	4.0	18.4	8.0	C 800	200	0.5

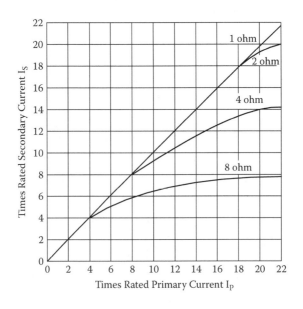

FIGURE 6.3 Overcurrent ratio curve for a T class CT

Instrument Transformers

CTs conforming to international standard (IEC 60044-1 [S8]) use a different method of classification, as discussed further in section 6.5.

6.2.5 Current Transformers Saturation

The saturation voltage (V_X) (defined as Vk in international standards, see section 6.5.1.1) is the voltage across the CT for which the peak induction just exceeds the saturation flux density. Though the burden of modern relays is small, system X/R ratio, CT winding resistance, and lead and contact resistance must be taken into consideration for saturation check during symmetrical and asymmetrical faults.

6.2.5.1 Saturation with Symmetrical Fault Current (No DC Component)

To avoid AC saturation,

$$V_X > I_S \times (Z_S + Z_B)$$

where

I_S = primary current divided by the CT turn ratio
Z_S = CT secondary winding resistance
Z_B = total burden (lead from CT terminal to devices + devices) connected to the secondary

To minimize the effect of saturation under steady-state fault condition, the saturation voltage must be less than 50% of the C voltage rating or C voltage rating greater than $2 \times V_X$.

6.2.5.2 Saturation with Asymmetrical Fault Current (with DC Component)

$$V_X > I_S \times (Z_S + Z_B) \times (1 + X/R)$$

where

X = primary system reactance up to the fault location
R = primary system resistance up to the fault location

The above selection criteria results in a large CT for systems with high fault current and high X/R ratio.

6.2.5.3 Example

1. Check the saturation voltage of current transformers for use on a 13.8 kV system
2. Primary system fault current, (I_f) = 15 kA, rms sym., X/R = 10
3. CT = 600:5 A, C200, CT standard burden = 2.0 ohm at $20 \times 5 = 100$ A
4. CT ratio = 600:5 A = 120
5. CT secondary winding resistance from manufacturer's data or fig. 6.2, Z_S = 0.31 ohm
6. Burden of lead + contact + relay and meters, Z_B = 0.3 ohm

- Saturation voltage for symmetrical fault current (no DC component):

$$V_X = I_S \times (Z_S + Z_B) = \frac{15000}{120} \times (0.31 + 0.3) = 76.25 \text{ V}$$
(no DC component)

< 50% of voltage rating of C200
CT will not saturate under the symmetrical fault conditions
- Saturation voltage for asymmetrical fault (with DC component):

$$V_X = I_S \times (Z_S + Z_B) \times \left(1 + \frac{X}{R}\right) = \frac{15000}{120} \times 0.61 \times (1 + 10) = 838.8 \text{ V}$$

CT will saturate for an asymmetrical fault.
- Change CT to 1200:5 A, C400, standard burden Z_B = 4.0 ohm
  CT ratio = 240
  CT secondary resistance, Z_S = 0.61 ohm
  Burden of lead + contact + relay and meter, Z_B = 0.3 ohm
- Saturation voltage for symmetrical fault current (no DC component)

$$V_X = I_S \times (Z_S + Z_B) = \frac{15000}{240} \times (0.61 + 0.3) = 56.88 \text{ V}$$
(no DC component)

<50% of voltage rating of C200
CT will not saturate under symmetrical fault condition
- Saturation voltage for asymmetrical fault current (with DC component)

$$V_X = I_S \times (Z_S + Z_B) \times \left(1 + \frac{X}{R}\right) = \frac{15000}{240} \times 0.91 \times (1 + 10) = 625 \text{ V}$$

CT will saturate for an asymmetrical fault

6.2.5.4 Recommendations

- Select the highest ratio that can be used within the relay setting range to reduce the secondary fault current. Modern microprocessor-based relays have a wide pickup range, and there is no point in selecting a low CT ratio.
- Specify the highest accuracy class CT available for the equipment in use. Bushing-type CTs used in power transformers and high-voltage circuit breakers can be purchased with high accuracy voltage. However, there is a problem with medium-voltage switchgear due to the limited space available for mounting the required number of current transformers. Tables 6.2 and 6.3 provide the data on current transformers that are available with maximum accuracy voltage.
- Use a higher CT ratio for instantaneous overcurrent relay elements. This will reduce the relay current, which results in a lower saturation voltage, and the relay will operate before the CT goes into saturation.

Instrument Transformers

TABLE 6.3
Typical Accuracy Voltage of Bushing Current Transformer

Ratio	110–250 kV BIL (15–48.3 kV)	350 kV BIL (72.5 kV)	Above 350 kV BIL (Above 72.5 kV)
50/5	C20	...	...
100/5	C50	C50	...
200/5	C100	C100	C200
600/5	C200	C200	C400
800/5	C200	C400	C400
1200/5	C200/C400	C400	C800
2000/5	C400	C800	C800
3000/5	C800/C400	C800	C800
4000/5	C800	C800	C800
5000/5	C800	C800	C800

6.2.5.5 Saturation Factor and Time to Saturate

Saturation factor, known as K_S, is the ratio of saturation voltage to the excitation voltage. K_S is used to calculate the time to saturate. The IEEE 1976 publication on the transient response of current transformers, 76-CH1130-PWR [S6], provides curves from which the time to saturation can be calculated. The time to saturation is given by the equation:

$$T_S = T_1 \times l_n \times \left[1 - \frac{K_S - 1}{\frac{X}{R}}\right]$$

where

T_s = time to saturate
$T_1 = \dfrac{X}{\omega \times R}$
l_n = natural log function
K_S = saturation factor (V_X/V_R)
$\omega = 2\pi f$
f = frequency
X, R = system reactance and resistance at the fault location

The following data is required to determine the time to saturate:

- System X/R ratio
- Maximum fault current as seen by the CT
- The remnant flux in the core

- CT secondary burden
- Saturation voltage

6.2.6 REMNANT FLUX

Remnant or trapped flux in a CT core is established when:

- DC current is applied for testing
- The primary current is interrupted while the CT is in a saturated stage

The CT produces an undistorted secondary current when the remnant flux is of the opposite polarity to the flux due to the transient component of the fault current. However, if the remnant flux is of the same polarity as the flux due to the transient component of the fault current, the wave form produced by the CT will be distorted.

The remnant flux remains in the CT core until it is demagnetized. This remnance can be reduced using power frequency voltage injection when the CT is de-energized. The following steps can then be used for demagnetizing the CT:

- With the primary winding open circuited, connect a variable voltage source across the secondary winding.
- Increase the voltage until the core enters the saturation region. The saturation point is detected by observing a disproportionate increase in exciting current.
- Reduce the voltage to zero over a period of 3 seconds.

The remnant flux from a CT that is in service and carrying current can be reduced using the following steps:

- Insert a variable resistor in the CT secondary circuit, using the usual precautions when handling live CT circuits.
- Increase the resistance to achieve core saturation.
- Reduce the resistance to zero.

6.2.7 ACCURACY VOLTAGE OF BUSHING CURRENT TRANSFORMERS

The typical maximum accuracy voltage available in a bushing current transformer is provided in table 6.3.

6.2.8 ACCURACY VOLTAGE OF CURRENT TRANSFORMERS USED IN SWITCHGEAR

The accuracy voltage of current transformers used in medium-voltage switchgear is provided in table 6.4. Typically, current transformers for switchgear application are designed for 1.0 V per turn or less. Hence, the manufacturer's data must be used before finalizing the design.

Instrument Transformers

TABLE 6.4
CT Accuracy Voltage Available in Medium-Voltage Switchgear

Ratio	Minimum Accuracy Class, C37.20.2-1999	Accuracy Class Normally Provided by the Manufacturer	Accuracy Class That Can Be Provided[a]
50:5	C10	C10	C20
75:5	C10	C10	C20
100:5	C10	C10	C20
150:5	C20	C20	C50
200:5	C20	C20	C50
300:5	C20	C50	C100
400:5	C50	C50	C100
600:5	C50	C100	C200
800:5	C50	C100	C200
1200:5	C100	C200	C400
1500:5	C100	C200	C400
2000:5	C100	C200	C400
3000:5	C100	C200	C400
4000:5	C100	C200	C400

[a] Requires a CT of increased width. This may limit the number of CTs that can be accommodated in the metal-clad switchgear.

6.2.9 CURRENT TRANSFORMERS IN SERIES

Although two identical current transformers with similar excitation characteristics can be connected in series to double the accuracy voltage, this connection is not recommended. If a primary fault should occur between the two current transformers, excessive overvoltage can occur in each CT, which may require overvoltage protection. It is better to use one CT with a higher ratio and good performance.

6.2.10 POLARITY OF CT: POLARITY MARKING

One primary and one secondary terminal of the current transformer are marked to indicate the relative instantaneous direction of the primary and secondary current (see fig. 6.4). When the current is flowing "in" at the marked primary terminal, the current is flowing "out" at the marked secondary terminal.

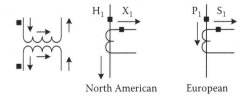

North American   European

FIGURE 6.4 CT polarity marking

6.2.11 Auxiliary Current Transformers

The addition of an auxiliary CT adds burden to the main CT. However, the effect on the main CT will be a decrease in burden when the current is stepped down and an increase when the current is stepped up. The apparent impedance to the main CT of the portion of the burden in the secondary of the auxiliary CT is that portion multiplied by the square of the auxiliary CT ratio. A 1.0-ohm burden connected to the auxiliary CT will appear as 0.25 ohm for a ratio of 2:1 (stepped down) and as 4.0 ohm when the ratio is 1:2 (stepped up). For this reason, the current-stepped-up applications shall be avoided. The internal burden of the auxiliary CT shall be added to the main CT burden.

6.2.12 Metering Current Transformers

Metering accuracy of a current transformer is stated in a different way, e.g., 0.3 B 0.5 for a North American CT, where:

0.3 = % accuracy (or 0.6 or 1.2 or 2.4)
B = burden in ohms = 0.5, one of a number of standard burdens in ohms, from 0.1 to 2.0 (or higher)

Standard burden designation and data for metering CT with 5 A secondary winding is provided in table 6.5 from IEEE C57.13-1999.

The relaying CT must perform with acceptable accuracy at (high) fault values of current, whereas a metering CT must perform well in the lower (i.e., load) current range. Thus, a relaying CT must be capable of substantial voltage output at high current, which can stress the meters connected to a protection CT. The method of avoiding this problem, if a shared CT has to be used, is to connect a metering-class interposing CT after the protection circuit and use that to feed the metering circuit.

6.3 VOLTAGE TRANSFORMERS

6.3.1 Inductive Voltage Transformer

A current transformer behaves like a short-circuited transformer, whereas a voltage transformer (VT) has a characteristic of a transformer at no load. The equivalent

TABLE 6.5
Standard Burden for Metering CT with 5 A Secondary Winding

Burden Designation	Resistance, Ω	Inductance, mH	Impedance, Ω	Terminal Voltage at 5 A	VA at 5 A	Power Factor
B-0.1	0.09	0.116	0.1	0.5	2.5	0.9
B-0.2	0.18	0.232	0.2	1.0	5.0	0.9
B-0.5	0.45	0.58	0.5	2.5	12.5	0.9
B-0.9	0.81	1.040	0.9	4.5	22.5	0.9
B-1.8	1.62	2.080	1.8	9.0	45.0	0.9

Source: IEEE C57.13-1999.

Instrument Transformers

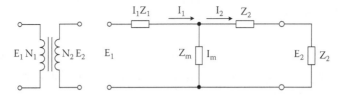

FIGURE 6.5 Equivalent circuit of a voltage transformer

circuit of a voltage transformer is shown in fig. 6.5. When the voltage drop caused by the magnetizing current in the primary winding is neglected. The following can be deduced:

$E_1/E_2 = N_1/N_2 = I_2/I_1$ (neglecting the drop in the two windings)
The voltage drop caused by the load current is the vector sum of:

$$\Delta E = (I_1 \times Z_1) + (I_0 \times Z_M) + (I_2 \times Z_2)$$

The ratio error is given by:

$$\frac{E_1 - \Delta E}{E_2}$$

The voltage drop caused by the magnetizing current is constant, whereas the voltage drop caused by the transformer impedance (Z_1 and Z_2) is proportional to the load current.

6.3.2 Capacitor Voltage Transformers

The physical size and cost of a voltage transformer increases with the increase in voltage rating. In general, a capacitor voltage transformer (CVT) becomes more economical compared with an electromagnetic voltage transformer at voltages above 120 kV. A CVT is a combination of capacitive voltage divider and an inductive voltage transformer rated at a lower voltage. The principle diagram of a CVT in a single-line form is shown in fig. 6.6.

The ratio of the capacitive voltage divider is given by the factor K_1:

$$K_1 = \frac{(C_1 + C_2)}{C_1} = \frac{E_1}{E_2}$$

The ratio of the electromagnetic transformer or C_2 is given by the factor K_2:

$$K_2 = E_2/E_1$$

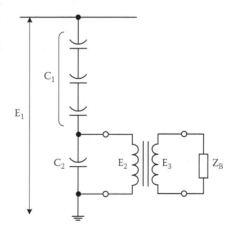

FIGURE 6.6 Capacitor voltage transformer

and the total ratio factor K is:

$$K = K_1 * K_2$$

Because a CVT contains capacitance and inductance, oscillation may occur. These oscillations are eliminated by adding resistance to the load circuit or into the secondary winding.

6.3.3 Standard Accuracy Class and Burden

The standard accuracy classes at the rated burden are 0.3, 0.6, and 1.2. For medium- and high-voltage systems, VT is selected with an accuracy class of 0.3 at the rated burden. The standard burden designations are W, X, M, Y, Z, and ZZ. The characteristics of a VT with standard burdens are given in table 6.6. The thermal burden is higher; however the standard accuracy limit is exceeded if the VT is loaded to this level.

For high-voltage systems such as 230 kV, VT can be supplied with an accuracy of 0.3% over a wide range, e.g., 0.3WXYZ and often ZZ, i.e., 12.5 VA, 25 VA, 75 VA, 200 VA, and 400 VA. VT at medium- and low-voltage systems is ordered for a narrow range. The accuracy class of voltage transformers built to IEEE [S1] and CSA [S7] standards is given in table 6.7

6.3.4 Ratings and Characteristics

Voltage transformers are divided into five groups, groups 1 to 5. The ratings and characteristics are given in the IEEE standard [S1] (tables 10 to 14), and typical con-

TABLE 6.6
Standard Burden Designation of Voltage Transformer

Designation	VA	Power Factor
W	12.5	0.1
X	25	0.7
M	35	0.2
Y	75	0.85
Z	200	0.85
ZZ	400	0.85

TABLE 6.7
Accuracy Class of Voltage Transformer

Class	Range of Burden, %	Range of Voltage, %	Load PF	Application
0.3	0–100	90–110	0.3	Revenue metering
0.6	0–100	90–110	0.6	Standard
1.2	0–100	90–110	1.2	Relaying

Instrument Transformers

TABLE 6.8
VT Groups and Their Applications

VT Group	Application	Table and Fig. No. in IEEE C57.13 [S1]
Group 1	VT is rated for 100% of primary voltage and is connected line-to-line or line-to-ground, e.g., VT rated at 24 kV can be connected line-to-line on a 2.4 kV system or line-to-ground on a 4.16 kV effectively grounded system	table 10 and figs. 6a, 6b
Group 2	VT is primarily for line-to-line service and may be applied line-to-ground or line-to-neutral at a winding voltage equal to the primary voltage rating divided by 3; the thermal burden will be reduced at the lower voltage	table 11 and figs. 6c, 6d
Group 3 for outdoor use	These are for outdoor use, rated for line-to-ground connections only, and have two secondary windings; they may have neutral fully insulated (two bushings) or neutral grounded (single bushing); the overvoltage rating is 110% continuous and 3 times the phase-to-neutral voltage for 1 min	table 12 and fig. 6e
Group 4 for outdoor use	These are further subdivided into two Groups, 4a and 4b; Group 4 transformers are used for line-to-ground connections only	table 13 and figs. 6f, 6g
Group 4a	Applied on effectively grounded systems, the voltage rating is the phase-to-neutral system voltage	
Group 4b	VT is insulated for line-to-line voltage and operates at approximately 58.7% of the rated voltage	
Group 5 for outdoor use	These are for line-to-ground connection only	table 14 and fig. 6h

nections are shown in figs. 6(a–h) of the standard. The application aspects are summarized in table 6.8.

For industrial power systems with low- or high-resistance grounded neutral, two single-phase VTs, connected in open Delta, are used for the majority of applications. However, for Group 4b applications where directional overcurrent or distance protection is used, three single-phase VTs, rated for line-to-line voltage, are connected line-to-ground.

6.3.5 PROTECTION OF VOLTAGE TRANSFORMERS

Voltage transformers rated for system voltage of up to 34.5 kV are protected by a current-limiting fuse. Above this voltage rating, it is difficult to find a fuse of adequate fault-interrupting rating.

The primary fuse current rating is selected to protect the VT against the bolted fault on the secondary terminals. The VT is designed to withstand the mechanical and thermal stresses resulting from a bolted fault at the secondary terminals for a duration of 1 s.

A miniature circuit breaker or a fuse on the secondary is used to protect the VT against overloads. The secondary protective device must carry the rated current at the thermal burden. The secondary protective device is located near the VT. If a number of loads are fed by the VT, each branch is protected by a molded-case circuit breaker or a fuse not higher than half the rating of the main protective device.

The primary or secondary fuse elements shall not be mechanically weak, as they can open the circuit due to vibration or if dropped accidentally. The primary fuse on the VT connected to an isolated phase bus in a power plant is subject to vibration. A current-limiting fuse rated 0.5 A for a 0.3Z VT may fail under such conditions. Discuss with the manufacturer and increase the fuse rating to 0.75 A or 1.0 A. However, ensure that the large fuse protects the VT against the secondary fault current.

6.3.6 Ferro-Resonance in Voltage Transformer

A line-to-ground-connected VT on an ungrounded system can experience overvoltage with very high exciting current, resulting in complete damage to the primary winding. Examination of a failed VT caused by ferro-resonance shows only primary winding roasted from end to end (see fig. 6.7). The system capacitance in parallel with the VT inductance can result in a parallel resonance. This condition can occur during a line-to-ground fault on an ungrounded system. The behavior of an ungrounded system during ground faults is explained in chapter 4.

To prevent the overvoltage caused by the ferro-resonance, a loading resistor is added to the VT secondary to damp the circuit. The damping resistor is connected across the open-corner Delta of the three single-phase voltage transformers, as shown in fig. 6.8. Typical values of the damping resistor ratings to prevent ferro-resonance for different system voltages, extracted from Westinghouse paper 44-060 [4], are given in Table 6.9.

6.4 GROUNDING OF SECONDARY

Grounding of the current and voltage transformer secondary circuits and secondary equipment cases is a must for safety reasons. To prevent circulating current caused

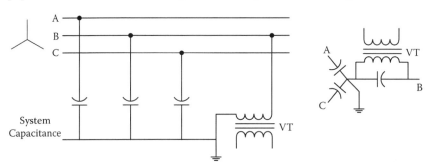

FIGURE 6.7 Ferro-resonance in a voltage transformer

Instrument Transformers

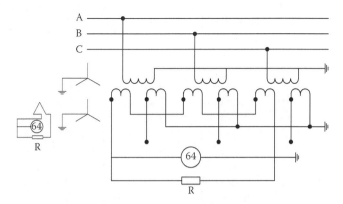

FIGURE 6.8 Damping resistor connected across open corner delta

TABLE 6.9
Damping Resistor in the Secondary Circuit to Prevent Ferro-Resonance

Primary System Voltage, V	Loading on Secondary, W	Equivalent Resistance, Ω	
		Per Phase at 120 V	Across Open-Corner Delta
Up to 600 V	200	72	216
6.9–15.0	500	30	90
25 kV and above	750	20	60

Source: Westinghouse, "Instrument Transformer," Technical Data 44-060, Westinghouse, Nov. 1971.

by the potential difference, connect the secondary circuit to the station ground at one point only. The recommended location and method of grounding is covered in detail in IEEE C57.13.3, "IEEE Guide for the Grounding of Instrument Transformer" [S4].

For voltage transformers, the connection to the station ground bus at the transformer location eliminates the possibility of accidentally removing the ground connection. However, for current transformers, the secondary circuit shall be grounded at the switchgear or protection/control panel. This arrangement facilitates the insulation testing of the CT.

6.5 EUROPEAN STANDARDS

Voltage and current transformers built to IEC standards are covered by the IEC 60044 series [S8–S11]. These replace the earlier IEC 185 and 186 standards.

6.5.1 CURRENT TRANSFORMERS

6.5.1.1 Protection-Class Current Transformers

These are defined as Class P or Class PX. P-class CT, intended for standard overcurrent protection, can be specified as 5P (5% error) or 10P (10% error). The CT is represented by VA, class, and accuracy limit factor, e.g., 30 VA, 5P20.

A class-PX current transformer is used with unit and other vital protection schemes where performance under transient conditions is important. For such transformers, knee-point voltage and secondary winding resistance is defined. These are similar to North American C-class CTs.

The current-transformer requirement for different protection applications is covered by Roberts et al. [3] and in chapter 11 of this book.

The knee-point voltage can be estimated by the following equation:

$$V_K = \frac{V_A \times A_{LF}}{I_N} + A_{LF} \times I_N \times R_{CT}$$

where
V_K = knee-point voltage
V_A = CT-rated burden in VA
A_{LF} = accuracy limit factor
I_N = CT-rated current, 5 A or 1 A
R_{CT} = CT secondary winding, ohm

6.5.1.2 Example

Current transformer 600/5 A, 2 5VA, 5P20, R_{CT} = 0.2 ohm

$$V_K = \frac{25 \times 20}{5} + 20 \times 5 \times 0.2 = 120 \; V$$

6.5.1.3 Measuring or Metering Current Transformers

The accuracy class of measuring or metering current transformers is: 0.1, 0.2, 0.5, 1.0, and 3.0. This is the ratio error in percent of the rated current and rated burden in VA.

6.5.2 VOLTAGE TRANSFORMERS

6.5.2.1 Standard Accuracy Class

The standard accuracy class (voltage ratio error in percent) of measuring voltage transformers is: 0.1, 0.2, 0.5, 1.0, and 3.0. This is the voltage ratio error in percent at 0.8–1.2 per-unit (P.U.)-rated voltage and 0.25–1.0 P.U.-rated burden. The accuracy class 0.1 is used in laboratory applications and 0.2 for revenue metering.

Instrument Transformers

TABLE 6.10
Voltage Factors

Voltage Factor (V_f)	Time Rating or Duration	Primary Winding, Connected/System Earthing
1.2	Continuous	Between phases or transformer neutral point and earth in any network
1.5	30 s	Between phases or transformer neutral point and earth in any network
1.2 / 1.9	Continuous / 30 s	Between phases and earth in any ineffectively earthed system with automatic earth-fault tripping system
1.2 / 1.9	Continuous / 8 h	Between phases and earth in isolated neutral without automatic earth-fault tripping

The accuracy class of protection voltage transformers is 3P and 6P. This is the voltage ratio error in percent at 0.25–1.0 P.U.-rated burden and 0.05 V_f (voltage factor) rated primary voltage.

6.5.2.2 Voltage Factor

The voltage factor (V_f) is the upper limit of operating voltage expressed in per-unit (P.U.) of operating voltage. Voltage factors and time rating or duration of maximum voltage are given in table 6.10. For unit-connected generators with high-resistance grounding, the V_f shall be 1.2 for continuous operation and 1.9 for 8 h.

REFERENCES

Standards

S1. IEEE C57.13-1993, IEEE Standard Requirements for Instrument Transformers, 1993.
S2. IEEE C57.13.1-1993, IEEE Guide for Field Testing of Relaying Current Transformers, 1993.
S3. ANSI C57.13.2-1992, Conformance Test Procedures for Instrument Transformers, 1992.
S4. IEEE C57.13.3-1983 (R1990), IEEE Guide for the Grounding of Instrument Transformer Secondary Circuits and Cases, 1990.
S5. IEEE C57.110-1996, IEEE Guide for the Application of Current Transformers Used for Protective Relaying Purposes, 1996.
S6. IEEE Special Publication 76-CH1130-PWR, Transient Response of Current Transformers, Jan. 1976.
S7. CSA CAN3-C13-M83, Instrument Transformers.
S8. IEC 60044-1, Instrument Transformers, Part 1: Current Transformers.
S9. IEC 60044-2, Instrument Transformers, Part 2: Inductive Voltage Transformers.
S10. IEC 60044-5, Instrument Transformers, Part 5: Capacitor Voltage Transformers.
S11. IEC 60044-6, Instrument Transformers, Part 6: Requirements for Protective Current Transformers for Transient Performance.

BIBLIOGRAPHY

1. Alstom T&D, *Network and Protection Guide*, Alstom (now Areva) T&D, chap. 6.
2. Alstom Technical Guide, Application Notes, MICOM P342/343, "Current Transformer Requirements," pp. 142–148.
3. J. Roberts, Stanley E. Zochol, and Gabriel Benmouyal, "Selecting CTs to Optimize Relay Performance," Western Protection Conference, Spokane, WA, Oct. 1996.
4. Westinghouse, "Instrument Transformer," Technical Data 44-060, Westinghouse, Nov. 1971.
5. R. F. Karlicek and E. R. Taylor, "Ferro-Resonance of Grounded Potential Transformer on Ungrounded Power Systems," *AIEEE Trans.* 78 (pt. 3A): 669 (1959).

7 Switchgear, Circuit Breakers, and Motor Control Center

7.1 LOW VOLTAGE

Low-voltage, metal-enclosed switchgear are designed for the control and protection of power circuits of fans, pumps, lighting, and machines at 220 V, 240 V, 380 V, 480 V, or 600 V. These switchgear are adapted for power centers, such as central-station auxiliary power circuits for fans, blowers, compressors, pumps, and lighting circuits; and industrial-plant power and lighting networks, power feeders, lighting feeders, power generation, and auxiliary power drives for machine tools and material-handling equipment.

The standards for low voltage in North America are ANSI C37.13, C37.16, C37.20, and C37.50, which cover metal-enclosed units with air circuit breakers and static trip elements.

7.1.1 Short Circuit

The short-circuit ratings given in standards and in manufacturers' data sheets are based on a short-circuit power factor of 15% or X/R ratio of 6.6. When the short-circuit power factor is less than 15% (X/R > 6.6), the computed fault current must be increased by a factor given in table 7.1.

7.1.2 Important Features to Be Specified

In addition to detailing the fault duty, the other important features to be specified when purchasing voltage circuit breakers are:

- Full compartmentalization, including barriers between bus and cable compartments, barriers between adjacent cable compartments, and insulated runback bus/conductors
- Trip units that are microprocessor based and include features such as self-checking, rms sensing, metering, remote communications, and adjustable protection settings
- Short-delay element that is selective (with no instantaneous override) up to the circuit-breaker interrupting rating
- Instantaneous element that can be switched off for coordination

TABLE 7.1
Selection of Multiplying Factors

System Short-Circuit PF, %	System X/R Ratio	Multiplying Factor	
		Nonfused CB	Fused CB
20	4.9	1.00	1.00
15	6.6	1.00-	1.07
12	8.27	1.04-	1.11
10	9.95	1.07-	1.15
8.5	11.72	1.09-	1.18
7	14.25	1.11-	1.21
5	20.00	1.14-	1.26

Source: ANSI/IEEE C37.012-1997, IEEE Application Guide for Capacitance Current Switching for AC High-Voltage Circuit Breakers Rated on a Symmetrical Current Basis, 1997.

7.2 MEDIUM VOLTAGE

For medium-voltage systems, the metal-clad and metal-enclosed switchgear provide centralized circuit control (fig. 7.1). A complete line of functional units is available for control and protection of motors, transformers, generators, distribution lines, rectifiers, and similar power equipment. Each standard self-contained unit utilizes basic-world reliable components, including vacuum circuit breakers, magnetic blast breakers, instruments, relays, meters, instrument transformers, and control devices.

The standards for medium voltage in North America are:

- For circuit breakers rated on a symmetrical basis: ANSI C37.04, C37.06, C37.09, C37.010, C37.011, C37.12
- For circuit breakers rated on a total current basis: ANSI C37.4, C37.6, C37.9, C37.10, C37.11, C37.12
- For switchgear: ANSI C37.20, NEMA SG4, NEMA SG5, CSA C22.2 No. 31, and EEMAC G8-2

FIGURE 7.1 Medium-voltage switchgear

Switchgear, Circuit Breakers, and Motor Control Center

The types are:

- Draw-out: up to 34.5 kV
- Vacuum: 5 kV to 34.5 kV
- SF_6: 15 kV to 72.5 kV

The preferred rating groups from ANSI C37.06-2000 for indoor circuit breakers are given in tables 7.2 and 7.3 for 4.76–38 kV and 15.5–72 kV, respectively. Similarly, tables 7.4 and 7.5 show the ratings for outdoor circuit breakers over the same range of voltages.

7.3 LOAD-INTERRUPTER SWITCHGEAR

Load-interrupter switchgear provide a reliable and safe method of switching as well as overcurrent protection for primary feeders (fig. 7.2). The use of load-interrupter switchgear is a cost-effective solution for distribution equipment when infrequent switching and protective relaying is not required. It can be used to interrupt load, sectionalize, isolate, and provide short-circuit protection via fuses. It is reliable, low maintenance, and economical for medium-voltage distribution applications such as main switchgear, a load-switching center, and unit substations.

Load-interrupter switchgear is generally used as secondary distribution equipment in an industrial or commercial setting, including:

- Industrial plant that requires power transformer
- Hospitals
- Airports

The standards for load-interrupter switchgear in North America are: ANSI/IEEE C37.20, ANSI C37.57, and ANSI C37.58.

Design features include:

- Reinforced-glass inspection windows to facilitate observation of switch position and general condition
- Split front door with mechanical interlock to ensure switch position
- Provisions for user padlock
- Switch position indicator, open-closed
- Removable phase barriers for easy switch maintenance
- Horizontal isolation barrier between switch and fuse compartments
- Rear compartment space for cable termination

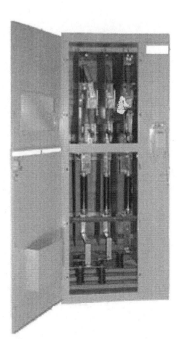

FIGURE 7.2 Load-interrupter switchgear

TABLE 7.2
Preferred Ratings for Indoor Circuit Breakers in the 4.76–38 kV Range

Line No.	Rated Max. Voltage (kV, rms) [a]	Rated Voltage Range Factor K	Rated Continuous Current (A, rms)	Rated Short-Circuit and Short-Time Current (kA, rms)	Rated Transient Recovery Voltage [b]		Rated Interrupting Time (ms) [d]	Rated Permissible Tripping Delay Y Time (s)	Rated Closing and Latching Current [e] (kA, Peak)
					Rated Voltage E_2 (kV, Peak)	Rated Time to Peak T_2 [c] (μs)			
1	4.76	1.0	1200, 2000	31.5	8.9	50	83	2	82
2	4.76	1.0	1200, 2000	40	8.9	50	83	2	104
3	4.76	1.0	1200, 2000, 3000	50	8.9	50	83	2	130
4	8.25	1.0	1200, 2000, 3000	40	15.5	60	83	2	104
5	15	1.0	1200, 2000	20	28	75	83	2	52
6	15	1.0	1200, 2000	25	28	75	83	2	65
7	15	1.0	1200, 2000	31.5	28	75	83	2	82
8	15	1.0	1200, 2000, 3000	40	28	75	83	2	104
9	15	1.0	1200, 2000, 3000	50	28	75	83	2	130
10	15	1.0	1200, 2000, 3000	63	28	75	83	2	164
11	27	1.0	1200	16	51	105	83	2	42
12	27	1.0	1200, 2000	25	51	105	83	2	65
13	38	1.0	1200	16	71	125	83	2	42
14	38	1.0	1200, 2000	25	71	125	83	2	65
15	38	1.0	1200, 2000, 3000	31.5	71	125	83	2	82
16	38	1.0	1200, 2000, 3000	40	71	125	83	2	104

Note: The preferred ratings are for 60 Hz systems. For other system frequencies, refer to ANSI/IEEE C37.010. Current values have been rounded off to the nearest kA, except that two significant digits are used for values below 10 kA.

a   The voltage rating is based on ANSI C84.1, where applicable, and is the maximum voltage for which the breaker is designed and the upper limit for operation.

b   The rated transient recovery voltage envelope is the "one-minus-cosine" (1 − cosine) shape, ANSI/IEEE C37.04.

c   If the source of power to a circuit breaker is a single transformer or a bank of transformers and there are no substantial capacitors or loaded feeders connected to the source side of the circuit breaker, the transient recovery voltage may be more severe than those covered in these tables. T_2 values for these applications are being developed.

d   The ratings in this column are the maximum time interval to be expected during a breaker opening operation between the instant of energizing the trip circuit and the interruption of the main circuit on the primary arcing contacts under certain specified conditions. The value may be exceeded under certain conditions, as specified in ANSI/IEEE C37.04, subclause covering "Rated Interrupting Time."

e   Rated closing and latching current (kA, peak) of the circuit breaker is 2.6 times the rated short-circuit. (If expressed in terms of kA, rms total current, the equivalent value is 1.55 times rated short-circuit current.)

Source: ANSI C37.06-2000.

TABLE 7.3
Preferred Capacitance Current-Switching Ratings for Indoor Circuit Breakers in the 4.76–38 kV Range

Line No.	Rated Max. Voltage (kV, rms)	Rated Continuous Current, (A, rms)	Rated Short-Circuit Current (kA, rms)	General Purpose CB [a,b]	Definite-Purpose CB [b,c] Back-to-Back Capacitor Switching					
				Rated Cable Charging Current (A, rms)	Rated Isolated Capacitor Bank Current (A, rms)	Rated Cable Charging Current (A, rms)	Rated Isolated Capacitor Bank Current (A, rms)	Rated Capacitor Bank Current (A, rms)	Rated Inrush Current [d]	
									(kA, Peak)	Frequency (Hz)
1	4.76	1200	31.5, 40, 50	10	400	10	630	630	15	2000
2	4.76	2000	31.5, 40, 50	10	400	10	1000	1000	15	1270
3	4.76	3000	50	10	400	10	1600	1600	25	1330
4	8.25	1200	40	10	250	10	630	630	15	2000
5	8.25	2000	40	10	250	10	1000	1000	15	1270
6	8.25	3000	40	10	250	10	1600	1600	25	1330
7	15	1200	20, 25, 31.5	25	250	25	630	630	15	2000
8	15	2000	20, 25 31.5	25	250	25	1000	1000	15	1270
9	15	1200	40, 50, 63	25	250	25	630	630	15	2000
10	15	2000	40, 50, 63	25	250	25	1000	1000	18	2400
11	15	3000	40, 50, 63	25	250	25	1600	1600	25	1330
12	27	1200, 2000	16, 25	31.5	160	31.5	400	400	20	4240
13	38	1200, 2000, 3000	16, 25, 31.5, 40	50	100	50	250	250	20	4240

Note: The preferred ratings are for 60-Hz systems. For other system frequencies, refer to ANSI/IEEE C37.010. Current values have been rounded off to the nearest kA, except that two significant digits are used for values below 10 kA.

a For general-purpose circuit breakers, no ratings for back-to-back capacitor switching application are established. The capacitor bank or cable shall be "isolated" as defined in ANSI/IEEE C37.04, clause "Rated Capacitance Switching Current."
b The circuit breaker shall be capable of switching any capacitive current within the range of 30–100% of the ratings listed at any voltage up to the rated maximum voltage. All circuit breakers shall meet the general-purpose circuit-breaker ratings values shown.
c If the circuit breaker is identified as a "definite purpose circuit breaker for capacitance current switching," it shall meet the specified ratings voltage.
d The rated transient inrush current peak is the highest magnitude of current that the circuit breaker shall be required to close at any voltage up to the rated maximum voltage and shall be determined by the system and unmodified by the circuit breaker. The rated transient inrush current frequency is the highest frequency that the circuit breaker shall be required to close at 100% rated back-to-back capacitor switching inrush current rating.

Source: ANSI C37.06-2000.

TABLE 7.4
Preferred Ratings for Outdoor Circuit Breakers in the 15.5–72.5 kV Range

Line No.	Rated Max. Voltage [a] (kV, rms)	Rated Voltage Range Factor K	Rated Continuous Current (A, rms)	Rated Short-Circuit and Short-Time Current (kA, rms)	Rated Transient Recovery Voltage [b]			Maximum Permissible Tripping Delay Y (s)	Rated Closing and Latching Current [e] (kA, Peak)
					Rated Voltage E_2 (kV, Peak)	Rated Time to Peak T_2 [c] (μs)	Rated Peek Interrupting Time [d] (ms)		
1	15.5	1.0	600, 1200	12.5	29	36	83	2	33
2	15.5	1.0	1200, 2000	20	29	36	83	2	52
3	15.5	1.0	1200, 2000	25	29	36	83	2	65
4	15.5	1.0	1200, 2000, 3000	40	29	36	83	2	104
5	25.8	1.0	1200, 2000	12.5	48.5	52	83	2	33
6	25.8	1.0	1200, 2000	25	48.5	52	83	2	65
7	38.0	1.0	1200, 2000	16	71	63	83	2	42
8	38.0	1.0	1200, 2000	20	71	63	83	2	52
9	38.0	1.0	1200, 2000	25	71	63	83	2	65
10	38.0	1.0	1200, 2000	31.5	71	63	83	2	82
11	38.0	1.0	1200, 2000, 3000	40	71	63	83	2	104
12	48.3	1.0	1200, 2000	20	91	80	83	2	52
13	48.3	1.0	1200, 2000	31.5	91	80	83	2	82
14	48.3	1.0	1200, 2000, 3000	40	91	80	83	2	104
15	72.5	1.0	1200, 2000	20	136	106	83	2	52
16	72.5	1.0	1200, 2000	31.5	136	106	83	2	82

Note: The preferred ratings are for 60-Hz systems. For other system frequencies, refer to ANSI/IEEE C37.010. Current values have been rounded off to the nearest kA, except that two significant digits are used for values below 10 kA.

a The voltage rating is based on ANSI C84.1, where applicable, and is the maximum voltage for which the breaker is designed and the upper limit for operation.

b The rated transient recovery voltage envelope is the "one-minus-cosine" (1 − cosine) shape, ANSI/IEEE C37.04.

c If the source of power to a circuit breaker is a single transformer or a bank of transformers and there are no substantial capacitors or loaded feeders connected to the source side of the circuit breaker, the transient recovery voltage may be more severe than those covered in these tables. T_2 values for these applications are being developed.

d The ratings in this column are the maximum time interval to be expected during a breaker opening operation between the instant of energizing the trip circuit and the interruption of the main circuit on the primary arcing contacts under certain specified conditions. The value may be exceeded under certain conditions, as specified in ANSI/IEEE C37.04, subclause covering "Rated Interrupting Time."

e Rated closing and latching current (kA, peak) of the circuit breaker is 2.6 times the rated short-circuit. (If expressed in terms of kA, rms total current, the equivalent value is 1.55 times rated short-circuit current.)

Source: ANSI C37.06-2000.

TABLE 7.5
Preferred Capacitance Current Switching Ratings for Outdoor Circuit Breakers in the 15.5–72.5 kV Range

Line No.	Rated Max. Voltage (kV, rms)	Rated Continuous Current (A, rms)	Rated Short Circuit (kA, rms)	General Purpose CB [a,b]		Definite Purpose CB [b,c]					
						Isolated		Back-to-Back			
				Rated Overhead Line Current (A, rms)	Rated Isolated Capacitor Bank or Cable (A)	Rated Capacitor Bank Current (A, rms)	Rated Overhead Line Current (A, rms)	Rated Capacitor Bank Current (A, rms)	Rated Inrush Current [d,e] (kA, Peak)	Frequency (Hz)
1	15.5	600, 1200	12.5	2	250	400	100	400	20	4240
2	15.5	1200, 2000	20	2	250	400	100	400	20	4240
3	15.5	1200, 2000	25	2	250	400	100	400	20	4240
4	15.5	1200, 2000, 3000	40	2	250	400	100	400	20	4240
5	25.8	1200, 2000	12.5	5	250	400	100	400	20	4240
6	25.8	1200, 2000	25	5	250	400	100	400	20	4240
7	38.0	1200, 2000	16	5	250	250	100	250	20	4240
8	38.0	1200, 2000	20	5	250	250	100	250	20	4240
9	38.0	1200, 2000	25	5	250	250	100	250	20	4240
10	38.0	1200, 2000	31.5	5	250	250	100	250	20	4240
11	38.0	1200, 2000, 3000	40	5	250	250	100	250	20	4240
12	48.3	1200, 2000	20	10	250	250	100	250	20	6800
13	48.3	1200, 2000	31.5	10	250	250	100	250	20	6800
14	48.3	1200, 2000, 3000	40	10	250	250	100	250	20	6800
15	72.5	1200, 2000	20	20	250	630	100	630	25	3360
16	72.5	1200, 2000	31.5	20	250	630	100	630	25	3360

Note: The preferred ratings are for 60-Hz systems. For other system frequencies, refer to ANSI/IEEE C37.010. Current values have been rounded off to the nearest kA, except that two significant digits are used for values below 10 kA.

a For general-purpose circuit breakers, no ratings for back-to-back capacitor switching application are established. The capacitor bank or cable shall be "isolated" as defined in ANSI/IEEE C37.04, clause "Rated Capacitance Switching Current."

b The circuit breaker shall be capable of switching any capacitive current within the range of 30–100% of the ratings listed at any voltage up to the rated maximum voltage. All circuit breakers shall meet the general-purpose circuit-breaker ratings values shown.

c If the circuit breaker is identified as a "definite purpose circuit breaker for capacitance current switching," it shall meet the specified ratings voltage.

d The rated transient inrush current peak is the highest magnitude of current that the circuit breaker shall be required to close at any voltage up to the rated maximum voltage and shall be determined by the system and unmodified by the circuit breaker. The rated transient inrush current frequency is the highest frequency that the circuit breaker shall be required to close at 100% rated back-to-back capacitor switching inrush current rating.

e The transient inrush current in circuit breakers applied in GIS substations has a very high equivalent frequency (up to the MHz range, depending on the bus length), with an initial peak current of several thousand amperes (depending on the surge impedance of the bus).

Source: ANSI C37.06-2000.

- Meters and instruments safely isolated from high-voltage equipment by a grounded steel barrier (optional)
- Epoxy insulators, with porcelain insulators offered as an option
- Tin-plated copper bus, with silver plating offered as an option
- Additional options, including lightning arrestors, voltage indicators, shunt trips, auxiliary contacts, blown-fuse trip indicator

7.4 POWER FUSE

Power fuses (fig. 7.3) are especially suited for protecting transformers, capacitor banks, and cables in outdoor distribution substations through 34.5 kV. They incorporate precision-engineered, nondamageable silver or nickel-chrome fusible elements with time-current characteristics that are precise and permanently accurate, assuring not only dependable performance, but also continued reliability of system coordination plans. With power fuses, source-side devices may be set for faster operation than practical with other power fuses or circuit breakers, thereby providing better system protection without compromising coordination.

The power fuses are offered with maximum continuous current ratings of 200 A, 300 A, 400 A, and 720 A in a variety of fault-interrupting ratings. They are available in a wide variety of ampere ratings, in three different speeds: standard, slow, and coordinating. This broad selection of ampere ratings and speeds permits close fusing to achieve maximum protection.

7.4.1 Advantages

- High interrupting capacity, up to 200 kA
- Reduces let-through energy and thus provides better protection for downstream elements
- Low initial cost

7.4.2 Disadvantages

- Not effective in resistance-grounded system (fuse will not see line-to-ground [L-G] faults); upstream breaker with ground-fault element will sense/trip, causing blackout in a larger area
- Source for single phasing (one blown fuse), which poses danger to electric motors
- High operating expenses

FIGURE 7.3 Power fuse

7.4.3 Fuse Size for Motor Branch Circuit

Manufacturers provide selection tables. Their selection is generally based on:

- 6.0-P.U. (per unit) motor inrush current
- 6 s starting time
- Two starts per hour

Fuse size selected on the above basis will not be suitable for high-inertia loads (blowers, etc.) or high-efficiency motors (starting inrush > 6.00 P.U.).

7.4.4 Application of Fuse Let-Through Charts

Fuse let-through charts, as shown in fig. 7.4, are used to determine the prospective fault current that will be available at the load side of the fuse. Thus, a lower rated device, i.e., molded-case circuit breaker, can be used in the circuit.

Prospective rms current should not be used to check the downstream equipment interrupting capacity. Compare $I^2 \times t$ value or ensure that the combination has been tested and certified.

7.5 MEDIUM- AND HIGH-VOLTAGE CIRCUIT BREAKER

Circuit breakers (fig. 7.5) are devices capable of interrupting fault currents and reclosing onto faults. While manufacturers' brochures and standards give extensive data on selection and performance of circuit breakers, the station designer should have a

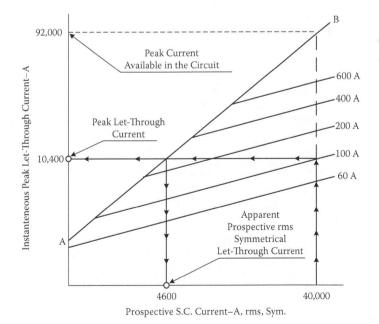

FIGURE 7.4 Fuse let-through chart

FIGURE 7.5 Circuit breakers

reasonable knowledge of the operating principles of the various types of breakers to assess their suitability for the application at hand. Assessment of station modifications and upgrading old equipment also requires knowledge of the specific characteristics of circuit breakers.

7.5.1 Overview of Medium- and High-Voltage Circuit-Breaker Standards

Standard design and performance requirements for HV (high voltage) and EHV (extremely high voltage) circuit breakers are given in the American standard IEEE C37 series and the international standard IEC 60056 [S10]. The IEEE C37 series comprises, among others, the following documents considered to be of principal importance in the specification of circuit breakers. There are some differences between the two standards, but in recent years there has been an effort to harmonize them, and for industrial installations in North America, it is acceptable to use the ANSI standards only.

IEEE C37.04 [S4] gives definitions and outlines underlying assumptions and criteria on which breaker performance is based. For the actual numerical values of the performance requirements, the standard refers to other standards, mainly C37.06 [S5] and C37.09 [S6]. Among other definitions, this standard defines the meaning of rated short-circuit current, rated interrupting time, rated transient recovery voltage, etc., and defines the required elements of the breaker interrupting performance.

ANSI C37.06 [S5] presents a series of tables giving the required design characteristics of the circuit breakers. The tables are entitled "Recommended Ratings ...," but all major manufacturers would comply with the recommendation. The tables are presented in three voltage ranges covering 4.16 kV through to 800 kV, including gas-insulated switchgear (GIS). The standard eliminates the voltage-range factor K that

Switchgear, Circuit Breakers, and Motor Control Center

was in the 1979 version of C37.06 for the 4.76–38 kV voltage range, but repeats the table for the old standard.

IEEE C37.09 [S6] outlines the test procedures for production, duty cycle, and routine acceptance testing of circuit breakers. It is for guidance in writing detailed technical specifications, and for engineers performing and witnessing acceptance tests.

IEEE C37.010 [S7] is not a standard but an application guide for the benefit of users. The guide, among others, elaborates on:

- Circuit-breaker ratings as affected by ambient temperature
- Emergency overload conditions
- Correct use of interrupting times
- Use of the E/X rating multipliers for high X/R ratios and tripping times shorter or longer than the standard range of tripping time for the breaker

ANSI/IEEE C37.011 [S8] is really a guide rather than a standard. It describes methods of calculating transient recovery voltage (TRV) and presents guidelines for parameters of different network components to be used in TRV calculations, whether done manually or by digital computers.

ANSI/IEEE C37.012 [S9] and C37.010 [S7] are also application guides for the calculation of inrush currents and transient recovery voltages of capacitor banks, cables, and long transmission lines.

7.5.2 Basic Performance Requirements: Medium- and High-Voltage Circuit-Breaker Standards

To be effective as an emergency shutdown device in a power system, a circuit breaker has to satisfy the following technical requirements:

- The breaker has to be capable of interrupting the highest value of symmetrical and asymmetrical power frequency fault current, which exists or is predicted to exist at the point of its installation in the network, during operation at highest system voltage, and stable operating conditions.
- The breaker has to withstand the highest rate of rise of TRV upon interrupting a current from a few amperes up to its maximum interrupting capability. For breakers where the TRV is likely to exceed the breaker capability, measures of TRV suppression across the breaker contacts should be implemented.
- The breaker should be capable of interrupting the out-of-phase switching current calculated for applicable conditions, or should be blocked if this current exceeds the standard limitation of the breaker for such a condition.
- The breaker should be able to withstand the electromechanical forces upon making the highest fault current and be able to latch itself closed.
- The breaker should withstand across the open contact without flashover the highest power frequency, switching impulse, and lightning impulse overvoltages it may be exposed to or have adequate overvoltage protection should the calculated values exceed the capabilities specified in the standards.

- The external insulation of live tank breakers and breaker bushings should withstand the power frequency and the switching and lightning impulse overvoltages specified in the relevant standards.

TRV in industrial installations is usually not a significant problem. There are, however, situations, particularly when involving vacuum breakers, where TRV should be checked.

7.5.3 CIRCUIT-BREAKER TYPES

Circuit breakers can be classified by the arc-quenching media they use. The following is a brief listing of the different types, in the order of their historical development.

7.5.3.1 Air-Magnetic Air Quenching

These types of breakers date back to the early 1900s and were used in the medium-voltage range to about 15 kV. They interrupt the current by drawing the arc into and extending it in a magnetic field and creating an increased air flow to extinguish it. This type of breaker was used in industrial and power installations to about the early 1960s. A well-known example is the GE or Westinghouse Magna-blast breaker, still in service in a number of installations. They are no longer readily available.

7.5.3.2 Oil Circuit Breakers

Oil circuit breakers (fig. 7.6) were the dominant type of breakers used in both the medium- and high-voltage range up to 345 kV until the early 1960s in the North American and British practice. The arc quenching is by the turbulent flow of oil originally by the effect of the arc and later by some improved design of the construction of the chamber surrounding the contacts. These breakers are distinguished by the relatively large amount of oil required for the effective and efficient arc interruption.

Oil Circuit in a 41 kV

Oil Circuit Breakers in a Circuit

FIGURE 7.6 Oil circuit breakers

Switchgear, Circuit Breakers, and Motor Control Center

Oil circuit breakers are used to switch circuits and equipment in and out of a system in a substation. They are oil filled to provide cooling and to prevent arcing when the switch is activated.

7.5.3.3 Air-Blast Breakers

Air-blast breakers (fig. 7.7) were used in Europe from the early 1930s in both the medium- and high-voltage range, but they were used in North America only from the late 1950s, and because of the advent of SF_6 gas circuit breakers, only for a relatively short period of time. All air-blast breakers are based on the principle of high-pressure air, on the order of 200 psi, forced through a nozzle and passed at high speed across the arc, partly to lengthen the arc and partly to carry away the arc/plasma product from in between the contacts. To this end, a central high-pressure storage or high-pressure individual breaker air storage is used.

Air-blast breakers have been manufactured for voltages up to 800 kV, and were used for the development of 500 kV systems. Because of their good functionality at low temperatures, they were frequently used in the northern latitudes.

7.5.3.4 SF_6 Gas Circuit Breakers

SF_6 breakers (fig. 7.8) were developed in the 1960s and quickly became the preferred choice for wide areas of medium- and high-voltage applications. These breakers utilize SF_6 gas for the extinction of arc, having dielectric properties superior to that of air, leading to a much more economical design in the construction and dimensions

FIGURE 7.7 Air-blast breakers

FIGURE 7.8 SF$_6$ gas circuit breakers

of the breaker. The voltage rating for a single interrupter head is mostly limited to 245 kV, hence more than one head in series is used for higher voltages. The initial development of the SF$_6$ breaker was the so-called double-pressure breaker mimicking the operating principle of an air-blast breaker.

This construction was replaced by the so-called puffer breakers. The SF$_6$ gas is passed to and held in the contact chamber at a gas pressure of about 50 to 70 psi (three to five times atmospheric pressure). Upon contact parting, a piston compresses the gas and blasts it through the arc, returning it to the closed SF$_6$ circulation loop in the breaker. SF$_6$ breakers are the breaker type predominantly used in the HV and EHV voltage range, and widely used for medium voltages.

7.5.3.5 Vacuum Circuit Breakers

Vacuum breakers (fig. 7.9) utilize the superior dielectric nature of the vacuum for arc extinction. They consist of a vacuum-tight "bottle" housing the fixed and moving Cu or Cr contacts. The contact movement and separation is very small, on the order of millimeters rather than centimeters, which makes the operation of these breakers and their ability of arc extinction very fast. Some problems associated with this phenomenon will be discussed later.

Vacuum breakers for fault-current interrupting capabilities similar to SF$_6$ breakers are available in the medium-voltage range up to about 35 kV and at 25-kV single phase (50-kV equivalent) as

FIGURE 7.9 Vacuum circuit breakers

Switchgear, Circuit Breakers, and Motor Control Center

dual bottle assemblies for electric traction. Vacuum switches as load-break switches consisting of a series of assembly of vacuum bottles are available up to 245 kV.

7.5.4 Circuit-Breaker Short-Circuit Rating and E/X Method

This method is a way of estimating the relationship between asymmetric breaking current and the computed steady-state fault current to avoid the need to calculate the actual asymmetrical current at contact parting when rating circuit breakers.

To illustrate the effect of asymmetry in a typical fault-current envelope, fig. 7.10 shows the plot of the short-circuit current with the time of contact parting marked:

- Curve A is the total peak value of the short-circuit current along the major current loops.
- Curve D is the rms value of the fault current.
- Curve C is the DC component, which reflects the asymmetry.

Figure 7.11 shows the DC component of the short-circuit current as a percentage of the symmetrical rms component. The magnitude of the DC component will occur within 45 ms of fault inception, corresponding to an X/R ratio of 17. This is the DC current for which the breaker has to be designed and tested. The value of X/R = 17 is a cutoff point in the IEEE standard, implying a design but not necessarily an actual limitation in assigned interrupting capability.

The marks on the curve show the percentage of DC current that a breaker with a given cycle classification (contact parting time) would experience over the time from fault inception to contact separation.

The E/X method is a unique feature of the ANSI/IEEE circuit-breaker standard and is described in detail in the standard C37.010 [S7]. A family of curves, precalculated in the standard, gives the relation between initial short-circuit current, system X/R ratio, contact parting time after fault inception, and the E/X multiplier. The required interrupting current is deemed to be the initial value of the fault current, three phase or single phase, multiplied by the E/X factor read from the curves. This

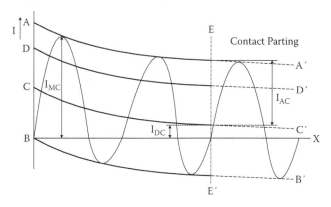

FIGURE 7.10   Short-circuit current

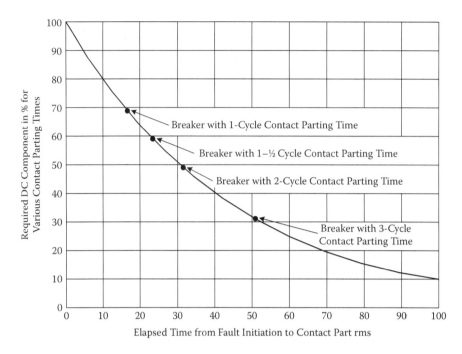

FIGURE 7.11 Required DC-component interrupting capability

factor includes the effect of the DC component at contact parting. A total of four sets of curves have been developed, including faults remote from or close to generation.

The E/X multiplier, therefore, can be thought of as a correction factor to account for the DC component. The use of the E/X correction can be omitted when the initial fault current is less than 80% of the rated interrupting capability of a breaker in the C37.06 rating tables and the X/R ratio of the system seen from the location of the breaker is less than 17.

The contact parting time of the breaker is entered in the curves as two, three, five, or eight cycles as the design operating speed given by the breaker manufacturer.

7.6 SF_6 GAS-INSULATED SWITCHGEAR

Gas-insulated switchgear (GIS) has come to be a major piece of substation equipment. Over that period, GIS has been improved by technological advancements that have increased the interrupting capacity of the circuit breaker (CB) and decreased the number of breaks per single CB without using an air condenser. The configuration of the GIS comes in several arrangements, such as the single bus, double bus, and ring-bus types. Because the arrangement is determined by a user's specifications, we must create a different design each time.

GIS offers high reliability, safety, and maintenance-free features in a much smaller space compared with conventional switchgear. The SF_6 GIS (fig. 7.12) guarantees outstanding advantages for the planning and operation of high-voltage supply

Switchgear, Circuit Breakers, and Motor Control Center

FIGURE 7.12 SF_6 gas-insulated switchgear

networks. For instance, the space requirement can be less than 10% of the space taken up by an equivalent conventional installation.

As all live parts are metal enclosed and hermetically sealed, the SF_6 GIS is completely immune to atmospheric pollution (fog, sand, salt, etc.) and industrial pollution (dust, smoke, gas, etc.). With SF_6 GIS, all foreign bodies (persons, animals, and tools) are prevented from coming into contact with live parts. This ensures maximum safety to personnel and increased continuity of operation.

7.6.1 Main Features

- *High reliability*: The live parts are sealed in metal enclosures filled with pure SF_6 gas. Thus, the switchgear is protected from contamination by smoke, chemical fumes, dust or salt-laden spray or by the ingress of small animals. This protection prevents the inside components from oxidation and rust caused by oxygen and moisture.
- *Space reduction*: A medium-voltage GIS can be installed into a space about half that required for conventional air-insulated metal-clad switchgear.
- *High safety*: All the live parts are fully enclosed in grounded metal enclosures. Thus, there is no danger of an electric shock. SF_6 is an inert, nonflammable, nontoxic, and odorless gas used as an insulation and arc-extinguishing medium. Therefore, it is safe for personnel and there is no fire hazard.
- *Short installation time*: A GIS of about four panels can be transported fully constructed. The GIS can be energized by connecting external power cables to the prefabricated plug-in type terminals, thus shortening installation time.

7.6.2 Design Concepts

Table 7.6 shows the design concepts of the standard SF_6 GIS. To minimize the installation volume, all enclosures are three-phase common type for a high-voltage GIS, and only the main bus enclosures are three-phase common for a 220 kV or larger GIS. Parts applicable to indoor and outdoor uses have been designed. For outdoor use, liquid packing is injected into the flanges of main circuits to keep the environmental resistance high.

TABLE 7.6
Design Concepts of Standard GIS

GIS	1. Reliability improved by reducing the number of parts and the size and weight of the GIS
	2. Same modular components applicable to every configuration
	3. Transportable structure as a fully assembled GIS
	4. CTs arranged at both sides of GCB
	5. Insulating spacers arranged vertically
	6. Applicable for both outdoor and indoor use
GCB[a]	1. No condenser between poles
	2. Three-phase common-type motor-spring-operated system for up to 245 kV GCB; single-phase hydraulic-operated system for 300 kV or larger
DS/GS[b]	1. Three-phase common buses are standard; isolated-phase buses also available without changing the structure of GIS
	2. Motor-spring-operated system for high-speed grounding switch for line
Lightning arrestors	1. Made compact through the use of zinc oxide elements having high-pressure resistance
	2. Reduction of insulation level made it possible to reinforce the insulation coordination
Main bus	1. Three-phase common buses are standard; isolated-phase buses also available without changing the structure of the GIS
	2. Bellows provided at each bay

[a] GCB: gas circuit breaker.
[b] DS/GS: disconnecting switch/grounding switch.

Improvements in reliability of assembly and installation works are achieved by assembling one bay each and transporting it as a fully assembled GIS. Insulating spacers are vertically arranged to eliminate the effect of particles, thereby increasing insulation reliability.

7.6.3 Specifications

The specifications and structure of transmission line circuits in GIS (e.g., 550 kV) are shown in fig. 7.13. The factors that determine the size of the tank are the designed gas pressure, the allowable particle length for dielectric characteristics, and the temperature-rise specifications for the current-carrying performance.

Switchgear, Circuit Breakers, and Motor Control Center

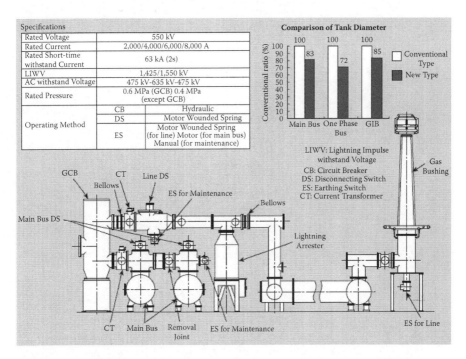

FIGURE 7.13 Example of specification and structure of line circuit of 550 kV GIS

7.7 LOW- AND MEDIUM-VOLTAGE MOTOR CONTROL CENTERS

7.7.1 Low-Voltage Motor Control Center

A motor control center (MCC) is a modular cabinet system for powering and controlling motors in a factory. Several low-voltage MCCs (fig. 7.14) in a factory may be powered from a main switchgear, which in turn gets its power from a transformer attached to the incoming line from the power company. MCCs are very common sights in factories with heavy machinery, although the purpose they serve is changing.

MCC will be three bus bars to carry three-phase, 600 V, 60 Hz electricity. The buckets are intended to contain a circuit breaker, a motor starter, and a control power transformer, although their actual contents will vary widely according to the factory's needs. The circuit breaker will have a handle that goes through the door to shut off the power without opening it.

Several years ago, MCCs housed almost all the control for motors in an industrial setting. The trend lately, however, has been to use the MCC only to provide points from which to distribute power to electrical panels. In this case, the MCC bucket will have only a circuit breaker and a handle to shut the circuit breaker off; the electrical panel would be a separate enclosure that houses all the motor control for a specific machine. The power and popularity of programmable logic controllers (PLC) for control and the size of high-horsepower variable-frequency drives have contributed to this trend, as they do not fit in a standard MCC very well.

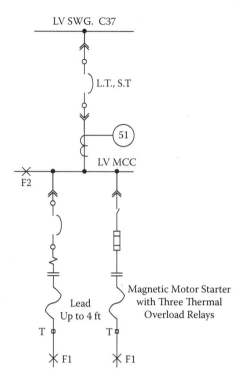

FIGURE 7.14 Low-voltage MCC

7.7.2 MEDIUM-VOLTAGE MOTOR CONTROL CENTER

A medium-voltage controller (fig. 7.15) for electrical equipment, such as motors, transformers, reactors, and capacitors, has an arc-resistant cabinet, swaged internal electrical connections, a one-piece self-aligning withdrawable finger cluster, a pull-out instrument compartment, a load discharge device, cast fuse holders, a disconnect switch, a switch illuminator, low-power current transformers, and an optical temperature measuring system.

Internal electrical connections between controller components are made with connectors and cables. The cable conductors are swaged or compression welded in the connector barrels, forming a cold-welded electrical and mechanical junction.

The pull-out instrument compartment includes a sliding rack to which a hinged instrument mounting panel is attached. The mounting panel slides out of the compartment and swings away from the opening in the cabinet, allowing access to the cabinet internals.

7.7.3 SHORT-CIRCUIT CAPACITY

Short-circuit rating, as defined in the NEMA standards, is based on a short circuit beyond (up to 4 ft) load terminals. This will considerably reduce the let-through fault current due to the impedance of various elements. These include bus bars, circuit breakers, fuses, overload heaters, leads, etc. Standards permit some damage to the equipment; however, the door must not open.

The bus bars (main and vertical) may be damaged by bus faults if they are not protected by instantaneous elements (50) or current-limiting fuses.

7.7.4 COORDINATED PROTECTION IN MEDIUM-VOLTAGE FUSED STARTERS

- Overload/stall relay must provide protection up to 12 × motor full-load current
- Minimum fuse rating of 1.3 × motor full-load current
- Fuse minimum melting curve is to:
  1. Intersect overload/stall relay curve at 110% of the motor locked-rotor current
  2. Intersect contactor drop-out time at a current below the contactor interrupting rating
- Fuse must be large enough to permit motor starts in hot conditions

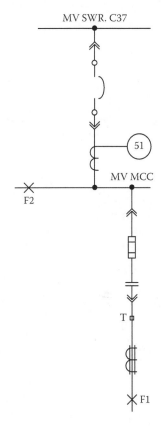

FIGURE 7.15 Medium-voltage MCC

7.7.5 MAIN AND VERTICAL BUS BARS

Main and vertical bus bars must be braced for short circuits against mechanical and thermal damage. Values given in the standards and manufacturers' catalogs are valid if the bus bars are protected by instantaneous elements. Specify as 22 kA, 42 kA, etc., for 0.5 s. This will permit the use of a short delay in the static-trip unit of the feeder breaker for coordination purposes without risking damage.

7.7.6 MINIMUM SIZE

Minimum size shall be as specified by NEMA/EEMAC 1. Derate for jogging or reversing duty. At upper limits, use one size larger for high-inertia loads or loads with long acceleration time. For example, for a 100 hp fan, use a size 5 starter instead of size 4.

7.7.7 IEC-RATED CONTACTORS

Manufacturers offer contactors to IEC standards. For each size, IEC has four categories (AC 1, AC 2, AC 3, and AC 4). The minimum equivalent to NEMA is AC 3. Compare current rating and duty cycle before selecting the category.

REFERENCES

Standards

S1. CNS3989-C1044, NEMA-ICS-2-322 JEM-1195, IEC-349-1-MCC.
S2. ANSI/IEEE C37.20, ANSI C37.57 and C37.58, NEMA SG-5, CSA C22.31.
S3. IEEE C37.71 (TM), C37.72 (TM), and C37.73 (TM) Load Interrupter Standards.
S4. IEEE C37.04-1999, Standard Rating Structure for AC High-Voltage Circuit Breakers Rated on Symmetrical Current Basis, 1999.
S5. ANSI C37.06-1997, AC High-Voltage Circuit Breakers Rated on a Symmetrical Current Basis: Preferred Ratings and Related Capabilities, 1997.
S6. IEEE C37.09-1999, Standard Test Procedure for AC High-Voltage Circuit Breakers Rated on a Symmetrical Current Basis, 1999.
S7. IEEE C37.010-1997, IEEE Application Guide for AC High-Voltage Circuit Breakers Rated on a Symmetrical Current Basis, 1997.
S8. ANSI/IEEE C37.011-1997, Transient Recovery Voltage for AC High-Voltage Circuit Breakers Rated on a Symmetrical Current Basis, 1997.
S9. ANSI/IEEE C37.012-1997, IEEE Application Guide for Capacitance Current Switching for AC High-Voltage Circuit Breakers Rated on a Symmetrical Current Basis, 1997.
S10. IEC 60056, High-Voltage Alternating-Current Circuit Breakers.
S11. IEEE C37.10.
S12. IEEE C37.13, 1990.
S13. NEMA/EEMAC IC1 ANSI and ANSI C19.
S14. CNS3989, NEMA-ICS-2-322, IEC-439-1.
S15. ANSI C37-06-2000.

8 Station Battery

8.1 INTRODUCTION

In many industrial processes, and particularly for the electrical power distribution systems within plants, highly reliable auxiliary power suppliers are required. The expansion of the use of computer control systems and digital electrical network protective systems have increased these requirements. Some systems, such as motor control centers (MCC), can have their control and protection circuits powered from the same AC power that feeds the MCC bus. In such cases, individual control transformers are used for each circuit. However, most of the more complex control and protection systems, particularly those that need to be functional even when the power to that part of the electrical network is interrupted, need more highly reliable power supplies. Such power supplies are then typically based on DC-backed systems, with batteries providing the energy storage required.

8.2 SYSTEM TYPES

Power supply systems that use battery systems to provide energy storage fall into two basic categories: (a) systems that feed DC power directly to the individual systems requiring continuous power and (b) systems that provide reliable uninterruptible AC power via uninterruptible power systems (UPS). The former systems require batteries, chargers, and a DC distribution system, while the UPS replaces the DC distribution by an AC distribution system that is fed by an inverter system (or systems) to provide the required AC power from the battery.

In addition, high-speed static switches provide fast transfer to an alternative source, for inverter or similar failures, hence providing the "uninterruptible" nature of the supply. Because supply systems based on direct DC distribution are fed directly from a battery system, they do not require complex switched backup or the complexity of an inverter. However, both DC and UPS AC-fed systems usually have switched transfer capability to allow for battery and switchgear maintenance.

Uninterruptible power supply systems have been often used to feed those computer systems and computer peripherals that require AC power. In the industrial context, it has also been practice to feed the control devices, such as programmable logic controllers (PLC), from AC power. However, the majority of electronic systems operate internally on DC, and most use, or have available, DC power supply options using AC/DC converters. Such equipment can therefore also be fed from a DC power distribution system.

As an alternative to the use of a UPS for those individual equipments that require AC input power, individual or grouped inverters can be provided to feed these devices from a DC distribution system.

Direct DC distribution systems do not usually provide high-speed transfer to an alternative power source; the individual devices fed from such systems are either duplicated or provided with back-up facilities. Most power system protective devices are usually configured in this way, although in many cases, such connections may only cover for individual power supply circuit failures. In more critical cases, duplicated systems may be fed from two independent DC-battery-backed systems, or individual equipment may be dual powered so that failure of one power feed does not shut down that equipment. However, dual-powered systems must be carefully designed so as not to cause "common-cause" hazards to both in-feeding systems.

The batteries used for the direct DC power supply system's battery systems have a number of different duty cycles and standby times, dependent on their function, and different designs of battery have been developed for such different duty cycles. UPS battery systems in particular are required to be operated continuously in standby mode, with the battery normally on float charge, and then provide full power for the designed back-up time after the AC power supply fails.

However, the DC systems feeding switchgear and switchyards typically carry heavy short-term loads, often in excess of the battery charger rating, and even when AC power is available, the peak power is drawn from the battery. The battery is then immediately recharged as soon as the load is removed. This cyclic loading is usually mixed with a lower level of steady-state load, and this steady load is usually fed from the battery charger at the same time as the charger provides a float charge to the battery.

During the loss of AC power to the charging system, the steady base load and the peak load must be fed. In those cases, the peak loads that occur during a power failure depend on the switching activity in the station by the battery. The switching sequence that is required to safely shut down and then to restore the AC power system are the most important loads that need to be considered in designing such systems. However, there may also be other loading conditions that occur during a power outage that combine the pulsed cyclic loads with different levels of steady load.

8.3 DC DISTRIBUTION SYSTEMS

The method of connection of the battery, battery charger, and distribution systems depends on the duty, the type or load, and whether the system needs to be duplicated or whether duplicate chargers are required. One typical example for a 125 V system is shown in fig. 8.1.

In this example, an alternative connection is shown for the battery-to-charger connection, and that alternative is preferred when sensitive digital systems are being fed or when there is a longer distance between the distribution panel and the batteries and charger. This alternative has the advantage that the battery capacitance acts as a filter and reduces the electrical noise on the DC bus and limits interference into associated electronic and communication systems.

Figure 8.1 also shows the possibility of cross connection to a second DC system, and fig. 8.2 shows a fully duplicated high-reliability system with manual transfer capabilities. In this configuration, the transfer (marked * in fig. 8.2) would be break-before-make, resulting in a short loss of power while the transfer is made. However, by adding block-

Station Battery

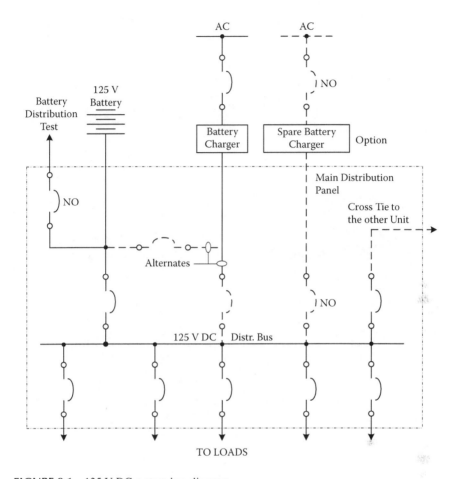

FIGURE 8.1 125 V DC system key diagram
All breakers normally closed except those marked "NO"
"----------" indicates optional or alternative features
Fuse may be substituted for circuit breakers for any or all circuits except where transfer switching is required

ing diodes into the crossover circuits, a short time parallel can be allowed without unduly hazarding both supplies. Similarly, in fig 8.1, a standby battery charger is shown with its circuit breaker normally open. Again, by providing blocking diodes on each charger feed and purchasing chargers designed to operate in parallel, both chargers could be operated simultaneously to share the load. An extension to such a system, which would be applicable when high-reliability DC supplies are necessary, is shown in fig. 8.2.

As in all dual, dual-supply, or redundant application cases, it is essential to minimize and, if possible, eliminate common-cause events that fail both systems. This aspect should always be considered when deciding whether to apply live or dead transfer and/or dually powered devices or deciding to provide totally independent redundant systems.

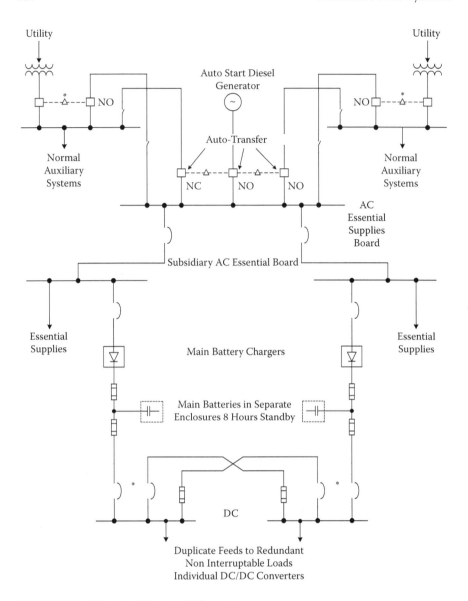

FIGURE 8.2 High-reliability dual-DC supply system

In the example shown in fig. 8.2, the AC system feed and the DC transfer connections use a crossover circuit rather than a single tie switch. Such a connection provides a simple transfer connection for operation and interlocking and allows each switch or circuit breaker in the transfer arrangements to be isolated for maintenance without the need to shut down both supply systems.

Battery protection fuses are shown. The application of such fuses to protect both the battery and the battery cables to the maximum extent possible are discussed further by Smith [5]. In this particular example, fuses are provided on both the charger

Station Battery

and feeder side to avoid the risk of a fuse operation disconnecting the battery while leaving the charger feeding the DC loads alone. This connection is required if the charger regulation and filtering is such that the resultant feed is unsuitable to feed sensitive electronic systems. In many cases, unless special-duty chargers are purchased, the charger can, if operated without a battery connection, cause premature power supply failure and unacceptable electrical noise levels in the distribution system.

For any battery-protection fuse arrangement, the fuses should be located as close as possible to the battery. However, for those batteries and battery configurations that can cause accumulation of explosive by-products, such as free-breathing and vented lead–acid batteries in battery rooms, the fuses need to be placed in a suitable protected enclosure, mounted just outside the battery room. In such cases, separate positive and negative connections should be run from the fuse enclosure to each battery terminal, thereby retaining the maximum possible separation. This needs to be done to minimize the risk of a fault between poles in the unprotected zone. However, from the fuses to the charger and the distribution boards, standard-two conductor cable could be used.

A third example is shown in fig. 8.3, which is an example of a dual supply and dual-battery AC and DC power supply arrangement. For a small hydroelectric generating station, AC and DC distribution supply is used. However, the transfer switch arrangements shown between the chargers and the two batteries in this example, which allow either charger to charge either battery, are less common.

In this arrangement, the battery protection fuse is a single fuse in the battery connection and would be suitable where the enhanced-performance chargers are used or if the downstream devices have wide power supply voltage range and have enhanced electrical noise withstand levels. Such battery fuse arrangements were also common when only electromechanical relay devices were used, noting that there were still risks in operation on a charger alone if emergency DC motors were fed from the same bus.

The single transfer switch arrangement to switch between the two battery systems would be typical for stations where operators were available to make the transfer if there were a failure on either battery system. For remotely controlled stations or for sites where automated transfer is more appropriate, the single transfer switch could be replaced by an interlocked autotransfer arrangement. This could also be provided with a short parallel (make before break) transfer by the addition of blocking diodes in each battery feed, noting that the provision of blocking diodes and protective circuit breakers would also make a continuous dual-feed arrangement feasible.

8.4 TYPES OF BATTERY

Historically, the early types of large rechargeable batteries for industrial and switchyard/power plant use were all lead–acid and, initially, open-top batteries were used, which could be easily refilled and their electrolyte levels and specific gravity simply checked. These batteries gave off significant levels of hydrogen and needed to be housed in separate ventilated rooms, and to minimize the explosion risk, explosion-proof lighting equipment was used in these battery rooms. Later developments produced the covered cell with a vented cap to allow the electrolyte to be checked and

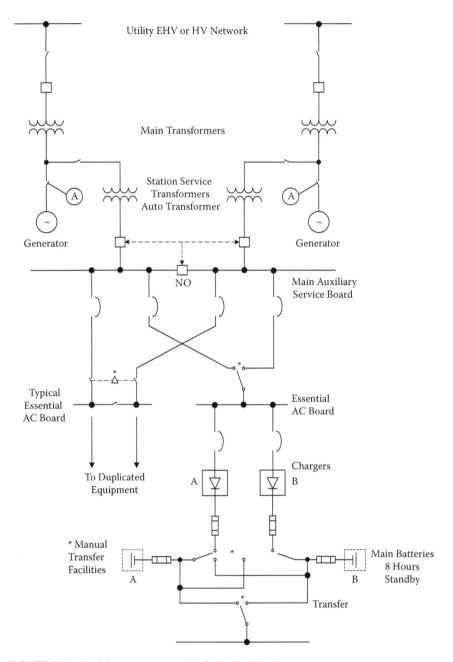

FIGURE 8.3 Dual-battery system with single distribution

topped up, while parallel developments with charger control significantly reduced the amount of gassing that occurred and, in many cases, limited this to occasions when the battery was being fast charged after a deep discharge. These developments also significantly reduced the boil off of electrolyte in the charging cycle, and charger settings were chosen to minimize the amount of loss and hence the requirements for frequent topping-up of the electrolyte.

More recent developments have produced the vented-valve-regulated lead–acid (VRLA) which is a lead–acid cell that is sealed, with the exception of a valve that opens to the atmosphere when the internal gas pressure in the cell exceeds atmospheric pressure by a preselected amount. VRLA cells provide a means for recombination of internally generated oxygen and the suppression of hydrogen gas evolution to limit water consumption. These come in two versions, the "absorptive glass-microfiber" (AGM) type, where the electrolyte is contained in the microfiber "sponge," and the gelled electrolyte "gel-cell" type. Parallel developments in other battery types have produced a design of a large nickel–cadmium cell that is suitable for large stationary rechargeable battery use. With these developments, the main choices for such batteries for industrial and switchyard use are:

- Lead–acid (antimony) VRLA
- Lead–calcium VRLA
- Nickel–cadmium
- Flooded-cell lead–acid (antimony or lead–calcium)

For industrial use, lead–acid batteries currently occupy over 90% of the market, although developments for other uses of more sophisticated battery chemical configurations may change this in future. Lead–acid batteries have a long life so long as the configuration of the cells and cell electrodes (plates) is correctly designed for the duty cycle. However, their performance at high and low temperature is compromised, and special precautions need to be taken outside the range 0–40°C. Although nickel–cadmium batteries are more costly, difficult to dispose of, and hence costly to recycle, they have application where extreme temperatures in the range −30°C to +50°C exist.

Nickel–cadmium batteries are not acid based and do not produce explosive gases during their charge–recharge cycles. They can therefore be placed more flexibly within a building and can be installed within cabinets, often in the same cabinets as the associated charger. They are therefore often used where a small amount of equipment needs to be supplied with DC, as that equipment and the associated battery system can be installed together in the same room. A typical application would be for local communications equipment. On the other hand, it should be noted that the environmental regulations are more stringent for nickel–cadmium batteries.

One of the advantages of lead–acid batteries is that the voltage change during the discharge cycle changes more gradually than nickel–cadmium, which makes charger control and alarming more sensitive, giving more warning before dangerous levels of discharge are reached.

With the VRLA battery cells, much lower levels of hydrogen are produced, and in most cases the internal pressure is kept within the limits of the regulating valves, and no hydrogen is emitted. Fewer precautions therefore need to be taken with their location. However the various safety codes [S6–S8] that require specific precautions in terms of restricting access to battery systems for personnel safety still have to be met, even if the risk of explosion and the need for ventilation are reduced. Both versions of the VRLA battery have their electrolyte immobilized in the spongy material or as a gel, and they are often configured with their cells on their side, which can provide more flexible location possibilities. This allows the installation flexibility, which used to be an advantage of the nickel–cadmium battery, to be combined with the advantages of lead–acid in terms of performance and cost.

The VRLA cell types are predicted to have shorter life cycles than the flooded electrolyte, breathing types, especially if overload or overcharge incidents cause them to vent. However, the actual long-life experience is not yet available, and better-than-predicted performance may be achieved if their environment, particularly temperature, can be controlled. Their virtually maintenance-free operation can also make their use economically attractive, even if the cells need to be replaced more frequently, noting that their life should still exceed 10–15 years. In both types of VRLA cells, recombination techniques are applied, and provided that the design duty and recharge cycles are met, problems with internal pressure rise should not occur and the risk of venting should be minimal.

The flooded-cell types, where the electrolyte level and specific gravity can be checked and corrected, may still have advantages, particularly where more extreme duty cycles and fast recharge are required, or where the duty cannot be easily controlled. However, the safety precautions appropriate to the handling of acids and for venting and explosion risk containment, including the need for eyewash and shower facilities, which were common with the older battery installations, will then have to be met.

8.5 BATTERY CHARGERS

Normally the charger operates continuously to supply the continuous load plus a small current to maintain the battery at full charge (float). Any excess load is supplied by the battery, which is automatically recharged when the intermittent load ceases. The battery carries the entire load of the DC system when the AC power supply to the charger is lost or the charger has failed. The charger capacity needs to be large enough to fully recharge the battery from fully discharged in a specific time, and this needs to be done while also feeding the normal standing load on the battery. In such a case, it is possible to compute the charger size using the following:

$$\text{Charger capacity (amp)} = L + \frac{1.15 \times C}{H}$$

where

L = continuous load, A
C = battery capacity, A·h
H = recharge time, usually between 8 and 10 h

In some cases, where the full-duty cycle is not well known or may well change with time, it is often considered that the continuous load is equivalent to the full battery rating divided by the design standby time. As this is typically 8 h for station batteries, the charger ampere rating is usually somewhat larger than twice the equivalent 8-h battery discharge rate.

Station Battery 189

Where the battery and charger are used in areas where temperature and altitude derating apply, the results obtained from the above sizing analysis should be increased by multiplying the calculated result by two additional factors, $1/C_1$ for temperature and $1/C_2$ for altitude. As discussed by Nelson and Bolin [1], the typical value for C_1 would be 0.8 at 55°C, while C_2 would be 0.9 at 1500 m.

Battery chargers need to be appropriate for their application, and if chargers are to be used in parallel then, as noted in section 8.3, both chargers should have similar characteristics and be specifically designed to operate in parallel. In such a case, the charger sizing rules can be modified to allow a longer design recharge time if one of the chargers is not available, rather than sizing each single charger to provide the full duty. Chargers must be designed to suit the battery design and charging cycle requirements of the particular battery design, especially if VRLA cells are used or in cases where low gassing during recharge is required.

Chargers used for communication equipment batteries need to be filtered to meet the psophometric (audible noise weighting) noise requirements of the DC system [S9], while those that could need to provide power with the battery disconnected need to have sufficient filtering and voltage control to keep the DC supply and superimposed DC ripple within the tolerances required of the load equipment.

Chargers can use either single- or three-phase power, dependent on the supply type and voltage available and their electrical load. As battery chargers are typically switched power devices, they can produce harmonics back into the power supply, and this needs to be controlled. This is always easier if three-phase connections are used and if a higher voltage level is available, and hence large battery chargers are usually three phase and fed from the 600 V or 480 V power level in North American applications.

8.6 APPLICATION CRITERIA

The types of loads that are typically provided with DC power have been discussed in some detail in the earlier sections of this chapter. One of the differences between UPS AC-fed battery-backed systems and those systems that usually are directly fed by DC is based on whether a load can withstand the ¼-cycle loss of voltage typically found when switching to UPS standby feeds without appreciable effect. In this respect, direct DC-fed power supplies will usually have fast circuit breakers and fuses, and hence the power supplies only have to withstand the voltage drop during fault clearing. Also, direct DC-fed power units usually can include capacitor storage to ride through such incidents.

8.6.1 Application Voltage Levels

Battery systems with DC distribution are typically rated at 24, 48, 110/125, and 220/250 V. Systems that use 24 V are typical of those used for industrial control, while 48 V is almost universally used for telecommunications. Because 48 V was also the voltage used for relay-based substation remote-control systems, the same techniques were used in electromechanical telephone switching.

The 110/125 V level is typically used for most switchyard and switchgear applications, with 125 V being common in North America, while 110 V is typical for international and European designs. Voltage levels of 220 V and 250 V have been used extensively in generating stations, particularly where DC motors or motorized valves are used, although as more digital systems with low energy requirements are used, separate lower voltage battery systems are often used to feed the digital devices in such stations. Many large UPS systems use DC voltages in the 200–300 V range, while the smaller individual units, such as those used to provide reliable power for individual computers, may have a 12 V battery. As the battery is internal to the UPS, the battery voltage does not have to match any standard voltage level for equipment, and therefore the actual voltage used can be chosen to match the duty cycle required of the UPS.

The basic voltage rating of a fully charged lead–acid design is a nominal 2 V per cell, and that of a fully charged nickel–cadmium battery is 1.2 V per cell. This leads to the typical lead–acid battery configurations of 12 cells for 24 V, 24 cells for 48 V, and 55 or 60 cells for 100 V or 125 V batteries, respectively. Although the North American battery voltage is often given as 125 V, it is actually 120 V nominal when 60 cells are used and, in many cases, is usually somewhat smaller to limit the service voltage when providing the battery with an equalizing charge. The respective cell numbers for nickel–cadmium are therefore 18–20 for 24 V, 36–40 for 48 V, and pro rata for the higher voltages, although these numbers tend to vary more than the lead–acid cell counts between particular applications.

8.6.2 Rating

Rating is given in ampere-hours (Ah) at a specific discharge rate for a fixed period. This Ah rating is the product of the nominal discharge current in amperes multiplied by a given time period (at initial temperature of 25°C). For the longer times, the rating is close to linear, but for shorter periods the linearity is lost, and manufacturers' tables are needed to obtain these shorter time ratings. Typically, a 300 Ah battery with a nominal 8 h discharge rate can also deliver, proportionally, 100 A for 3 hours, whereas it cannot deliver 300 A for 1 h.

Most batteries are almost always made up of a series connection of single cells. Consequently, the Ah capacity of a battery is the same as the Ah capacity of each of the single cells, depending on the dimensions and number of plates and the specific design of the cell.

8.6.3 Load Classification

The loads that emergency power supply systems have to supply, whether UPS AC systems or direct DC distribution systems, will vary in time, depending on the type of load they have to feed once the main AC power feed has been lost. The main classifications used in calculating the battery size required to provide the specified standby load are:

- *Continuous*: lighting, indication lamps, communication equipment, power supply converters, etc.
- *Noncontinuous*: emergency pump motors, fire protection system, valve operations, etc.

- *Momentary*: switchgear, circuit breaker, disconnect and load-break switch operation, synchronous generator field flashing, motor-starting inrush currents, etc.

It is usual to classify loads as noncontinuous if the power is taken for 1 minute or longer, while momentary loads are taken for shorter times.

Continuous loads are a significant part of the load cycle when sizing the battery or fixing the standby time that is necessary. In some cases, particularly when long standby times are required, it is sometimes necessary to consider whether all loads are of the same level of importance and, in some cases, shed some of the continuous load to safeguard the more essential loads. Typically, for switchyard and substation applications, where reswitching the network is of particular importance, emergency lighting can be cut after a sufficient period to evacuate an area and only turned back on, usually by reinitiating the timer control of the lighting, when the lighting is actually required.

The other alternative is to use a separate battery or UPS for the lighting or to use self-contained battery-backed emergency lighting units to provide the necessary escape and lead lighting. These individual units also carry the advantage that the failure of one such individual unit should not remove all the lighting in an area, whereas if a central supply unit fails, all emergency lighting can be lost.

8.7 BATTERY SIZING

Batteries need to be sized correctly to be able to feed the required load for the required time, and a number of factors need to be decided to be able to optimize the battery for the duty expected. Some of these factors are fixed within the chemistry of each type of cell and, in some cases, the physical structure of the plates that make up the cell. The performance is also influenced by the temperature and other location factors, and as an optimal combination of cells is needed to provide the required performance, the following factors need to be considered:

- Maximum system voltage
- Minimum system voltage
- Correction factor
- Duty cycle

8.7.1 System Voltage

The cells that make up any battery have a limited voltage range specific to the type of cell being used. In the case of lead–acid batteries, the cell nominal voltage, which is the voltage of a fully charged cell without any input charge or load, is 2 V. On the other hand, the minimum voltage that a battery cell can safely supply a load without damage is typically 1.7 V, although to give a margin of safety, it is more normal to use 1.75 as the operating minimum. Similarly, to be able to charge a battery, the voltage across each cell must be more than the nominal 2 V, and to keep a battery fully charged, each cell typically needs to be kept energized at 2.2 to 2.25 V, dependent on cell construction.

This is the float-charge voltage. Because individual cells in a battery can develop higher impedance than others when floated for a significant time, or after they are discharged, batteries only charged on "float" can result in some cells being less charged than others. To overcome this condition, it is necessary to subject the battery to a higher voltage, the equalize charge voltage, which could be up to 2.7 V per cell. Although the higher voltage would allow a faster recharge and would even up the charge on individual cells more quickly, this level of cell voltage would make the battery voltage range exceed the rating of most equipment that uses DC supplies. It is therefore usual practice to keep the equalize charge in the range of 2.33 to 2.5 V per cell and extend the time required to equalize the battery. Based on the above, the common battery size for a 125 V North American battery uses 60 cells with a battery voltage range of 105 to 140 V DC. This range is computed as follows:

- Equalizing voltage = 2.33 V per cell
- Maximum battery voltage under equalized charge = 60 V × 2.33 = 140 V
- Minimum volts per cell = 1.75 V
- Minimum battery voltage = 60 V × 1.75 = 105 V

Because the equipment fed by such a battery must also be operable with a level of voltage drop in the associated distribution cables, the operating range should cover the range 100–140 V. For international use, the typical lead–acid battery consists of 55 cells having a battery voltage range of 96–128 V, resulting in a required equipment voltage range of 91–128 V. Some earlier equipment, particularly the incandescent lights used for display indications, could not cover this range easily, and hence there is some use of batteries with a smaller number of cells and some use of end-cell switching when equalizing a battery. It should be noted that, when using a smaller number of cells and a fixed minimum equipment operating voltage, the actual Ah rating needs to be increased to match the lower voltage range available unless the minimum equipment voltage rating can also be reduced.

The above examples are based on lead–acid battery designs. For nickel–cadmium and for other battery types, a similar series of voltage levels, and hence the number of cells used for a particular battery rating, can be established. In this respect, as nickel–cadmium is the most common, the associated voltages over and above their 1.2 V per cell are 1.4–1.47 V float charge, 1.50–1.65 V equalize, and 0.95–1.0 V for the minimum for discharge, typically leading to the use of 100 cells for a North American battery design and 92–94 cells internationally. It should be noted that the voltage range for the nickel–cadmium battery system is larger than for the lead–acid-based system, and hence precautions may need to be taken to protect sensitive equipment when nickel–cadmium batteries are used. However, for many modern digital systems, wide-range power units are used, and the large range of battery systems is less of a problem.

8.7.2 Correction Factor

The capacity of all batteries changes with temperature, and for lead–acid batteries, more change is found, particularly at the lower temperatures. The battery therefore has to be sized to provide the required standby time even under the worst applicable temperature conditions. As each particular cell type has its own particular characteristics, the design curves for the particular cell type should be used in calculating the appropriate derating factor. Batteries also age with time, and it is usual to add a margin of 25% to cover that factor. Also, as loads may increase, even during plant design, a design margin of about 10–15% would be appropriate.

In new installations, the initial capacity is usually less than 100% (about 90%), and will only reach 100% after a few equalizing charges.

8.7.3 Duty Cycle

It is necessary to detail the amount of power required for each function over the design discharge period. Typically, the various classes of load considered are as follows:

1. Continuous load (indicating lamps, relays, etc.): 8 h
2. Communications (UPS, etc.): 3 h
3. Emergency light: ½–3 h
4. Intermittent or momentary (CB close and trip): 1 min

General-purpose battery systems typically include both load types 1 and 4, as noted above, while the other types, particularly UPS, would be provided by dedicated batteries and require sufficient size to feed a fixed load level for a fixed time. For mixed-use batteries, the loading with time and the method recommended to calculate the battery size required in power houses and switchyards is detailed in IEEE standards 485 [S1] and 1115 [S2], and are equally applicable to industrial situations. In such a case, the worst-case loading needs to consider a significant amount of momentary switching load both at the start and end of the duty cycle, with a few random load events throughout the discharge cycle. When this duty is added up, the peak and total load can be calculated and then, using the specific battery-type design figures, the battery-plate configuration and amp-hour rating can be computed. The battery size given by each supplier depends on the minimum voltage at the end of the cycle which, for a lead–acid battery with an 8-h standby time, should be not less than 1.75 V per cell.

Although actual switchgear operating times are short, the standards recommend using a 1-min value for a sum of the current taken by all devices simultaneously, with the circuit breakers tripping at the beginning and closing at the end (8 h) of the discharge cycle.

REFERENCES

Standards

S1. IEEE 485-2000, IEEE Recommended Practice for Sizing Larger Lead Storage Batteries for Generating Stations and Substations, 2000.

S2. IEEE 1115-2000, IEEE Recommended Practice for Sizing Nickel Cadmium Batteries for Generating Stations and Substations, 2000.
S3. IEEE 446-1987, IEEE Recommended Practice for Emergency and Standby Power Systems, 1987.
S4. IEEE 450-1995, IEEE Recommended Practice for Maintenance, Testing and Replacement of Vented Lead-Acid Batteries for Stationary Application, 1995.
S5. IEEE 485-1997, IEEE Recommended Practice for Maintenance, Testing and Replacement of Nickel Cadmium Batteries for Generating Stations and Substations, 1997.
S6. ANSI/IEEE C2-1999, National Electrical Safety Code (NESC), 1999.
S7. ANSI/NFPA 70-2005, National Electrical Code (NEC), 2005.
S8. Canadian Standard CSA C22.1-2002, Canadian Electrical Code, 2002.
S9. International Telecommunications Union, Telecommunications Standards Bureau (ITU-T), CCIF-1951, 1951.
S10. IEEE 1375-1998, IEEE Guide for the Protection of Stationary Battery Systems, 1998.
S11. IEEE 1188-1996, IEEE Recommended Practice for Maintenance, Testing and Replacement of Valve-Regulated Lead-Acid (VRLA) Batteries for Stationary Application, 1996.
S12. IEEE 1187-1996, IEEE Recommended Practice for Installation Design and Installation of Valve-Regulated Lead-Acid (VRLA) Batteries for Stationary Application, 1996.
S13. IEEE 484-1996, IEEE Recommended Practice for Installation Design and Installation of Vented Lead-Acid Batteries for Stationary Application, 1996.

BIBLIOGRAPHY

1. J. P. Nelson and W. D. Bolin, "Basics and Advances in Battery Systems," *IEEE Trans. IAS* 31 (2): (1995).
2. Exide, "Stationary Lead-Acid Battery Systems for Industrial and Utility Operations," document C 50.00, ESB Exide.
3. Gould, "Stationary Battery Presentation: Questions and Answers," document GBC-33 2/79 3C/7, Gould Manufacturing of Canada Ltd., Industrial Battery Division, 1979.
4. G. D. Gregory, "Applying Low-Voltage Circuit Breakers in Direct Current Systems," *IEEE Trans. IAS* 31 (4): (1995).
5. R. L. Smith, Jr., "Control Batteries: Power System Life Savers," *IEEE Ind. Applic. Mag.* Nov./Dec. (1995): 18–25.
6. A. R. Mann, "Nickle-Cadmium Batteries for Standby Power Systems," *IEEE Electron. Power* July/Aug. (1981): 661–663.
7. A. I. Harrison, "Lead-Acid Stand-By Power Batteries in Telecommunications," *IEEE Electron. Power* July/Aug. (1981): 551–553.
8. J. Marks, "Batteries Need More than Specific Gravity Tests," *Electrical World* Feb. (1995): 27–29.

9 Application and Protection of Medium-Voltage Motors

9.1 INTRODUCTION AND OVERVIEW

The application of a correct motor to a given machine is a joint effort between the electrical and mechanical engineers. The motor characteristics must match or exceed the driven load requirement. Electric motors generally fall into three groups: small, medium, and large. Small-frame motors are low-voltage (120–575 V) machines rated up to 250 hp (187 kW). Medium-size motors are medium-voltage (2.3–4.0 kV) machines rated up to 250–1500 hp (1120 kW). Large motors are medium-voltage (4.0–13.8 kV) machines rated 1500 hp (1120 kW) and up to 3000 hp. In this chapter, more coverage is provided for the application of medium- and large-sized motors.

9.2 LOAD CHARACTERISTICS

The majority of the driven equipment falls into two categories, constant torque and torque as a function of speed, and these are shown in figs. 9.1 and 9.2, respectively. Steady-state or running conditions (full-load torque, power) can be readily calculated to determine the motor rating. However, for some applications, transient conditions related to acceleration (starting) of the motor and load combination are more critical than the steady-state performance.

9.2.1 Load Categories

9.2.1.1 Constant-Torque Drives

See fig. 9.1. Torque is constant for any given load condition; power is then proportional to the speed.

$$P = \omega \times T$$

$$\omega = \frac{2 \times \pi \times N}{60}$$

where

P = power in kW
ω = rotational speed in rad/s
N = rotational speed in rpm
T = torque in newton-meters

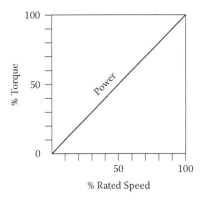 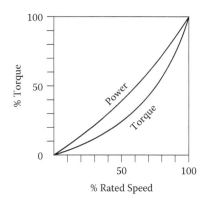

FIGURE 9.1 Constant-torque drives   **FIGURE 9.2** Torque vs. speed

Examples of constant-torque drives are grinding mills, reciprocating compressors, conveyors, etc.

9.2.1.2 Torque as a Function of Speed Drives

See fig. 9.2. This type of machine has a torque requirement that depends on speed (for any given load condition).

$$T = C_3 * N^2$$

$$P = C_4 * N \times T = C_5 * N^3$$

Examples of these types of drives are fan pumps, blowers, centrifugal compressors, etc.

9.2.2 Steady-State Power

Steady-state power at the motor shaft can be calculated using the following equations. For output power at operating running speed,

$$P_r \text{ (hp)} = \frac{N \times T}{5252}$$

$$P_r \text{ (kW)} = \frac{N \times T_m}{974}$$

and for power drawn from the system,

$$P_{IN} = \frac{P_r}{\eta}$$

where
- N = running or operating speed in rpm
- N_S = synchronous speed in rpm

- T = torque in lb-ft
- T_m = torque in mkg
- η = motor efficiency in per-unit
- P_r = output power in hp or in kW
- P_{IN} = power drawn from the system

9.2.3 Average Starting Torque and Acceleration

Figure 9.3 shows the load and motor torques during acceleration. The motor must provide sufficient torque to accelerate the drive within its thermal rating:

- T_L = load torque
- T_M = motor torque
- $T_A = (T_M - T_L)$ accelerating torque
- N = operating speed in rpm
- N_S = synchronous speed in rpm

Transient conditions associated with starting and acceleration are outlined in the following sections.

9.2.3.1 Acceleration Torque

The torque required to accelerate the drive train at any point on the speed–torque curve is the difference between the motor torque and the load torque. Because the motor torque varies with the voltage squared, it must be adjusted for voltage drop during starting. If the acceleration torque produced by the motor is insufficient to bring the machine to its operating speed within the permissible time, the motor may suffer from rotor damage. Some drives are designed for unloaded start using air clutch or hydraulic coupling.

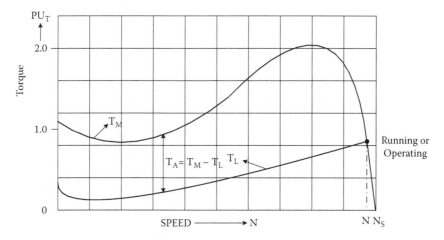

FIGURE 9.3 Speed–torque curve

- *Air Clutch*: This device is used primarily on low-speed applications. The motor is started and accelerated to full speed (and synchronized if it is a synchronous motor), and then the clutch is engaged to accelerate the driven equipment. In this case, the motor is completely unloaded and has to accelerate its own friction and windage loss plus accelerate its own inertia. The motor must, however, have sufficient breakdown or pull-out torque to transmit the load and the accelerating torque transmitted by the clutch during closing.
- *Hydraulic coupling*: This behaves in a similar manner to the air clutch, and is more often applied to high-speed applications.

9.2.3.2 Inertia

The inertia of a load and its speed–torque characteristic determine the acceleration torque required to bring the load up to speed. The acceleration time is directly proportional to the inertia (load plus motor) and inversely proportional to the acceleration torque. If the acceleration torque produced by the motor is insufficient to bring the machine (load) to its operating speed within the permissible time, the motor may stall or even may suffer damage to the rotor.

For the calculation of starting performance, the load should be adjusted by adding the friction and winding losses in the motor. The inertia or moment of inertia is expressed as:

- Wk^2 or WR^2, lb-ft^2 or J-kg-m^2
- W = weight, lb
- k or R = radius of gyration, ft

When the motor and the driven load run at different speeds, the load inertia at motor speed can be calculated using the following formula:

$$WR^2_{\text{Load}} \text{ at motor rpm} = WR^2_{\text{Load}} \times \left(\frac{\text{Load rpm}}{\text{Motor rpm}} \right)^2$$

9.2.3.3 Friction

Friction is higher during starting than under running condition, particularly just before movement starts. It is noticeable in equipment such as heavy machines with sleeve bearings without forced oil lubrication, where conveyor belts become stiff under low-temperature conditions. The motor locked-rotor or starting torque must be sufficient to overcome this breakaway torque.

9.2.3.4 Frequency of Starting

This poses a limiting factor because of heating of the motor, particularly of the rotor in large induction and all synchronous motors. When ascertaining the frequency of starting, consideration should be given to the number of starts for operational as well as maintenance purposes. When a large fan requires a rebalancing, the number of starts may exceed four. This will require a specially designed motor or controls arranged to limit the number of starts within the standards.

9.2.3.5 Induction-Motor Equipment Network

It is customary to derive induction-motor performance based on the equivalent network shown in fig. 9.4 which, with the equivalent diagram for a simple cage induction motor, becomes

$$\text{Motor slip} = \frac{N_S - N}{N_S} \text{ per unit}$$

Stator reactance = X_1 and stator resistance = R_1
Rotor reactance = X_2 and rotor resistance = R_2
Magnetizing reactance = X_m and magnetizing = R_m

It can be shown that all the significant operational characteristics of the induction motor can be derived from the above equivalent diagram. For details of the analysis of the equivalent diagram, refer to [14].

9.3 SQUIRREL-CAGE INDUCTION MOTORS

The squirrel-cage induction motor is the most commonly used type in the industry. The rotor (rotating part of the motor) has copper or aluminum bars that are welded to the end rings. This configuration is similar to that of a squirrel cage; hence it has acquired the name "squirrel cage" induction motor. The NEMA standard [S1] has defined a set of motor characteristics, dimensions, and performance for motors rated up to 500 hp (375 kW).

9.3.1 TORQUE CHARACTERISTICS

A typical speed–torque curve during starting (rest to full-load speed) is shown in fig. 9.5. Different types of torques developed by the motor are defined as follows:

- *Locked-rotor or breakaway torque (point "a")*: Locked-rotor torque of a motor is the minimum torque that it develops at rest for all angular positions of the rotor with rated voltage applied at rated frequency.

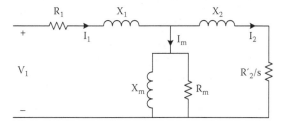

FIGURE 9.4 Induction-motor equivalent network

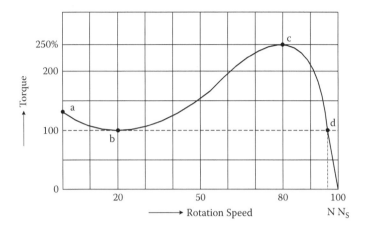

FIGURE 9.5 Speed–torque characteristic of an induction motor

- *Pull-up torque (point "b")*: The pull-up torque of an AC motor is the minimum torque developed by the motor during the period of acceleration from rest to the speed at which breakdown torque occurs.
- *Breakdown torque (point "c")*: The breakdown torque of a motor is the maximum torque that it develops with a rated voltage applied at a rated frequency, without an abrupt drop in speed.
- *Full-load torque (point "d")*: The full-load torque of a motor is the torque necessary to produce its rated power at full-load speed.

The operating point is the intersection of the motor and load torque curves. Slip is the difference between the motor synchronous speed and the operating speed.

NEMA MG-1 [S1] has defined the minimum torques (a, b, c) and maximum locked-rotor currents for motors rated 1–500 hp (375 kW). Minimum torques developed by motors rated >500 hp (375 kW) are 60% locked rotor, 60% pull-up, and 175% breakdown.

9.3.2 NEMA Design Letters

NEMA has assigned design letters for motors rated up to and including 500 hp (375 kW) that define the torque characteristics. NEMA design letters shown in fig. 9.6 do not apply to motors rated above 500 hp (375 kW). The design letters and hp ranges are:

- Designs A, B, and E: for motors rated 1–500 hp (375 kW)
- Design C: for motors rated 1–200 hp (150 kW)
- Design D: for motors rated 1–500 hp (375 kW)

Design E was introduced to accommodate the demand for a better than high-efficiency design, which resulted in a higher locked-rotor current and lower torques compared with design B motors.

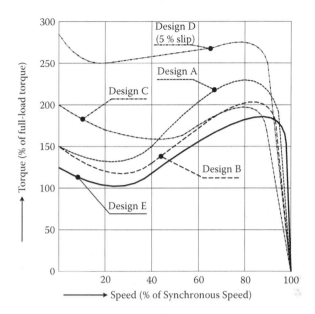

FIGURE 9.6 Typical speed–torque curve for NEMA design letter motors

9.3.3 NEMA Code Letter

NEMA has designated a set of code letters to define locked-rotor kVA per horsepower (kVA_{LR}/hp), and this letter appears on the motor nameplate. Each letter gives a range of kVA/hp, e.g., the letter "G" standard for 5.6–6.3.

9.3.4 Inertia

High-inertia applications create thermal and mechanical problems. Any machine that has inertia (referred to the motor shaft) greater than five times that of the motor rotor or that exceeds the value given in table 20-I of [S1] is considered to be a high-inertia load. The normal load inertia given in the standard is based on the formula:

$$\text{Normal load } WK^2 = A \times \left[\frac{hp^{0.95}}{\left(\frac{rpm}{1000}\right)^{2.4}} \right] - 0.0685 \times \left[\frac{hp^{1.5}}{\left(\frac{rpm}{1000}\right)^{1.8}} \right]$$

where

A = 24 for 300–1800 rpm motors
A = 27 for 3600 rpm motors

For motors driving loads with high inertia or load torques, copper or copper-alloy rotor bars should be used. Double-cage rotors or rotors with deep rectangular or other shapes of bar design are used when high locked-rotor torque, low full-load slip, and low starting current are desired. The deep rotor-bar design used in two- and four-pole motor design has an added advantage over double-cage rotors for heat dissipation during starting and acceleration. Typical high-inertia rotors are shown in figs. 9.7 and 9.8.

9.3.5 Service Factor

The service factor (SF) of an electric motor gives the measure of its overload capacity within defined temperature limits. The SF, defined by NEMA standards, is:

- Open drip-proof enclosures: 1.15 for motors rated 1–200 hp (150 kW)/3600 rpm and 1–500 hp (375 kW)/1800 rpm
- All other enclosures: 1.0

Manufacturers can supply motors with a service factor of 1.15; however, the temperature rise will be 10°C higher. A better approach is to purchase the motors

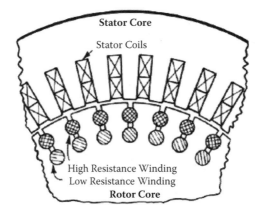

FIGURE 9.7 Double-cage rotor

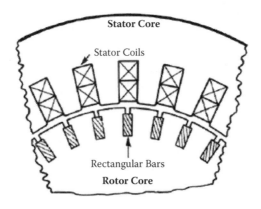

FIGURE 9.8 Rotor with deep bar

with Class 155(F) insulation systems, but limit the temperature rise to Class 130(B) systems (80°C) above an ambient of 40°C.

9.3.6 INSULATION SYSTEMS AND TEMPERATURE RISE

The Class 155(F) insulation system has become the industry standard. It is a good practice to purchase motors with Class 155(F) insulating systems but with a temperature rise of Class B (80°C) above an ambient of 40°C. This will reduce the losses, increase the motor insulation life, and permit overloads (up to 15%) without exceeding the temperature limits. The recommended temperature rise above an ambient temperature of 40°C shall not exceed the following:

- Motors rated 500 hp (375 kW) and less: 80°C by resistance
- Motors rated above 500 hp (375 kW): 80°C by an embedded resistance-temperature detector (RTD)

Experience has shown that an increase of 10°C in operating temperature above the limit will reduce the insulation system life by 50%.

Motor hp or kW rating is based on an ambient temperature of 40°C and an altitude (elevation) of 1000 m (3300 ft). The motor capacity is reduced if the ambient temperature is higher. The reduction in capacity will be 1% for each degree Celsius above 40°C ambient or each 100 m higher elevation. The motor capacity will be higher in the same order for lower ambient temperature or lower altitude. However, this margin is not used in practice.

9.3.7 SURGE-WITHSTAND CAPABILITY

Because the vacuum contactors and vacuum circuit breakers are used for switching the motor circuits, the insulation system of the stator winding shall be designed for higher surge-withstand capability. The recommended value is 3.5 P.U. (per-unit) at a rise time of 0.1 to 0.2 µs and 5.0 P.U. at 1.2 µs, where

$$1 \text{ P.U.} = \sqrt{\frac{2}{3}} \times V_{L-L}$$

9.3.8 POWER SUPPLY VOLTAGE AND FREQUENCY

Motors are designed to operate successfully at rated load with a variation of:

±10% of rated voltage with rated frequency or
±5% of rated frequency with rated voltage or
±10% combined variation of voltage and frequency provide the frequency variation does not exceed +5%

The torque developed by the motor at any speed is approximately proportional to the voltage squared and inversely proportional to the frequency squared. A voltage dip of 30% or higher for a few cycles could stall a fully loaded motor.

For unbalance in supply voltage, the motor current will be on the order of 6 to 10 times the voltage unbalance. The motor output must be reduced using the derating factors given in [S1]. Because the motor-protection relays are calibrated for current unbalance of negative-sequence current, a proper setting based on voltage unbalance becomes a problem.

9.3.9 Starting or Locked-Rotor Power Factor

Data on motor-starting power factor is required for the evaluation of motor performance during acceleration and for harmonic analysis. The suggested values are provided in table 9.1, which can be used for rough calculations if the vendor data is not available. The starting power factor will be higher with low starting current.

9.3.10 Enclosures

In North America, enclosure and cooling are defined by a single heading, whereas in Europe, the IEC standard separates enclosure/protection and cooling.

NEMA frame motors (1–500 hp [375 kW]) are generally drip proof (DP) or totally enclosed fan-cooled (TEFC) type, depending on the environment. Large motors (>500 hp) have several options, depending on the environment. The most common enclosures used in industrial plants with their NEMA and IEC designations are given in table 9.2.

A brief description of enclosures used in industrial plants is provided here:

- *Totally enclosed fan cooled (TEFC)*: A TEFC motor is a totally enclosed machine, cooled by a fan integral with the motor but external to the enclosure. TEFC motors are preferred for ratings up to 500 hp (375 kW). They are available in ratings up to 1000 hp (750 kW); however, considerable savings can be realized by specifying alternative enclosure for ratings above 500 hp (375 kW).
- *Weather-protected Type I (WPI)*: WPI is an open machine with the ventilated openings constructed to minimize the entrance of rain, snow, and airborne particles to the electrical parts.

TABLE 9.1
Starting Power Factor

Motor Rating			Motor Rating		
hp	kW	Starting PF	hp	kW	Starting PF
≤50	≤37	0.4	1000	750	0.15
100	75	0.3	2500–7000	1875–5250	0.13
500	375	0.2	>7000	5250	0.1

TABLE 9.2
Motor Enclosures

NEMA Designation	Description	IEC Designation Protection	Cooling Simplified Code
DPG	Drip-proof guarded	IP 22	IC 01
TENV	Totally enclosed nonventilated	IP 44	IC 410
TEFC	Totally enclosed fan cooled	IP 44/54	IC 411
WP I	Weather protected, type I	IP 23	IC 01
WP II	Weather protected, type II	IP W24	IC 01
TEAAC	Totally enclosed air-to-air cooled	IP 44/54	IC 611/616
TEWAC	Totally enclosed water-to-air cooled	IP 44/54	IC 81W
TEPV or TEFV	Totally enclosed pipe ventilated or totally enclosed forced ventilated	IP 44	IC 31/37

- *Weather-protected Type II (WPII)*: In addition to the protection for WPI machines, the ventilation openings at intake and discharge are so arranged that high-velocity air particles blown into the motor by storm or high winds can be discharged without entering the internal ventilating passages leading directly to the electric parts. This is the most popular enclosure being used for outdoor applications. More frequent changing or cleaning of filters is required if the environment is damp and dirty.
- *Totally enclosed air-to-air cooled (TEAAC)*: A TEAAC motor is a totally enclosed machine cooled by circulating the internal air through a heat exchanger, which is cooled by the external air. The design incorporates an air-to-air heat exchanger and is used when the conventional TEFC motor design is not available or its cost is very high.
- *Totally enclosed water-air cooled (TEWAC)*: A TEWAC motor is a totally enclosed machine that is cooled by circulating air, which in turn is cooled by circulating water. A TEWAC enclosure is more economical when cooling water is readily available. The use of a sealed insulation system eliminates the need for double-walled coolers.
- *Totally enclosed pipe or forced ventilated (TEPV/TEFV)*: A TEPV or TEFV motor is a totally enclosed machine, except that the inlet and outlet openings are provided for connection to pipes or ducts for cooling air. The cooling air may be brought from outside.

9.3.11 MOTORS FOR CLASSIFIED (HAZARDOUS) AREAS

Motors and their enclosures and accessories for classified areas must be approved for the location and shall comply with the electrical codes having jurisdiction over them. References [S8] and [S9], prepared by the American Petroleum Institute (API), have covered requirements for squirrel-cage induction and synchronous motors, respectively.

TABLE 9.3
Maximum Surface Temperature Classification

T Rating, NEC/IEC	Max. Temp. (°C)	T Rating, NEC/IEC	Max. Temp. (°C)
T1	450	T4	135
T2	300	T5	100
T3	200	T6	85

The surface temperature of the enclosure or any part in the motor shall not exceed the temperature classification or code rating known as the "T" rating, defined by NEC and IEC standards. The T ratings, listed in table 9.3, are used to represent the maximum operating temperature on the surface of the equipment. This temperature must not exceed the ignition temperature of the surrounding atmosphere. Temperature-sensitive devices such as RTDs do not pose any problem. Space heaters provided in the motor to prevent condensation may be acceptable, provided that the maximum surface temperature does not exceed the T rating.

"P"-type protection is used for motors operating in an explosive environment, where there is a risk of explosion. In a P-type protection, flammable gas is prevented from entering the enclosure by purging and pressurizing the motor enclosure. Before energizing the motor, the enclosure is pressurized above the atmosphere, purged to remove any trace of gas, and then a positive pressure is maintained to prevent the entry of the gas. Reference [5] explains the application of P-type motors used in classified areas that meet the requirements of NEC and IEC standards.

9.3.12 BEARING AND SHAFT CURRENTS

This section applies to all types of motors. The types of motor bearings used for motor application are sleeve, antifriction, and thrust.

9.3.12.1 Sleeve Bearings

This type of bearing is suitable for operating at any speed up to 3600 rpm and is used for two-pole (3600 rpm) motors rated above 1000 hp (750 kW) and four-pole (1800 rpm) above 3000 hp (2250 kW). Sleeve bearings theoretically have infinite life; however, they are more expensive compared with antifriction bearings.

Sleeve bearings are self-lubricated or flood lubricated. Self-lubricated bearings employ an oil ring, which rests on top of the shaft and is submerged in the oil reservoir. When the shaft rotates, the oil ring also rotates and delivers oil from the reservoir to the bearing. Flood lubrication is employed when the heat generated from axial and radial loads cannot be dissipated by convection or radiation. An auxiliary lube oil system is required for flood lubrication. The system consists of an AC pump, DC back-up pump, reservoir, and cooler. In most applications, the auxiliary lube system for the driven (mechanical) equipment also supplies the lubrication to the motor bearings.

9.3.12.2 Antifriction Bearing

The motor can experience severe vibration running through the critical speed when accelerating up to and coasting down from the operating speed. Because antifriction bearings have very little or no damping, they are not suitable for applications that have the rotor critical speed below the operating speed.

The life of an antifriction bearing is defined as the number of revolutions, which is equal to the operating hours multiplied by the speed in rpm. The "10" in the B_{10} life, or L_{10} life, means that 90% of a large group of identical bearings can be expected to reach or exceed the specified life, and 10% will not. The manufacturers use the term "basic load rating," known as "C," as a measure of the load-carrying capacity.

9.3.12.3 Thrust Bearing

Thrust bearings are used for vertical motors. These are designed to carry axial thrust in addition to the radial load.

9.3.12.4 Bearing Protection Against Failures

Temperature sensors such as thermocouples, thermistors, and resistance-temperature detectors (RTD) are employed to monitor the bearing temperature. The common settings based on a maximum temperature rise of 50°C over an ambient of 40°C are 85°C for alarm and 90°C for trip. The temperature of an antifriction bearing does not change significantly until the failure mode has started.

The rate of temperature rise in a failing bearing is usually so rapid that the sensors are not fast enough to prevent the failure. However, a fast trip will minimize the motor rotor and stator damage. Bearing protection based on the rate of change in temperature is better than sensing the total temperature. Initiation of bearing failure causes high vibration and noise. Accelerometers mounted on the bearing housing can sense this change and provide adequate warning.

9.3.12.5 Supply Voltage Unbalance

Induction motors are designed to operate satisfactorily without rotor overheating for a supply voltage unbalance of <1%. The rotor heating rapidly increases for voltage unbalance, causing a rotor current component with a reverse rotation. It has been found that an unbalance voltage with 3% negative-sequence component would cause a motor derating of 15–20%. Several incidents of rotor failures have been observed due to excessive voltage unbalance.

9.3.12.6 Shaft Current

Shaft current or stray current in motors is usually caused by dissymmetry in the magnetic circuit or by an uneven air gap. These currents are present in large machines and may cause damage to the motor bearing. The oil film in sleeve bearings acts as an insulator and prevents the flow of shaft current when the induced voltage is small. The following methods are used to eliminate the shaft current:

- *For pedestal bearings*: One or both pedestals are insulated from the base plate.
- *For vertical motors*: Insulation is applied between thrust and guide bearings and the upper supporting structure.
- *For motors with bracket bearings*: Bearing housings are insulated from the end brackets.
- *For engine-type motors*: Consult the manufacturer when the bearing is a part of the driven compressor.

Any path that will bypass the insulation shall be checked and corrective action taken. These include pipe connections for lubrication and coolers, wiring conduits and metal connectors, cable sheath, and mounting bolts in contact with grounded reinforcing steel bars.

9.3.13 Accessories

The following accessories are recommended for induction and synchronous motors:

- *Resistance-temperature detectors (RTD)*: For medium-voltage motors of all ratings. A minimum of two RTDs per phase is embedded and equally spaced around the stator windings, with one RTD in each bearing. The RTD shall be of platinum-type 100-ohm nominal at 0°C.
- *Surge protection*: For motors rated 1500 hp (1120 kW) and larger or for motors connected to an open overhead line. The equipment is installed in the main terminal box. The protection shall include surge capacitors with damping resistors and surge arrestors.
- *Current transformers for differential protection*: For motors rated 1000 hp (750 kW) and larger. Three current transformers (CT) rated 100/5 A with minimum accuracy of C30 (accuracy maintained to 30 V secondary voltage) are required for flux balance differential protection. However, for a full-differential scheme, the protection will include a motor feeder, and three CTs of the same ratio and accuracy as provided at the motor starter shall be installed at the neutral end of the winding.
- *Space heaters*: To prevent condensation when the motor is not in operation. The heaters are wired to a separate terminal box.
- *Grounding pads:* are fittings which connect grounding cable to flat steel surface to have low resistance ground path.
- *Lifting devices*: For installation and removal of switchgear equipment such as breakers.

9.3.14 Motor Terminal Box

The terminal box must be of ample size to accommodate the number of conductors entering the box, providing sufficient space for rearranging the multi- or parallel conductors for connection to the motor terminals and stress cones, surge arrestors and capacitors, and damping resistors if used. The terminal box and the component must be designed and built to withstand the maximum three-phase fault level. The fault level is considerably reduced due to the current-limiting action of the fuse in a NEMA E2 controller. However, a rupture disc is required in the terminal box when the motor is controlled by a circuit breaker. The disc prevents the terminal box from explosion due to the pressure buildup during the short circuits. Separate terminal boxes are required for resistance-temperature leads and space-heater leads.

9.4 WOUND ROTOR (SLIP RING) INDUCTION MOTORS

This type of motor is identical to a squirrel-cage induction motor except for the rotor, which consists of a symmetrical, three-phase winding. The three terminals are connected to three slip rings. Three external variable resistors are connected to the three slip rings via carbon brushes. Motor and resistor connections are shown in fig. 9.9.

Carbon brush wear is considerably higher in a humid environment or when the current density is lower than the recommended value. The carbon dust deposits on the slip ring and adjacent metal parts can cause a ground fault that may initiate a flashover and fire. A slip-ring flashover protection that will open the motor controller upon detecting a ground fault is required to prevent a fire in the housing. For large motors, a liquid rheostat is used to dissipate the generated heat more efficiently.

Wound-rotor or slip-ring induction motors are used where high starting torque or small variation in speed is required. The most common applications are:

- *Conveyor drives*: Higher starting torque is required for loaded systems.
- *Cranes*: For lifting heavy loads and changing speeds.
- *Crushers, ball and sag mills*: High starting torque is required.

However, the wound-rotor induction motor with resistor banks for starting is not in demand and is being replaced by other types for the following reasons:

- Higher initial cost compared with squirrel-cage induction motors
- Higher maintenance cost due to slip rings, carbon brushes, and liquid resistor

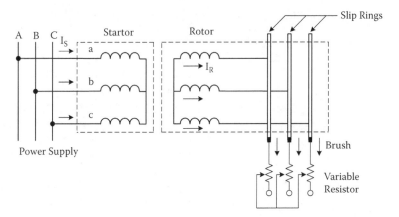

FIGURE 9.9 Wound-rotor or slip-ring induction motor

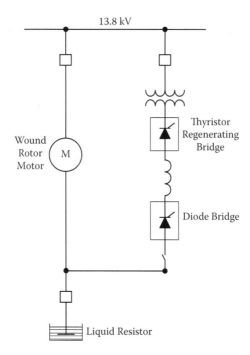

FIGURE 9.10 Wound rotor with slip-recovery scheme

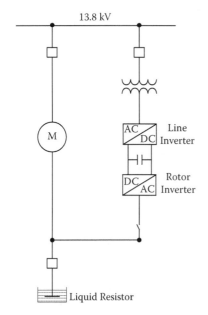

FIGURE 9.11 Wound-rotor induction motor

- Low efficiency (high power loss in external resistor) when slip energy recovery is not used

Two schemes, "slip energy recovery drive" and "doubly fed drive," are used for wound-rotor induction motors when variation in speed is required. These are shown in figs. 9.10 and 9.11, respectively. The slip-energy recovery scheme utilizes a diode bridge and a thyristor regenerative bridge.

In a doubly fed wound-rotor induction motor scheme, pulse-width-modulated (PWM) voltage source inverters utilizing insulated-gate bipolar transistors (IGBT) are used. This scheme has some advantages over the slip recovery. These include reduced harmonic current, higher power factor (>90%), and motoring or regeneration with no dynamic braking.

Application and Protection of Medium-Voltage Motors 211

9.5 SYNCHRONOUS MOTORS

9.5.1 An Overview

The stator of a synchronous motor consists of an iron core and winding similar to a squirrel-cage induction motor. The rotor consists of salient poles, field pole winding, and amortisseur or damper windings. Infrequently used high speed (3600 rpm) synchronous motors may use cylindrical rotors. The amortisseur winding consists of cage bars embedded in the pole faces and short circuited at each end by the end rings. Field poles are magnetized by DC from the rotating exciter mounted on the same shaft.

The motor starts as an induction motor, and the shape of the speed–torque curve depends on the resistance and reactance of the cage or damper winding. DC excitation is applied to the field winding at a speed known as pull-in speed when the motor torque equals the load torque. The motor will pull the connected load into synchronism upon the application of the DC field. During starting, the field winding is short circuited through a field-discharge resistor to limit the induced voltage.

Synchronous motors are classified as high-speed machines when rpm is >500 and low speed when rpm is <500. In certain applications, the synchronous motors are preferred over squirrel-cage induction motors for the following reasons:

- They can be made to operate at a leading power factor and thereby improve the overall power factor of the plant power system.
- They are less costly in certain horsepower and speed ranges, especially at speeds less than 600 rpm. Synchronous motors are cost effective when hp/rpm is >5 or kW/rpm is >6.7.
- They can be constructed with wider air gaps than induction motors, which make them better mechanically and facilitates ease of maintenance.
- The efficiency is higher, on the order of 1.5–2%.

9.5.2 Power Supply Voltage and Frequency

Motors are designed to operate successfully at rated load with a variation of voltage and frequency as described for induction motors, covered in section 9.3.8.

The torque developed by the synchronous motor is approximately proportional to the voltage or voltage squared, depending on the excitation source. Torque is proportional to the voltage squared ($T \cong V^2$) during starting or running when the excitation power is taken from the same bus, and it is proportional to the voltage ($T \cong V$) during running when the excitation power is from a different source.

For unbalance in supply voltage, the motor current will be on the order of 4 to 10 times the voltage unbalance. The motor output must be reduced using the derating factors given in [S1]. Because the motor-protection relays are calibrated for current unbalance or negative-sequence current, a proper setting based on voltage unbalance becomes a problem.

9.5.3 Insulation System and Temperature Rise

The insulation system shall be Class F (155) for armature and Class F or H (180) for field winding. However, the temperature rise at full load shall not exceed:

- *Armature winding*: 80°C by embedded detector (RTD)
- *Field winding*: salient pole 80°C by resistance
- *Field winding*: cylindrical rotor 80°C by resistance

9.5.4 Torque Characteristics

The different types of torques developed by a synchronous motor are as follows:

- *Starting or breakaway torque*: Torque developed at the instant of starting at zero speed with rated voltage applied at rated frequency.
- *Pull-up torque*: Minimum torque developed by the motor during acceleration from rest to the pull-in speed.
- *Pull-in torque*: is defined as the maximum load torque under which the motor will pull its connected load inertia (WR^2) into synchronism at rated voltage and frequency when excitation is applied. At the pull-in point, the motor torque just equals the load torque. The torque developed by the amortisseur and field windings becomes zero at synchronous speed and cannot pull the motor into step. The torque developed by a synchronous motor (when operating as an induction motor) at 95% speed (5% slip) is defined as "nominal pull-in torque" and is used as a characteristic value of the motor. The slip from which the motor will pull into step on the application of DC excitation can be expressed by the formula [2]:

$$\% \, Slip \, (s) = \frac{C}{N_S} \times \frac{HP}{WR^2}$$

where
  WR^2 = inertia of motor rotor and load
  N_S = synchronous speed in rpm
  C = constant

  The above equation shows that a higher speed is required to synchronize the motor for a load with high inertia.
- *Synchronous torque*: Steady-state (load dependent) torque developed during operation.
- *Pullout torque*: Maximum steady-state torque developed by the motor for 1 minute before it pulls out of step due to overload.
- *Maximum torque*: The peak of the torque–slip curve at rated voltage. The stable operating range of the motor is between the slip corresponding to the maximum torque and the synchronous speed. If, during transient changes in load, the slip increases beyond this point, the motor may stall.

9.5.5 Inertia (Wk^2)

The NEMA standard [S1] has established normal load inertia (Wk^2) based on the formula:

$$\text{Normal load Wk}^2 = \frac{3.75 \times HP^{1.15}}{(rpm/1000)^2}$$

The calculated values for different speed and horsepower are provided in the cited standard.

9.5.6 Excitation System

Brushless excitation has become the industry standard. A typical brushless excitation system is shown in fig. 9.12. The exciter is a brushless AC generator fitted with a rotating three-phase bridge rectifier to give a controlled DC output. The exciter is flange mounted on the nondrive end of the motor shaft. During starting, the field winding is shorted through a discharge resistor to block the DC current until the rotor is near full speed. The power required by the brushless exciter field is on the order of 100 V DC, with current ranging from 5 to 15 A.

The exciter field winding is energized upon starting the synchronous motor. As the motor speed increases, AC voltage is induced in the exciter armature winding. The thyristor is in the off state and prevents the flow of DC current to the motor field winding. Motor field winding develops high voltage during acceleration, which is discharged through the resistor connected in parallel. Upon sensing the optimum phase angle and the rotating speed approaching synchronous speed, the thyristor is turned on and the path to the resistor is turned off. DC current is applied to the field winding, and the motor pulls in to the synchronism.

The excitation system is provided with a manual/auto control and two selectable operating modes, constant power factor and constant current or constant VAR. In a constant power factor mode, the field current is regulated as the load changes to maintain constant power factor. The constant power factor mode is used in systems such as TMP (thermomechanical pulp) plants in paper mills, where large synchronous motors provide sufficient VAR to the system. However, for plants with predominantly induction motors or inductive load, excitation control in a constant-

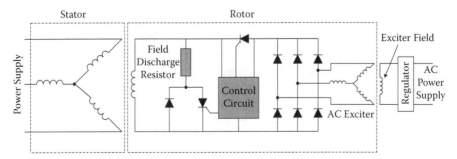

FIGURE 9.12 Brushless exciter

current mode is required to get more VAR from the motor during reduced-load operations. The following features are recommended for the excitation system control and protection:

- Free-standing excitation control panel
- Selector switch for VAR/PF and current controls
- Constant voltage control transformer for ride through during disturbances
- Excitation current and VAR/PF controllers with autotracking
- Maximum and minimum field current limit
- Field application by speed sensor or switch
- PLC with self-diagnostic, event logging, and monitoring features
- Redundant feedback
- Incomplete sequence, out of step or pull out, and under/overexcitation protection
- Motor field ground and temperature indication, alarm, and trip
- Brushless exciter diode monitoring
- Rotor protection and monitoring

9.5.7 Rotor Protection and Monitoring

A motor rotor protection and monitoring system consists of:

- Transmitter mounted on the rotor
- Receiver module mounted on the machine
- Serial data cable and PLC for processing mounted in the excitation control panel

The system offers the following features:

- Thermal protection during starts
- Temperature monitoring and protection of amortisseur and field windings
- Rotor ground-fault monitoring and protection

9.5.8 System Power Factor Improvement

Squirrel-cage induction motors consume about 0.3–0.6 kVA per hp for magnetizing, whereas synchronous motors can deliver capacitive kVA (VAR) if the rated excitation is maintained. The additional VAR will improve the overall plant power factor or can be used to reduce the voltage flicker in the plant power system. Figure 9.13 can be used to estimate the available capacitive kVA or VAR per hp when the motor is operating at a reduced load. These curves are based on reduced excitation at overload to maintain rated stator current. In paper mills, an unloaded tertiary refiner motor can supply additional VAR to reduce the voltage dip upon starting the large primary or secondary refiner.

9.5.9 Torque Pulsations during Acceleration

During starting and acceleration as an induction motor, the synchronous motor develops a pulsating or oscillating torque superimposed on the average torque (fig. 9.14). The frequency of this torque is equal to twice the slip frequency. For a 60 Hz system, the oscillating torque will go through all frequencies from 120 Hz to zero. The

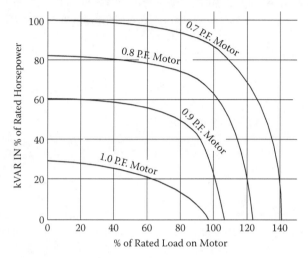

FIGURE 9.13 System power factor improvement (From EMC Publication Synchronizer 200-SYN-42, "The ABC of Synchronous Motors." With permission.)

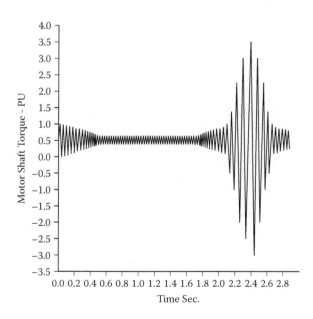

FIGURE 9.14 Torque pulsation

system with inertias has one or more natural torsional frequencies, and any frequency at or below twice the line frequency will be excited. When this condition occurs, there is a possibility of torsional resonance occurring with torque amplifications.

A steady-state and transient torque analysis is required for the complete drive train (including motor, coupling, gear, and driven equipment) to determine the magnitude of the torque amplification. The mechanical or electrical equipment vendor or both shall be asked to carry out this analysis. If excessive torque amplification is identified, then necessary steps must be taken to reduce its magnitude to an acceptable level. The corrective action may include:

- Change motor or driven equipment shaft stiffness to shift the resonant frequency
- Limit torque amplification by employing a self-resetting shear-type coupling
- Adjust damping by inserting an electrometric coupling

9.5.10 CURRENT PULSATIONS

Low-speed synchronous motors are used for driving reciprocating compressors, which operate at low speeds on the order of 300–600 rpm. The compressor cylinders pass through suction/pressure strokes, so that the shaft torque pulsates regularly through each revolution. The periodic torque pulsation at the motor shaft is reflected in a similar variation in pulsating motor current drawn from the line. The amount of variation depends on the compressor configuration, weight of the unbalanced parts, and dissymmetry caused by the part-load operation.

The current pulsation is the difference between the maximum and minimum values, expressed in percent of full-load current, which can be calculated from the motor design characteristics and harmonic analysis of the torque variation produced by the compressor. A higher inertia in the drive train (compressor, motor, etc.) helps to reduce the pulsation by supplying the torque peaks from the stored energy. The frequency of pulsation depends on the motor speed. For example, for a 360-rpm machine connected to a 60 Hz system, the frequency will be six pulsations per second.

The standards allow a maximum current pulsation of 66% of motor full-load current. The compressor vendor shall be asked to carry out a transient analysis and determine the required inertia to limit current pulsation that can be tolerated by the system. The current pulsation will cause voltage flicker, which must be checked against the flicker criteria, and corrective measures taken to ensure that the limits are not exceeded. A lower value of current pulsation (40% or 25%) can be specified, or the system configuration can be changed to circumvent the flicker problem. An envelope of the motor current drawn from the system is shown in fig. 9.15. "A – B" is the current pulsation encountered in each revolution.

Current pulsation = $A - B$, ampere, peak

$$= \frac{A - B}{\sqrt{2}}, \text{rms ampere}$$

Application and Protection of Medium-Voltage Motors

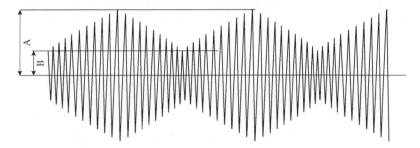

FIGURE 9.15 Current pulsation

$$\text{Percent current pulsation} = \frac{(A-B) \times 100}{\sqrt{2} \times \text{Motor Full Load Current}}$$

9.5.11 APPLICATION

Synchronous motors are used on a wide range of applications, particularly when the power output is high or the speed is low. Common applications are refiners, grinders, fans, metal rolling mills, marine propulsion, extruders, and reciprocating compressors.

9.6 ELECTRIC MOTORS FOR VARIABLE FREQUENCY DRIVES

The introduction of PWM (pulse-width modulation) drives caused a substantial increase in premature failure of low voltage (≤600 V) motors. The causes of their failures were attributed to the following:

- A majority of the users were not aware of the impact of the PWM output voltage with higher peaks and shorter rise times on the motor insulation.
- To minimize the spares, some of the users applied NEMA standard design squirrel-cage induction motors for drive applications.

Part 31 of "NEMA Standards Publication — Motors and Generators" [S1], "Definite Purpose Inverter Fed Motors," covers the design and application aspects, including related issues. The user must follow this standard for all applications where the drive is not equipped with an output filter to convert the pulses to a sine wave. The following recommendations are made for the application of definite-purpose electric motors:

- Service factor: 1.0
- Insulation class and temperature rise
  Motors rated 500 hp (375 kW) or less: 80°C rise by resistance
  Motors rated above 500 hp (375 kW): 80°C rise by embedded detector (RTD)
- Motor insulation system voltage rating and rise time (see table 9.4)
  - Neutral shift: The voltage difference between the source neutral and the motor neutral. The magnitude of the shift can be as high as 2.3 P.U.,

TABLE 9.4
Motor Voltage Rating and Rise Time

Motor Rated Voltage	Withstand Voltage	Rise Time
600 V or less	1.9 kV	≥0.1 μs
Above 600 V	2.5 kV (NEMA) 3.5 kV (IEEE)	≥1.0 μs

with resulting motor line-to-ground voltages of up to 3.3 P.U., where

$$1.0\ PU = \sqrt{\frac{2}{3}} V_{LL}$$

The magnitude of the neutral voltage can be reduced if the inverter is connected to an ungrounded power source by using an isolating transformer with ungrounded secondary, or by using separate reactors in both the positive and negative DC links. Application of an isolating transformer is highly recommended to circumvent the system neutral grounding problem associated with drives connected to the common grounded system.

- *Torsional considerations*: When an induction motor is operated from an inverter, torque ripple at various frequencies may exist over the operating speed range. Identify the frequency and amplitude of the torques and determine the possible effect on the motor and driven equipment. Operating or running the motor at or near the critical speed can cause excessive vibration that may result in bearing failure.
- *Shaft voltages and bearing currents*: Additional shaft voltages occur from voltage and current peaks, which are superimposed on the symmetrical phase quantities during inverter operation. These voltages will cause currents to flow through the bearings. It is recommended that the motor be equipped with an insulated bearing or that the shaft be suitably grounded if shaft voltages higher than 300 mV (peak) are present when tested to IEEE 112.
- *Motor feeder*: Use a symmetrical cable when a filter to convert a pulse to a sine wave is not provided at the drive output. A symmetrical cable consists of three conductors with three symmetrically placed ground conductors and heavy-gauge aluminum sheath.

A suggested arrangement for power supply to low- and medium-voltage drives is shown in fig. 9.16.

9.7 VOLTAGE DROP AND ACCELERATION TIME

9.7.1 Guidelines

Synchronous and squirrel-cage induction motors started on full voltage may draw four to eight times their rated full-load current. This excessive current may cause higher voltage drop and long acceleration time, which could result in the following:

Application and Protection of Medium-Voltage Motors

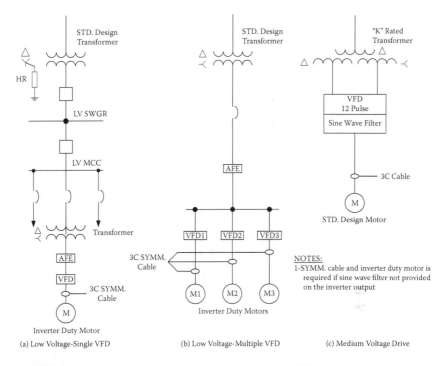

FIGURE 9.16 Electric motors for variable-frequency drives (VFD)

- For starting motors: reduced starting torque (torque varies as the voltage squared), which might not be sufficient to accelerate the load or might damage the rotor due to heating.
- For running motors: load torque might exceed the motor breakdown torque for squirrel-cage induction motors and pullout torque for synchronous motors.
- Voltage-sensitive devices such as drives will drop out if voltage drop exceeds 15%.

The following steps are recommended during the initial design stage:

- Perform voltage-drop calculations or a system study if one of the following conditions is present:
  1. Motor-starting kVA is greater than 150% of the transformer bank rating
  2. Fault level at the motor bus is less than six times motor-starting inrush
  3. Motor horsepower exceeds 10–15% of the generator rating.
- Evaluate the impact on voltage-sensitive loads if the voltage drop due to motor starting exceeds 12–15%.
- Perform detailed calculations, taking into account the resistance components and initial load if the voltage drop is 15% or higher. This should be carried out as a part of the system study, using power system analysis software.

- Check motor performance during acceleration for motors driving high-inertia loads or if acceleration time is close to or exceeds the motor hot-stall time. Motor performance curves can be generated by the motor designer and shall be specified in the request for bids.

The voltage drop and acceleration time can be computed using simplified hand calculations or power system analysis software. The following section deals with the hand calculations.

9.7.2 Voltage Drop Using Hand Calculations

9.7.2.1 Short-Circuit Method

Per-unit voltage drop

$$V_d = \frac{kVA_{ms}}{(kVA_{ms} + kVA_{sc})}$$

where

- kVA_{ms} = motor-starting kVA at 1.0 P.U. voltage
- kVA_{sc} = short-circuit kVA at the location

9.7.2.2 Per-Unit Method

Per-unit voltage drop at the motor:

$$V_d = 1 - \frac{Z_m}{Z_{tot}}$$

where

- Z_m = motor lock rotor impedance in P.U. at 1.0 P.U. voltage
- Z_{tot} = total circuit impedance in P.U.

For a simplified approach, resistance components can be neglected:

$$V_d = 1 - \frac{X_m}{X_{tot}}$$

Note:

- Use normal bus voltage as base kV and interpret the results accordingly.
- System voltage shall be minimum; assume 95% if it is not known.
- System impedance shall be based on the minimum short-circuit level.
- Include voltage drop caused by running motors for all detailed calculations.
- For transformer impedance, include plus (+) standard tolerance (7.5% for two-winding and 10% for three-winding).

Application and Protection of Medium-Voltage Motors

9.7.2.3 Motor Representation

A lightly loaded motor will tend to draw constant current over a wide range of terminal voltage. A fully loaded motor will tend to draw constant kVA. Motor kW remains relatively constant as the voltage drops, but the input current rises:

- *Running motor*: represent as a constant-kVA load. For hand calculations, convert these kVA values to impedance for an assumed voltage condition.
- *Starting motor*: represent as a shunt corresponding to the locked-rotor impedance.

9.7.2.4 Voltage Drop Caused by Running Load

For hand calculations, this can be considered in two ways:

- Add transformer voltage regulation due to the load prior to starting the motor to the voltage drop calculated without initial load.
- Assume running load as a constant kVA load, convert to equivalent impedance, and change its magnitude in proportion to the bus voltage during motor start. Combine the running-load impedance with the starting motor impedance. Ratio of this impedance to the total impedance will determine the bus voltage. Repeat this process if there is a large difference (5% or more) between the calculated bus voltage and the assumed value.

9.7.2.5 Commonly Used Formulae and Equations

Base MVA, MVA_{base} = 10.00 (for industrial systems)
= 100,000 (for utility or large power systems)
Base kV, (kV_{base}) = line-to-line voltage

$$\text{Base current } (I_{base}) = \frac{MVA_{base}}{\sqrt{3} \times kV_{base}}$$

$$\text{Base impedance (ohm), } Z_{base} = \frac{kV_{base}^2}{MVA_{base}}$$

$$\text{Base impedance (P.U.), } Z_{PU} = \frac{Z(Ohm) \times (base\ kVA)}{(base\ kV)^2 \times 1000}$$

$$Z_{PU} = \frac{Zpu\ (given\ base) \times New\ base\ VA}{(new\ base\ KVA) \times given\ base\ KVA}$$

$$Z_{PU}\ (\text{new voltage base}) = Z(given\ base\ voltage) \times \left[\frac{given\ base\ voltage}{New\ base\ voltage}\right]^2$$

Motor data and impedance
- Full-load kVA:

$$kVA_{FL} = \frac{hp \times 0.746}{Eff. \times PF_{FL}} \text{ at motor rated voltage}$$

$\approx hp$ for low voltage motors

- Locked-rotor kVA:

$$kVA_{LR} = \frac{I_{LR}}{I_{FL}} \times kVA_{LR} \text{ at motor rated voltage}$$

$$kVA_{LR} = \left[\frac{V_{BUS}}{V_{motor}}\right]^2 \times \frac{I_{LR}}{I_{FL}} \times kVA_{FL} \text{ at bus voltage}$$

$\frac{I_{LR}}{I_{FL}} \approx 6.5$ Or value from motor data sheet

$$Z_{SM} \text{ (starting motor)} = \frac{base\ kVA}{kVA_{LR}} \text{ P.U.}$$

$$Z_{RM} \text{ (running motor)} = \frac{base\ kVA}{kVA_{FL}} \text{ P.U.}$$

PF (starting motor) = value from motor vendor or from table 9.1.

9.7.2.6 Example Voltage Drop Due to Motor Starting

Figure 9.17 shows a 5000 hp motor connected to the 4.16 kV bus as calculated using the voltage drop upon starting the motor. The voltage drop at the 34.5 kV bus and 4.16 kV bus shall not exceed 5% and 20%, respectively.

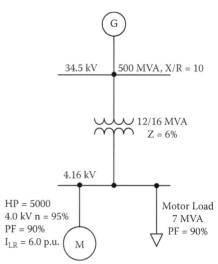

FIGURE 9.17 Example Single-line diagram

Application and Protection of Medium-Voltage Motors

Preliminary check

$$\text{Motor: } MVA_{FL} = \frac{HP \times 0.746}{\eta \times PF} = \frac{5000 \times 0.746}{0.95 \times 0.9} \times 10^{-3} = 4.36$$

where η = efficiency

$$MVA_{LR} = I_{RL} \times MVA_{FL} = 6 \times 4.36 = 26.16$$

$$\frac{MVA_{LR}}{MVA_{TR}} = \frac{26.16}{12} = 2.18 > 150\%$$

Short-circuit method

$$\text{Voltage drop at 34.5 kV: } \Delta V_{34.5} \frac{MVA_{LR}}{MVA_F + MVA_{LR}} = \frac{26.16}{500 + 26.16} = 5\%$$

Per-unit method (neglect resistanc component) (see fig. 9.18)
Base: MVA = 10; kV = 34.5 or 4.16

$$P.U. \ X_{SYS} = \frac{10}{500} = 0.02$$

$$P.U. \ X_{TR} = \frac{0.06}{12} \times 10 = 0.05$$

$$P.U. \ X_{SM} = \frac{10}{26.12} \times \left(\frac{4.0}{4.16}\right)^2 = 0.354$$

$$P.U. \ X_{RM} = \frac{10}{4.36} \times \left(\frac{4.0}{4.16}\right)^2 = 2.121$$

$$P.U. \ X_{LOAD} = \frac{10}{7.0} \times \left(\frac{4.0}{4.16}\right)^2 = 1.321$$

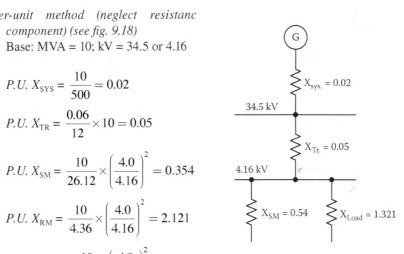

FIGURE 9.18 Impedance diagram

Voltage drops
Neglect the voltage drop caused by the initial load.

$$\Delta V_{34.5} = \frac{X_{SYS}}{X_{SYS} + X_{TR} + X_{SM}} = \frac{0.02}{0.02 + 0.05 + 0.354} = 0.047 P.U. = 4.7\%$$

$$\Delta V_{4.16} = \frac{X_{SYS} + X_{TR}}{X_{SYS} + X_{TR} + X_{SM}} = \frac{0.02 + 0.05}{0.02 + 0.05 + 0.354} = 0.165 P.U. = 16.5\%$$

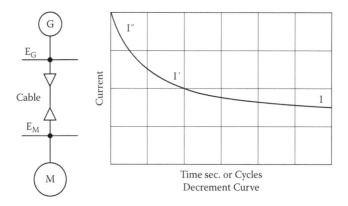

FIGURE 9.19 Load connected to a small generator

9.7.3 Voltage Dip When a Sudden Load Is Applied to a Small Generator

Figure 9.19 shows an induction motor connected to a small generator and generator decrement curve. The generator decrement curve shows the change in current with time when the generator terminals are short circuited or a motor-inrush current is drawn from the machine during starting. The change in current upon the application of a suddenly applied load is given by:

Initial or subtransient current:

$$I'' = \frac{E_g}{X''_d + X_e} \quad P.U.$$

Transient current:

$$I' = \frac{E_g}{X'_d + X_e} \quad P.U.$$

Sustained current:

$$I = \frac{E_g}{X_d + X_e} \quad P.U.$$

where
E_g = generator voltage behind the reactance or generator internal voltage
X_d'' = generator subtransient reactance (saturated)
X_d' = generator transient reactance
X_d = generator synchronous reactance
I'', I', I = subtransient, transient, and sustained current, respectively
X_e = equivalent reactance of the starting motor and interconnecting cable

The initial voltage at the motor terminals is given by:

$$E_{M1} = E_g \frac{X_e}{X''_d + X_e} \quad \text{at } t = 0$$

$$E_{M2} = E_g \frac{X_e}{X'_d + X_e} \quad \text{use this value for estimate}$$

$$E = E_g \frac{X_e}{X_d + X_e} \quad \text{no voltage regulator}$$

The actual voltage value will be between E_{M1} and E_{M2}, depending on the type of voltage regulator.

Example:

Estimate the voltage drop upon starting of a 200 hp SCIM connected to a 500 kVA generator.

Generator: 500 kVA, PF = 0.85, $X_d'' = 15\%$, $X_d' = 22\%$

Motor: 200 hp, 575 V, $\eta = 90\%$, $I_{LR} = 6.5$ P.U., PF = 0.85%

$$\text{Motor full-load kVA, kVA}_{FL} = \frac{200 \times 0.746}{0.9 \times 0.85} = 195.03 \text{ at 575 V}$$

$$195.03 \times \left(\frac{600}{575}\right)^2 = 212.36$$
$$= \quad \text{at 600 V}$$

Motor locked-rotor kVA, $\text{kVA}_{LR} = 6.5 \times 212.36 = 1380.33$ at 600 V

Convert motor reactance to generator base: $X_{LR} = \dfrac{500}{1380.326} = 0.362$
Initial voltage at start:

$$\text{Subtransient (X'')} \quad E_{T1} = \frac{X_{LR}}{X_{LR} + X_G''} = \frac{0.326}{0.326 + 0.15} = 70.7\%$$

$$\text{Transient (X')} \quad E_{T2} = \frac{0.326}{0.326 + 0.22} = 62.2\%$$

Note: A 200 hp motor with 6.5 P.U. I_{LR} is large for a 300 kVA generator with high X_d''. The following options can be used to reduce the voltage dip:

- Specify lower locked-rotor current, on the order of 5.0 P.U.
- Reduce the generator reactance and specify an excitation system with high ceiling voltage and fast response.
- Increase the generator kVA rating.

9.7.4 VOLTAGE DROP AND ACCELERATION TIME USING COMPUTER SOFTWARE

The data required for a load-flow or short-circuit study is practically the same as that required for a motor-starting study. For better accuracy, speed versus current and speed versus power factor characteristics are included in the model. Bus voltages and voltage dip can be determined using a conventional load-flow program. The starting motor is modeled as a constant-impedance load, and the load-flow solutions yield the bus voltages during starting.

For a system with a limited generation, the internal impedance of the generator connected to the system is ignored, and it is assumed that the generator voltage regulator had time to increase field excitation to maintain the desired generator terminal voltage. The impact of generator transient behavior cannot be ignored during motor starting. To model the transient behavior, the transient reactance (X_d') shall be connected in series with other source reactance.

9.7.5 Estimate of Acceleration Time Using Hand Calculations

The acceleration time can be computed using the equation:

$$\text{Acceleration time in seconds} = \frac{WR^2 \times rpm \text{ change}}{308 \times T_A}$$

- Draw motor and load speed–torque curves; convert the load torque (T_L) to motor speed if they are different.
- Draw vertical lines at 5 or 10% intervals from 5% to 95% of full-load speed. Using 5% intervals above 70% speed will give better results.
- Calculate the acceleration torque and time at each interval.

Total acceleration time:
T_A = net PU acceleration torque = $(T_M - T_L)$
T_M = motor rated torque, lb-ft
T_L = load torque, lb-ft
$WR^2 = WR^2_{MR}$ (motor rotor) + WR^2_{Load} (referred to motor rpm), lb-ft²

$$WR^2_{Load} \text{ at motor rpm} = WR^2_{Load} * \left(\frac{Load\ rpm}{Motor\ rpm}\right)^2 \text{ lb-ft}^2$$

$$\text{Rated motor torque } (T_M) = \frac{HP \times 5250}{Motor\ rpm} \text{ lb-ft}$$

$$\text{Load torque } (T_L) = \frac{HP \times 5250}{Load\ rpm} \text{ lb-ft}$$

$$T_L \text{ at motor rpm} = T_L \times \frac{Load\ rpm}{Motor\ rpm} \text{ lb-ft}$$

$$\text{Acceleration time in seconds for each speed change} = \frac{0.1\ WR^2}{308 \times T_A} = \frac{WR^2}{3080}\left[\frac{1}{T_A}\right]$$

$$\text{Total acceleration time in seconds} = \frac{WR^2}{3080}\left[\frac{1}{T_{A1}} + \frac{1}{T_{A2}} + \ldots + \frac{1}{T_{A10}}\right]$$

Application and Protection of Medium-Voltage Motors

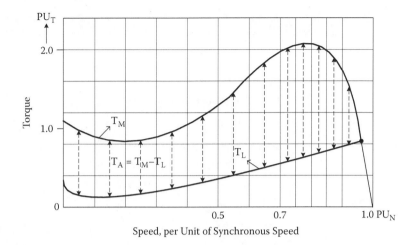

FIGURE 9.20 Acceleration torque vs. rotational speed

where $T_{A1}, T_{A2}, \ldots, T_{A10}$ are the calculated time segments between motor and load torque, as shown in fig. 9.20.

9.7.6 Estimate of Deceleration Time

In the case of a motor and a fan or a pump operating at rated speed and load on a temporary loss of power supply to the motor, the motor slowdown time can be calculated in the same manner as the acceleration time. Because the motor voltage would be zero, the motor output torque is considered to be zero. The fan or pump torque is then a decelerating torque having values that can be inserted into the time-interval equation for T_L.

9.7.7 Simplified Method for Estimating the Acceleration Time of Centrifugal Drives (Fans, Blowers)

A rough estimate of the acceleration time for centrifugal drives can be made using the average value of motor and load torques.

Average motor torque, $T_M = \dfrac{(3 \times T_{LR} + T_{BDN})}{4}$ lb-ft

Average load torque, $T_L = \dfrac{T_L}{3}$ lb-ft, at load speed

Acceleration time = $WR^2 \times \text{rpm}/308 \times \text{avg. } T_A$

where
  T_{LR} = motor locked-rotor torque, lb-ft
  T_L = load torque at load speed, lb-ft
  T_{BDN} = motor breakdown torque, lb-ft

Average acceleration torque, T_A = avg. T_M − avg. T_L lb-ft

Note: Motor torque (T_M) varies as the voltage squared (V^2); adjust the motor torque accordingly. Convert the load torque T_L and load WR^2 to motor speed using the following equations:

T_L at motor rpm = T_L × load rpm/motor rpm
WR^2_{Load} at motor rpm = WR^2_{Load} × (load rpm/motor rpm)^2

Example: Blower Driven by a Squirrel-Cage Induction Motor
Blower data: hp = 2050, rpm = 5730, WR^2 = 2628.5 lb-ft^2
Motor data: hp = 2100, rpm = 1790, I_{FL} = 265 A, I_{LR} = 7.05 PU, T_{LR} = 110%, T_{BDN} = 250% WR^2 = 3335.7 lb-ft^2

Blower torque (T_L) = $\dfrac{2050 \times 5250}{5730}$ = 1878.2 lb-ft at blower speed

Blower torque (T_L) at motor rpm = $1878.2 \times \dfrac{5730}{1790}$ = 6012 lb-ft

Blower inertia (WR^2) at motor rpm = $2628.5 \left(\dfrac{5730}{1790}\right)^2$ = 26,935 lb-ft^2

Motor torque (T_M) = $\dfrac{2100 \times 5250}{1790}$ = 6159 lb-ft

Average motor torque, T_M = $\dfrac{6159 \times (3 \times 1.1 + 2.5)}{4}$ = 8930.6 lb-ft

Average load torque, T_L = $\dfrac{6012}{3}$ = 2004 lb-ft

Average acceleration torque, T_A = 8930.6 − 2004 = 6926.6 lb-ft

Total WR^2 = 3335.7 + 26935 = 30270.7 lb-ft^2

Acceleration time in seconds: $Accl.Time = \dfrac{30270.7 \times 1790}{308 \times 6926.6}$ = 25.4 s

Note: Motor torque is based on 1.0 P.U. voltage $T_M \cong V^2$.

9.8 MOTOR CONTROLLERS AND STARTING METHODS

Large AC motors can be switched or controlled by medium-voltage fused starters rated 2.4–7.2 kV or by switchgear equipped with circuit breakers rated 2.4–34.5 kV.

9.8.1 Fused Starter

These are metal-enclosed units, built to NEMA standards and known as NEMA Class E2 controllers. These controllers comprise medium-voltage vacuum contactors and a HRC current-limiting fuse. The fuse interrupts the circuit at currents above the

Application and Protection of Medium-Voltage Motors

contactor interrupting rating. The contactors are available in two ratings (open), 400 A and 800 A, with an interrupting rating of 25 or 50 MVA. The continuous current rating is derated to 360 A and 720 A, respectively, to compensate for higher ambient temperature in the enclosure.

The complete controller is rated for 200 MVA at 2.4 kV and 400 MVA at 4.16 kV. The controller short-circuit rating is based on the fuse interrupting rating. The controllers for large AC motors shall be equipped with a contactor rated for 50 MVA (at 4.16 kV) or higher. The fuse rating is assigned "R" numbers (2R, 4R, ..., 24R). For example, the nominal current rating of 18 R fuse is 390 A. The "R" rated fuse will melt in a range of 5 to 35 s at a value of current equal to 100 times the "R" number. The following guidelines are recommended for a coordinated protection:

- The minimum fuse rating shall be 130% of the motor full-load current. The fuse continuous rating may need a further increase if the number of starts per hour exceeds four.
- Overload/stall relay must provide protection and open the contactor at up to 12 times the motor full-load current.
- Fuse minimum melting curve must intersect overload/stall relay curve at 10% or higher in excess of motor locked-rotor current and intersect the contactor drop-out time at a current below the contactor interrupting rating.

NEMA E2 controllers are compact, permit a greater number of switching operations before maintenance, and are more economical than the switchgear-type controller. However, they have the following inherent limitations:

- Bus bars and connections are noninsulated, and power-frequency-withstand voltage is low compared with the switchgear rating. This requires periodic inspection and maintenance to avoid tracking.
- The short-circuit rating of the controller is based on the fuse rating and the fault occurring on the load side of the terminal with a lead length of up to 4 ft. The let-through current is substantially reduced due to the current-limiting feature of the fuse. However, a fault on the bus or between the bus and the fuse may damage the entire bus-work, unless extra bus bracing is specified in the technical specification.
- For motors rated above 1500 hp (1120 kW) at 4.0 kV, it is difficult to coordinate with the upstream protective devices.

Insulated bus bars including connections and higher bracing are available as an option. It is highly recommended that the starter assembly be purchased with the following features:

- Complete bus-work insulated and braced to withstand the rated short-circuit current for a duration of one second
- Current transformers of "C" class and accuracy that will not saturate at 12 times motor full load or fuse let-through current.

9.8.2 Switchgear-Type Controllers

These are metal-clad units built to ANSI/IEEE switchgear standards. A vacuum or SF_6 circuit breaker is used as a switching device. Some of the safety features such as fail-safe operation, which are inherent with the contactor-type controllers, shall be implemented. These include AC and DC undervoltage release to open the circuit breaker upon loss of AC or DC control voltage.

9.8.3 Starting Methods

This section outlines some of the available motor-starting methods that are used in the industry. A one-line diagram of each type is shown in fig. 9.21.

9.8.3.1 Direct-on Line Start

See fig. 9.21(a). A direct-on line (DOL) start is the simplest, most common, and least expensive method of starting squirrel-cage induction and synchronous motors. Direct-on line starting offers high-acceleration torque and reduced acceleration time. However, the following criteria must be checked for large motors:

- The power system must be stiff, and the voltage drop caused by direct-on line starting must not exceed flicker limits at all voltage levels.
- With synchronous motors, high accelerating and oscillating (pulsating) torque may be a problem for the driven equipment during starting. A torsional analysis is required for such applications.
- With squirrel-cage induction motors, high starting and breakdown torques may cause a shock for some types of driven equipment.

9.8.3.2 Reactor Start

See fig. 9.21(b). A reactor is connected in the motor circuit, either in the line or at the neutral end, during start of the motor. The reactor is bypassed by switching a contactor or a circuit breaker when the motor has attained the rated speed. The starting current is reduced linearly, and the torque is reduced by the square of the voltage at the motor terminals. One advantage of the reactor starting is that the motor torque increases with the speed as the starting power factor improves. This increase in torque is an added boost for synchronous motors, where the motor torque drops off at the end of the acceleration.

The reactor ohmic value must be selected to reduce the motor-starting current drawn from the system to a value that will cause acceptable voltage flicker and at the same time provide sufficient voltage at the motor terminals to accelerate the load.

9.8.3.3 Autotransformer Start

See fig. 9.21(c). This method is similar to the reactor start, except that an autotransformer is switched into the motor circuit during starting and bypassed at the end of the start. Two switching arrangements, open transition and closed transition, are used.

Application and Protection of Medium-Voltage Motors

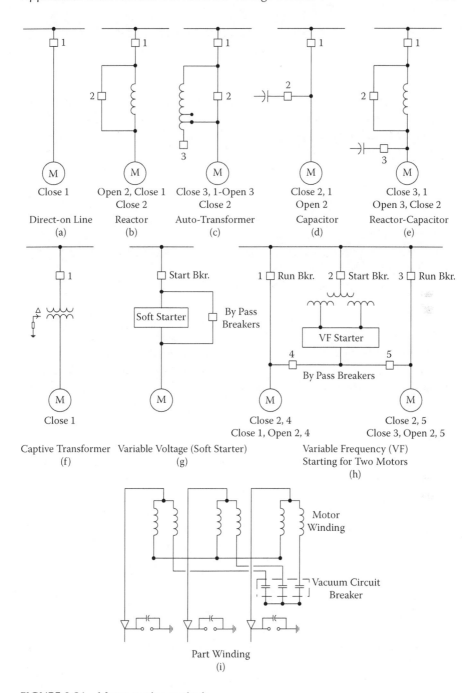

FIGURE 9.21 Motor-starting methods

In the open-transition scheme, the switching sequence is: close 1 and 3, open 1 and 3, close 2. This scheme provides open-circuit transition from reduced voltage to full voltage and is not recommended.

In the closed-transition scheme known as "Kordorfer," the switching sequence is: close 1 and 3, open 3, close 2, open 1. This switching arrangement introduces a portion of the autotransformer as a series reactor in the motor circuit during the transfer to full-motor voltage, thus reducing the transient current and torque.

The salient features are:

- Usually provided with 8%, 65%, and 50% taps. This permits the adjustment of the motor terminal voltage.
- Two additional contactors or circuit breakers and an autotransformer are required.
- The starting current as seen in the network is reduced as the square of the transformer ratio.
- Because the motor current is greater than that in the line with an autotransformer starter, the starter produces more torque per ampere of line current.

There have been many autotransformer failures caused by switching transients. A switching transient analysis using an electromagnetic transient program (EMTP) is highly recommended. The required insulation level and a need for surge-protective devices at the line and neutral end of the transformer can be determined from the study results.

9.8.3.4 Capacitor Start

See fig. 9.21(d). A capacitor is switched with the motor, which compensates part of the VAR drawn by the starting motor. The bank size is usually selected to provide about 50% of the motor-starting VAR. The capacitor is switched off at about 95% of the motor rated speed, and the motor current drops to about full load. The salient features of this method are:

- Because $C \approx V^2$, a smaller bank size at a lower voltage can be used to match with the motor-starting duty.
- The method improves the acceleration torque and reduces acceleration time, which makes it suitable for high-inertia or high-starting-torque loads.

A resonance check shall be made to ensure that the selected capacitor bank does not create a system problem.

9.8.3.5 Reactor-Capacitor Start

See fig. 9.21(e). This method is a combination of reactor and capacitor starting, as described earlier. This method can be used where the network is weak and reactive VAR is needed to improve motor performance during acceleration.

9.8.3.6 Captive Transformer Start

See fig. 9.21(f). The motor is energized through a two-winding transformer. The switching and protection are provided at the primary only. The transformer must be sized and built for motor-starting duty (impact loading). The salient features are:

- Can be used for large motors where the switching is done at a higher voltage than the motor rated voltage, e.g., a 50,000 hp, 13.8 kV motor can be switched at 34.5 kV through a captive transformer.
- Reduces short-circuit currents from the motor to the primary switchgear and from the system to the motor.
- High-resistance grounding can be provided for the motor, thus reducing iron damage.

9.8.3.7 Variable-Voltage Start

See fig. 9.21(g). This method is known as a "soft start" and is basically a reduced-voltage starting method similar to reactor start. In this method, a rectifier-inverter using IGBT is applied to vary the voltage at the motor terminals to reduce the starting current. Because the frequency remains the same, the motor-starting torque is also reduced in proportion to the voltage squared. The soft-start controller is bypassed by switching a contactor or circuit breaker when the motor has attained the rated speed. Because the output waveform from the controller is not a sine wave, there is some reduction in the motor torque compared with the reactor start. The controller will also inject current harmonics into the power system during the starting period. This method is not suitable for high-inertia and high-torque loads.

9.8.3.8 Variable-Frequency Start

See fig. 9.21(h). In this method, the ratio of voltage to frequency (volt/Hz) is maintained constant during the acceleration period or operating speed range. The voltage and frequency of the power supply to the motor is reduced to a low value to increase the ratio of the motor torque to the motor-starting current. At reduced frequency, the applied voltage and starting current are lower.

The motor is accelerated through a frequency converter, and upon reaching the system frequency, the motor is transferred to the network. The motor-starting torque can be shaped to suit the load characteristics. Two types of drives — load commuted inverter (LCI) and pulse-width modulation (PWM) — are used. LCI is a current-source inverter and is common with synchronous motors. The PWM drive is a voltage-source inverter, using switching devices such as IGBT, IGCT (integrated gate commutated thyristor), or IEGT (injection enhanced gate transistor) to chop the DC into pulses. The salient features are:

- Smooth acceleration and a negligible voltage drop in the network.
- No transient torques.
- High cost and requires a large space.
- One starter can be used for more than one motor.

- Harmonic generation, with the order depending upon the number of pulses. Filters may be required.

9.8.3.9 Part Winding Start

See fig. 9.21(i). This method can be used for applications where a synchronous motor is started unloaded. The motor is wound with two sets of Wye-connected stator windings, with the neutral end of one winding connected through a vacuum contactor or a circuit breaker. Upon starting, the contactor at the neutral end is open, and only one winding is in operation. The contactor is closed just after the motor has reached the synchronous speed, and the motor excitation is turned on. With this arrangement, the motor-starting (inrush) current is on the order of 65% to 75% of the normal starting current.

This method offers a limited reduction in starting current, relatively low starting torque, and is not suitable for all ratings and speeds.

REFERENCES

Standards

S1. NEMA MG1-2003, NEMA Standards Publication — Motors and Generators, 2003
S2. NEMA MG2–2001, Safety Standard and Guide for Selection, Installation, and Use of Electric Motors and Generators, 2001.
S3. IEEE 841-2001, IEEE Standard for Petroleum and Chemical Industry-Severe Duty Totally Enclosed Fan-Cooled (TEFC) Squirrel Cage Induction Motors — up to and Including 370 kW (500 hp), 2001.
S4. IEEE 1255-2000, IEEE Guide for the Evaluation of Torque Pulsations during Starting of Synchronous Motors, 2000.
S5. IEEE 1434-2000, IEEE Guide to Measurement of Partial Discharges in Rotating Machinery, 2000.
S6. IEEE 112-1996, IEEE Standard Test Procedure for Polyphase Induction Motors and Generators, 1996.
S7. IEEE 115-1995, IEEE Guide: Test Procedures for Synchronous Machines, 1995.
S8. API 541-April 1995, Form Wound Squirrel Cage Induction Motors — 250 Horsepower and Larger, 1995.
S9. API 546-June 1997, Brushless Synchronous Machines — 500 kVA and Larger, 1997.
S10. IEC 60034 (series), Rotating Electrical Machines.
S11. ANSI/NFPA 70, National Electrical Code, Articles 500–503.

BIBLIOGRAPHY

1. EMC Publication Synchronizer 200-SYN-28, "The ABC Induction Motors."
2. EMC Publication Synchronizer 200-SYN-42, "The ABC of Synchronous Motors."
3. EMC Publication Synchronizer 200-SYN-55, "The ABC of Motor Mechanics."
4. EMC Publication Synchronizer 200-SYN-71, "The ABC of Large Motors and Controls, Part 2: Applications."
5. EMC Publication Synchronizer 200-SYN-97, "The ABC of 'P' Type Motors."
6. General Electric Co., *Industrial Power Systems Data Book*, GE publication, Schenectady, NY: General Electric.
7. A. N. Eliasen, "High Inertia Drive Motors and Their Starting Characteristics," *IEEE Trans. PAS* 99 (4): 1472–1482 (1980).

8. James H. Diamond, "Stall Time, Acceleration Time, Frequency of Starting: The Myths and the Facts," *IEEE Trans. IAS* 29 (1): 42–51 (1993).
9. Shoaib Khan and Roland Brown, IEEE/IAS-PPIC 1992 Conference Record, "System Problems Associated with Large Synchronous Motors in TMP Plants," paper CH3142-7/92, 1992, pp. 23–25.
10. TAPPI Electrical Engineering Committee, "Selection of Chipper Drives," Atlanta: TAPPI.
11. Joseph Nevelsteen and Humberto Aragon, "Starting of Large Motors: Methods and Economics," *IEEE Trans. IAS* 25 (6): 1012–1018 (1989).
12. John H. Stout, "Capacitor Starting of Large Motors," *IEEE IAS Trans.* 14 (3): 209–212 (1978).
13. J. A. Zimmerman and W. O. Richardson, "Motor Starting with Capacitors," *IEEE IAS Trans.* 5 (6): 759–765 (1969).
14. C. J. Rogers and D. Shirmohamed, "Induction Machine Modelling for Electromagnetic Transient Program," *IEEE Trans. Energy Conversion* 2 (4): (1987).

10 Power and Control Cables

10.1 INTRODUCTION AND OVERVIEW

The electrical power distribution networks for industrial systems normally rely on cable feeders to distribute the necessary power to the various industrial processes. Some overhead lines are used, but these are normally only for the longer-length and higher-voltage circuits. Cables are similarly used to interconnect the various control and protection systems for both the electrical network and for the process. Each industrial plant has different physical constraints that result in different physical configurations for process and hence for the primary and secondary power distributions. Typical industrial distribution networks are illustrated and the design aspects quoted are discussed further in chapter 3.

The distances involved and the general configuration of the plant set the method of installation of the cables that can be used. Cables can be either installed in cable trays, in concrete or metallic ducts, or be buried directly in the soil. Each of these methods of cable installation, and the proximity of other load-carrying cables, affect the heat dissipation from the cable which, together with the ambient temperature, set the current-carrying capability or the ampacity of the cables. In any cable route, it is the most restricted part of the route, both in terms of the proximity of other cables and the type of installation, such as a duct bank running under a roadway, that will set the hot spot and will dominate the rating of the circuit. It is therefore necessary to identify any such restrictions before rating any particular feeder.

With the exception of the cables that distribute the auxiliary control power, most control and protection cables are not sized for their current rating, although for longer runs, the voltage drop may require a size change.

Auxiliary service cables will also have the same restrictions on their installation as power cables and need to be designed for the conditions where they are installed and for the installation methods used. Control cables need to meet similar requirements but also need to be designed and installed to minimize electromagnetic interference both from and into their service conductors, and this will also influence both their physical design and their installation in terms of the location and separation from other power or control cables.

10.2 CABLE SELECTION CRITERIA

The main factors that influence the choice of cable to be employed for any particular application are the cable size and whether single or three-core cables or more than one cable per phase needs to be used in any particular feeder. The type and size of the cable are selected based on four criteria. The first three, on which the main choice of cable size should be based, are the maximum loading of the feeder, the maximum voltage drop that can be allowed, and the fault current to which the cable is exposed. The fourth criterion is the level of insulation used in the cable, over and above that applicable to the voltage rating that has been chosen for the network, and this characteristic is based on the method of neutral grounding used in the system.

10.2.1 LOAD CURRENT CRITERIA

The primary choice of the cable size used is based on the load current for the particular circuit being considered. This is influenced by the type of cable, the method of installation, and the ambient temperature. Although these factors can all be calculated and allowed for, it is more convenient to use the various ampacity tables published in the cable manufacturers' data and, for cables rated at 5 kV and above, the IEEE/ICEA tables [S4]. For lower-voltage cables, the design codes or the regulatory authorities apply (NEC or Canadian code), and those tables give the cable sizes that are mandated for the loading levels of each particular circuit and the method of calculating that loading. These ampacity tables cover most normal applications but, for special conditions, detailed calculations need to be made and, for such calculations, a number of software tools are also available. These aspects are discussed in more detail in section 3.9.2 in chapter 3.

In any particular industrial system it is convenient to use the smallest number of different cable sizes possible to ease installation and reduce spares holding. It is therefore a design objective, if the plant configuration and loading distribution allows, to balance the circuit loading across a network so that a small number of standard cable sizes can be used. In this case, the loading of some circuits may have higher safety margins than others, but the savings noted above may outweigh the provision of more exactly rated cables.

10.2.2 VOLTAGE-DROP LIMITATIONS

Conductor size also influences cable impedance and hence the voltage drop along a feeder due to the load current being taken. It is therefore necessary to check that the drop of voltage along the cable route does not exceed the design criteria for the network or the operating voltage range of the equipment being fed. This review needs to take into account both the continuous and noncontinuous loads and any emergency overload that the cable will be required to carry. Voltage drop is a secondary factor and usually only occurs in very large installations with long cable runs. The review should check the voltage drop of the longest circuit, using the conductor size and the current capacity indicated. Manufacturers' data sheets also often include voltage-drop tables that can be used for a quick parametric check. Where such data are not available, it is necessary to calculate the voltage drop using the methods discussed in chapter 2.

It is normal practice to set the acceptable levels of voltage drop, or regulation, in a network when the design parameters for a network are defined. For medium-voltage circuits, 3–5% regulation is generally tolerable with reasonable regulation on the system as a whole. However, the voltage drop in any particular circuit should not be large enough to cause faulty operation of any industrial process supplied by that circuit. If any doubt exists, the manufacturer of that equipment needs to be contacted to determine the applicable voltage tolerances.

Power and Control Cables

10.2.3 Fault-Current Criteria

In addition to their continuous current rating, all cables also have a corresponding short-time rating. This is rarely a basis for the sizing of larger cables, but it must always be checked, when setting the protection or choosing feeder fusing, to ensure that the protection adequately protects the cable during faults. These calculations are considered in section 10.4.1, while the associated aspects are dealt with in more detail in chapter 2 (System Planning) and chapter 11 (Protection). In plants with high fault levels, the maximum available short-circuit current may dictate the smallest conductor size that can safely be used, as there are limits in the speed possible in the associated protective devices.

10.2.4 Insulation Levels and System Grounding

In North America, it is customary to design cables for three different overvoltage ratings, for each operating voltage, dependent on the type of neutral grounding system used. For solidly grounded systems, where healthy-phase voltage rise during ground faults is minimal, a basic cable, known as 100% insulation class, is used. For high-resistance grounded (HR) systems or for ungrounded systems, where the healthy-phase voltage rise can reach the full phase-to-phase voltage, a phase-voltage-rated cable, known as a 173% class, is available. There is also an intermediate class, known as 133% class, for use in conditions where an intermediate limited grounding current is allowed. As the cable's ability to withstand stress is also related to the time of exposure, there are intermediate usages if the ground-fault conditions are cleared quickly. A more detailed definition of the cable insulation requirements for the applicable grounding conditions is presented in section 10.5.3 of this chapter.

IEC standard cables designed for international usage are similarly classed as category A, B, and C, which effectively match the 100%, 133%, and 173% classes but with somewhat different withstand times. However, as the insulation requirements for category A and category B are virtually the same, all cables in these categories are effectively 133% rated.

10.3 CABLE SHIELDING

It is important, for optimum cable design, to control the stress levels in the various layers of insulation applied in the cable and to limit the external fields, particularly for the higher application voltages. Internal and external shielding is used to control these internal stresses.

10.3.1 Inner Shield

The conductor or inner shield usually consists of a semiconducting material applied over the conductor circumference to even out the conductor contours. This shield prevents the dielectric field "lines" from being distorted by the shape of the outer strands of the conductor and eliminates the peak stresses near the conductor. This layer also provides a smooth and compatible surface for the application of the next layer of insulation. Figure 10.1 illustrates the typical construction of medium-voltage shielded power cable.

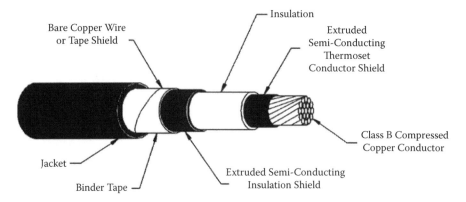

FIGURE 10.1 Power cable shield construction

10.3.2 Outer Shield

The outer shield forms the outer layer of the insulation grading and is connected to ground to fix the voltage gradient across the insulation. This outer copper shield, when correctly grounded, together with any necessary adjacent semiconducting shield, will also minimize the effect of the electric field outside the cable. The functions of this shield can therefore be summarized as follows:

- To confine electric field within the cable
- To equalize voltage stress within insulation and minimize any surface discharge
- To protect the cable from induced potential
- To limit electromagnetic or electrostatic interference (control systems, mobile radio, etc.) on adjacent systems
- To reduce any shock hazard (when properly grounded) in the vicinity of the cable

10.3.3 Outer or Overall Shield Size

In electrical networks, the most important element of cable sizing is the current-carrying capacity that is required to serve the load. However, it is the circulating currents and the short-time currents produced by ground faults in the vicinity of the cable that define the sheath or outer shield current rating. Although some cables also have a metallic armor, which also provides a path for unbalance and ground fault current, the outer shield is designed to have a low impedance to permit the flow of line-to-ground (L-G) fault currents to ensure the operation of the ground protective relays.

It is therefore the outer metallic shield that needs to be designed to withstand the short-circuit ground return current for internal cable faults, even if single-point sheath grounding is used. This is an important factor to ensure that the whole cable is not damaged by faults in the cable, so the cable needs to be repaired or resplicated at the point of fault only. The short-circuit criterion for the outer metallic sheaths withstand, expressed as its I^2t rating, needs to be greater than the current available from

Power and Control Cables

the system for the time that the ground-fault current flows. This can be expressed as:

$$(I_{GF})^2 \times 1 \text{ second}$$

where I_{GF} is the available ground fault current and the 1 second is the maximum time delay expected for most ground-fault protection systems.

10.3.4 Shield and Outer Sheath Grounding

Cable shields and the metallic sheath/armor should be solidly grounded at one or more points so that they operate at or near ground voltage at all times. Accidental removal of the shield ground can disturb the voltage distribution in the cable insulation, which can lead to a cable failure. This can also cause a hazard to personnel in the vicinity of the cable.

The simplest option for cable installation is to ground the sheath and the metallic armor in the cable termination at each end of the cable. However, this method of connection, grounding at both ends, results in circulating currents that may require cable derating, depending on the cable length and construction. The derating of cable ampacity due to such multiple-point shield grounding has a negligible effect for three-phase circuits in the following cases:

- Three-conductor cables encased by a common shield or metallic sheath
- Single-conductor shielded cables, with 500-kcmil copper conductors or smaller, installed together in a common duct
- Triplexes or three-conductor cables with individually shielded cables, with 500-kcmil copper conductors, or smaller
- Single-conductor lead-sheathed cables, with 250-kcmil copper conductors or smaller, installed in a common duct

Short cable runs of other designs of cable, such as cables to interconnect adjacent panels, do not result in significant circulating currents, but for longer runs, single-end grounding should be used. However, grounding at one end results in standing voltages which, unless fully insulated from external access, need to be limited, and the safe limit for accessible sheaths is 25 V between the sheath and the local ground. In any case, this voltage rise needs to be limited to less than the outer sheath insulation rating. The length of the cable should therefore be limited by the acceptable voltage rise of the shield when the shield is grounded at only one point. An example of the maximum length for single-point sheath grounding is given in table 10.1.

10.3.5 Recommended Practice for Industrial Cable System

The additional notes on shielding in the standard reference for table 10.2 give two options:

1. Multiconductor cables should be considered in the 2–5 kV range where any of the following conditions exist:
   - Transition from moist to dry environment

TABLE 10.1
Maximum Lengths for Single-Conductor Cables Operating at Rated Ampacity with Single-Point Shield Grounding

Conductor Size	One Cable per Duct [a]				Three Cables per Duct [b]			
	Copper		Aluminum		Copper		Aluminum	
	Ampacity	Duct Length, ft (m)	Ampacity	Duct Length, ft (m)	Ampacity	Duct Length, ft (m)	Ampacity	Duct Length, ft (m)
1/0	249	1465 (446.5)	194	1875 (571.5)	214	4965 (1513)	167	6355 (1937)
4/0	371	1055 (321.5)	290	1350 (411.5)	278	3530 (1076)	248	4480 (1365)
350	496	820 (249.9)	387	1050 (320)	418	2610 (795.5)	329	3310 (1009)
500	608	695 (211.8)	472	890 (271.3)	504	2200 (670.6)	400	2770 (844.3)
750	762	595 (181.4)	601	750 (228.6)	626	1800 (548.6)	497	2260 (688.8)
1000	890	565 (172.2)	707	710 (216.4)	...	...	...	...
2000	1237	420 (128)	1022	508 (158.6)	...	...	...	...

[a] Single 15 kV cables in ducts on 7.5 in. (19.1 cm) centers operating at 75% load factor.
[b] Three single-conductor 15 kV cables in one duct operating at 75% load factor.

Source: ANSI/IEEE 525-1992, Guide for Selection and Installation of Control and Low-Voltage Cable Systems in Substations, 1992.

Power and Control Cables

- Transition from conducting to nonconducting environment
- Dry soil, such as desert
- Damp conduits or ducts
- Connection to overhead lines
- Locations where the cable surface collects conducting materials, such as soot or salt deposits
- Surface electrostatic discharges that are sufficient in magnitude to interfere with control and instrumentation circuit functions
- Safety to personnel is involved
- Long underground cables
- Direct earth burial

2. Both-end grounding shall be applied:
For the both-end grounding conditions, the cable rating always needs to be checked. In very high voltage cable installations, the alternatives are to use surge arrestors connected to the sheath at the ungrounded end of the cable or to use cross-bonded cable installation where the sheaths are transposed between the phase cables to balance out the induced voltages in the cable sheaths, and hence reduce the sheath voltage at the ungrounded end.

More details of the methods of sheath bonding and grounding and the associated calculations are given in IEEE Standard 575 [S3].

10.4 ADDITIONAL APPLICATION CONSIDERATIONS

10.4.1 SHORT-CIRCUIT CONDUCTOR HEATING

The short-circuit withstand capability of the cable main conductor can be determined by the following formula or from charts (graphs) published by ICEA [S4] or by the relevant cable manufacturers:

For copper conductors: $\left(\dfrac{I}{A}\right)^2 \times t = 0.0297 \times Log \dfrac{[T_2 + 234]}{[T_1 + 234]}$

For aluminum conductors: $\left(\dfrac{I}{A}\right)^2 \times t = 0.0125 \times Log \dfrac{[T_2 + 228]}{[T_1 + 228]}$

where

I = short-circuit current (A, rms) duration the entire interval of current flow
t = duration of short-circuit current, s
A = conductor cross section in circular mils
T_1 = initial conductor temperature, °C
T_2 = final conductor temperature, °C

- For XLPE/EPR insulation: $T_1 = 90$, $T_2 = 250$
- For low-voltage circuit breakers and noncurrent-limiting fuse: $t = 1$ cycle
- For current-limiting fuses, the conductor withstand ($I^2 t$) needs to be greater than that of the total fuse ($I^2 t$), also known as the let-through current

At T_2 temperatures exceeding the design limit temperatures for particular types of cable insulation, quantities of combustible vapors can be expelled, increasing the risk of fire and explosion.

The calculated short-circuit current used in the above thermal-withstand calculation has to be increased by a correction factor (K_0) to account for the asymmetry in the fault current. This can be expressed in terms of the system X/R ratio and fault duration, as shown in fig. 10.2 for 60 Hz systems.

10.4.2 Application of Power Cables for Variable-Frequency Drives

10.4.2.1 Feeder Cable to the Drive Cubicle

The power cable feeder to any variable-frequency drive (VFD) drive carries 60 Hz fundamental or sinusoidal current plus the harmonic currents produced by the drive. The selected feeder size needs to be based on the heating from the total rms current (fundamental plus harmonics) and the skin effect of the higher order harmonics. The cable therefore has to be derated to compensate for additional heat caused by the harmonic currents and the associated skin effect. Skin effect also depends on the conductor size and, hence, large conductor sizes should be avoided.

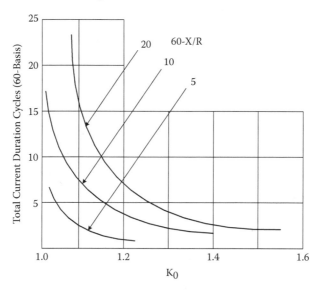

FIGURE 10.2 Correction factor K_0 for DC components of current

Figure 10.3 shows the cable derating factors plotted against percentage harmonic current with the harmonic mix associated with a typical six-pulse VFD. Due to the skin effect, more derating is required for large conductors.

A paper by Hiranandani [1] has covered the method of performing cable ampacity calculations to include the effects of harmonics. Some cable manufacturers have in-house computer software and can assist the users if the specific harmonic data is available for the drives under consideration.

In terms of the application of cables for variable-speed drives, or other significant harmonic sources, the following recommendations can be made:

- Use three-conductor and not single-conductor cables.
- Avoid large conductors to minimize the losses due to skin effect. The conductor size should not exceed 350 kcmil.
- Use shielded cables for equipment rated above 600 V.

10.4.2.2 Feeder from the Drive Cubicle to the Motor

Pulse-width modulated (PWM) inverters have an output voltage that has higher peaks and shorter rise times than a standard AC wave shape. The drive motor feeder is therefore subjected to higher voltages due to voltage reflection and also carries high-frequency currents. The high-frequency components of this fast rise time can induce bearing currents in the motor due to the inductive and capacitive coupling

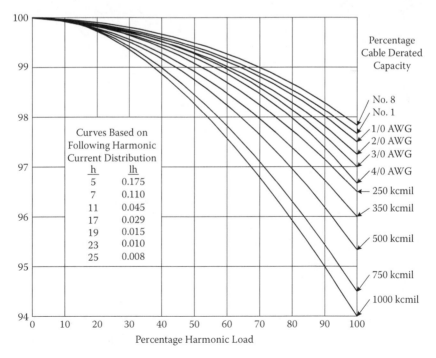

FIGURE 10.3 Cable derating vs. harmonics with six-pulse harmonic current distribution

between the stator and the rotor. The resultant high-frequency common-mode current returning to the PWM inverter through the motor bearings can exceed the safe level and damage the bearing surfaces. The selection criteria for the feeder cable from the drive to the motor need to include the following:

- Higher voltages due to voltage reflection
- High-frequency currents
- Common-mode currents that return through the cable

A paper published in the IEEE IAS magazine [3] has covered tests carried out with different types of cables available on the market and their influence on the motor shaft voltages and bearing currents. The type of cable recommended where a sine wave filter is not provided at the drive output is a three-conductor cable with three symmetrically placed ground conductors and a heavy-gauge aluminum sheath as armor. Teck type (steel tape) cables are not acceptable.

10.5 CABLE INSULATION

A very important parameter in cable selection is the insulation type. Insulation selection should be based on service life, dielectric characteristics, resistance to flame, mechanical strength and flexibility, temperature capability, moisture resistance, and the type of location where the cable is to be installed. Common insulation types applicable to medium-voltage cables are:

- Ethylene propylene rubber (EPR)
- Cross-linked polyethylene (XLPE)
- Tree-retardant cross-linked polyethylene (TR-XLPE)

These insulation materials have replaced the impregnated-paper designs that may still be found in some older installations.

10.5.1 STANDARDS

Cable standards are somewhat fragmented in both North America and internationally, with significant work being undertaken to harmonize the cable standards in Europe under the EU standards and in North America. It is likely to be some time before these two groups of standards are brought together due to the variations in operating voltages between the two areas [4].

At present there are two groups of standards for cables used in North America, one group produced by the Association of Edison Illuminating Companies (AEIC), historically representing the U.S. electrical utilities, and the other by the Insulated Cable Engineers Association (ICEA), which basically standardizes cables for general and industrial use. The ICEA is representative of the cable manufacturers and is the cable manufacturing arm of the U.S. National Electrical Manufacturers Association (NEMA). Both these groups issue voluntary standards that do not fully come under the ANSI standards.

Power and Control Cables

The IEEE, through its Insulated Conductors Committee, monitors the ICEA and AEIC work and has issued some standards in this area. It also monitors the international standards issued by the International Electrotechnical Commission (IEC) on behalf of the U.S. standards organizations. Cables used for low-voltage domestic and commercial installations need to meet the standards issued by the Underwriters Laboratory (UL) [S8] and need to be UL approved, under the U.S. National Electrical Code (NEC), for use in the United States. Similar approvals are issued by the Canadian Standards Authority (CSA) under the Canadian Electrical Code (CEC) for use in Canada.

10.5.2 INSULATION CHARACTERISTICS

The type of cable insulation used limits the maximum conductor temperature and hence the steady-state and short-time ratings of the cable. Typical temperature values for the types of extruded cable insulation are given in table 10.2.

The last two types of extruded insulation given in the table tend to suffer from permanent deformation following short-time temperature excursions, and this can lead to premature failure. Such cables are therefore given significantly lower short-time temperature rise limits. The older forms of impregnated-paper insulation also have correspondingly lower operating temperature limits and are now rarely used except at the highest voltage ratings and for submarine and extremely high voltage DC cable systems. In this respect, XLPE-insulated cables are now available for voltage levels up to 500 kV AC.

10.5.3 INSULATION LEVELS

The voltage rating of a cable is based, in part, on the thickness of the insulation and the type of the electrical system to which it is connected. These general system categories have been defined by the Association of Edison Illuminating Companies (AEIC) and apply to the three insulation levels used in North America as follows:

- *100%*: Cables in this category may be applied where the system is provided with protection such that ground faults will be cleared as rapidly as possible, but in any case within 1 minute. While these cables are applicable to the great majority of cable installations on grounded systems, they may also

TABLE 10.2
Maximum Conductor Temperature, °C

Thermosetting	Operating	Overload	Short Circuit
XLPE	90	130	250
EPR	90 (105)	130	250
PVC	75/85	105	150
Polyethylene	75	95	150

Source: ANSI/IEEE 525-1992, Guide for Selection and Installation of Control and Low-Voltage Cable Systems in Substations, 1992.

be used on other systems for which the application of cables is acceptable, provided that the above clearing requirements are met when completely de-energizing the faulted section.
- *133%*: This insulation level corresponds to that formerly designated for ungrounded systems. Cables in this category may be applied in situations where the clearing time requirements of the 100% level category cannot be met, and yet there is adequate assurance that the faulted section will be de-energized in 1 hour or less. They may also be used when additional insulation thickness over the 100% level category is desirable.
- *173%*: Cables in this category should be applied on systems where the time required to de-energize a grounded section is indefinite. Their use is also recommended for resonant grounded systems. It may be necessary to consult the cable manufacturer for the insulation thickness that applies in this case.

An example of how these levels are applied is given in table 10.3.

10.5.4 INSULATION RELIABILITY

Cable insulation is designed to provide a long service life while balancing the cost of manufacturing with the risk of premature failure in service. There is therefore always a statistical risk of failure, particularly if a particular cable is operated near its design rating for a significant part of its operational service. For more critical and

TABLE 10.3
Insulation Thickness for XLPE-Insulated Cables

System Voltage	Insulation Level, %	Insulation Thickness		Recommended for Grounding Type
		ICEA	AEIC	
600 V and 480 V	100	30	...	...
	133	45	...	...
	173	45–90	...	...
4.16 kV	100	90	90	Solid
	133	90	115	HR < 1 h
	173	...[a]	...[a]	HR > 1 h
8 kV	100	115	115	Solid or 4.16 kV HR (1 h)
	133	140	140	Resistance or 4.16 kV HR
15 kV	100	175	175	Solid
	133	215	220	Resistance

[a] For a 4.16-kV system and HR > 1 h, use 8 kV, 133% insulation with 140-mil insulation thickness.

for high-rated applications, the cable life can be extended using the following recommendations for longer life:

- Increase the insulation thickness to reduce the voltage stress (less volt/mil).
- Use the 133% insulation level for 5 kV and above.
- Lower the maximum operating temperature.
- Use cables manufactured using dry curing.
- Use EPR or TRXLP for wet locations (duct or directly buried).

All extruded dielectric cables risk treeing in service, i.e., the development of treelike tracks within the insulation, and XLPE insulation suffers from the accelerated development of such trees in wet environments. These factors influence the recommendations given above and those in the following section covering testing.

10.6 TESTING

10.6.1 HIGH-VOLTAGE DC TESTING (DC HI POT) OF MEDIUM-VOLTAGE CABLE INSULATION

It has been established that high-voltage DC testing can reduce the life of both new and aged medium-voltage cross-linked polyethylene (XLPE) cable insulation by accelerating treeing growth during the tests. It has similarly been assessed that high-voltage DC testing of completed cables at the factory, in addition to the standard AC withstand and corona tests, has no useful purpose and has been eliminated for both XLPE and EPR cable insulation. This is because these AC factory tests and the associated measurements taken during those tests are much more sensitive to detect incipient failure than the DC tests.

Site testing with DC has the advantage that such tests can be done with much smaller equipment than the equivalent AC test equipment and has therefore been retained for new cables and for initial follow-up testing. It is, however, recognized that such tests can only detect major damage or incorrect installation in splices or terminations and not the general health of the cable insulation, although monitoring the change in current with time can give some indication of less severe problems.

The following changes have therefore been made in AEIC Standard CS5 [S12] for XLPE-insulated cable:

- Factory DC testing was eliminated.
- DC proof testing levels during the first 5 years of service were reduced, and the time of application was reduced to 5 minutes.
- DC testing is not to be performed after 5 years.

The revised DC test voltage levels for XLPE cable from AEIC Standard CS5 is shown in table 10.4.

10.6.2 EPR INSULATION TESTING

EPR-insulated cables are less susceptible to aging by AC in-service stress and can be subjected to routine DC maintenance testing programs without fear of significantly

TABLE 10.4
AEIC CS5 XLPE Field Acceptance and Proof Test Values

Rated kV	Insulated Thickness (100/133% mil)	During Installation	First 5 Years
5	90/115	28/36	9/11
8	115/140	36/44	11/14
15	175/220	54/64	18/20
25	260/320	80/96	25/30
28	280/345	84/100	26/31
35	345/420	100/124	31/39
46	445/580	132/172	41/54

Note: All test voltages are 5 min duration, maximum.

TABLE 10.5
AEIC CS6 EPR Field Acceptance and Proof Test Values

Rated kV	Insulated Thickness (100/133% mil)	During Installation	First 5 Years
5	90/115	28/36	22/29
8	115/140	36/44	29/35
15	175/220	54/64	45/51
25	260/320	80/96	64/77
28	280/345	84/100	67/80
35	345/420	100/124	80/99
46	445/580	132/172	106/138

Note: All test voltages are 5 min duration, maximum.

degrading or failing the insulation system during the test. The efficacy of such tests has also been brought into question, and the revised DC test voltage level recommendations for EPR cables from AEIC Standard CS6 [S1] are shown in table 10.5.

10.7 CONTROL CABLES

Control cables are required for most protection and control applications within an industrial system. Although there are significant developments in the use of digital solutions in both the control and protection areas, and the consequential use of fiber-optic cables for data transmission and control interconnections, there will be a continued usage of physical control cables for local functions.

Control cables are usually multicore, with conductors built up to meet the requirements of their functions, with four-core cables used for substation current transformer (CT) and voltage transformer (VT) connections, and larger arrangements for more complex ancillary interconnections. Although there are physical advantages to using a full-layer approach to cable configuration, which leads to cables made up of 7, 11, 24, and 36 cores, it is common in North American practice to use plastic fillers to make up any conductor arrangement required by the purchaser.

Conductor sizes are usually in the range of 12 and 14 AWG stranded conductors for general control functions, with larger stranded conductors, such as 8, 9, or 10 AWG being used for 5 amp CT cable connections, depending on length. As discussed in the earlier sections for power cables, the voltage-drop calculations, and the equivalent burden calculations for CT and VT connections, need to be made to ensure that the associated control and protections function correctly. If these calculations indicate that it is necessary to reduce the voltage drop or burden, then the cable cross section will need to be increased.

Control cables are available as tray class without screening or armor and with copper or aluminum tape screens and, for North America, Teck steel tape armor. For most industrial applications, the Teck type is the most flexible in that it can be applied in protected and nonprotected areas. In many cases, this armor provides an adequate screen for most general purpose applications, and when more screening is required, an internal copper screen is added.

As there are no internally induced shield or armor currents, it is possible to operate with the shield and armor grounded at both ends for typical lengths of up to 1600 ft (500 m) and when the ground grids to which the cable terminal panels are connected are strongly interconnected. Where unarmored shielded cable is used, it is usually necessary to ensure that there is also a strong ground interconnection cable running along the same route to enable the shield to be grounded at both ends.

For special low-energy signaling, twisted-pair cables are typically used, with overall and individual pair screening used as applicable to the duty and length of that particular cable function. Where double screening is applied, there is an opportunity to achieve the maximum screening by connecting the outer screen to ground at both the cable terminations and grounding the inner screen at one end only. In all cases where single-end screen ground connections are used, it is important to tape back the screen at the ungrounded end to avoid inadvertent grounding at that end, which would defeat the screening improvement when only one end is connected. As noted for power cable screens, the screening effect is lost if the ground(s) become disconnected, and hence these should always be securely applied and be easily checkable.

Physically, control cables follow many of the same characteristics as the power cables discussed earlier in this chapter, and these aspects, such as location and routing, need to be applied with similar caution as noted for the power cables.

REFERENCES

Standards
S1. IEEE 141-1986, Recommended Practice for Electric Power Distribution for Industrial Plants, 1986.
S2. IEEE 404-2000, Cable Joints for Use with Extruded Dielectric Cable Rated 5,000 through 46,000 Volts, and Cable Joints for Use with Laminated Dielectric Cable Rated 2,500 through 500,000 Volts, 2000.

S3. IEEE 575-1988, Guide for the Application of Sheath-Bonding Methods for Single-Conductor Cables and the Calculation of Induced Voltages and Currents in Cable Sheaths, 1988.
S4. ICEA P-46-426, ICEA Power Cable Ampacities Copper and Aluminum Conductors.
S5. ICEA S-19-81, Rubber-Insulated Wire and Cable for the Transmission and Distribution of Electrical Energy.
S6. ICEA S-56-434, PE Insulated Communication Cables.
S7. ICEA S-73-532/NEMA WC57, Standard for Control Cables.
S8. UL 1072, Medium-Voltage Power Cables.
S9. ANSI/IEEE 383-2003, Type Test of Class 1E Electric Cables, Field Splices and Connections for Nuclear Power Stations, 2003.
S10. ANSI/IEEE 422-1986, Design & Installation of Cable Systems in Power Generating Stations, 1986.
S11. ANSI/IEEE 525-1992, Guide for Selection and Installation of Control and Low-Voltage Cable Systems in Substations, 1992.
S12. AEIC CS5-94, Specification for Cross-Linked Polyethylene Insulated Shielded Power Cables Rated 5 through 46 kV.
S13. AEIC CS6-96, Specification for Ethylene Propylene Rubber Shielded Power Cables Rated 5-69 kV.
S14. AEIC CS7-95, Specification for HV Cable Systems, Currently Covering XLPE Cables from 69–138 kV but under Revision to Be Extended to 345 kV.
S15. IEC 502, Extruded Solid Dielectric Insulated Power Cables for Rated Voltages from 1 kV up to 30 kV, now reissued as IEC 60502, Pt. 1, Covering 1 kV and 3 kV, and as IEC 60502, Pt. 2, Covering from 6 kV up to 30 kV.
S16. IEEE 400-1991, IEEE Guide for Making Direct-Voltage Tests on Power Cable Systems, 1991.

BIBLIOGRAPHY

1. Ajit Hiranandani, "Calculation of Cable Ampacities Including the Effects of Harmonics," *IEEE Ind. Appl. Mag.* 4 (2): 42–51 (1998).
2. J. M. Bentley and P. J. Link, "Evaluation of Motor Power Cables for PWM AC Drives," *IEEE Trans. IAS* 33 (2): 342–358 (1997).
3. "Characteristics of Shaft Voltage and Bearing Currents," *IEEE Ind. Applic. Mag.* Nov./Dec. (1997): 21–32.
4. F. H. Rocchio, H. R. Stewart, and D. A. Voltz, "Comparing U.S. and International Power and Control Cable Standards," *IEEE Ind. Applic. Mag.* Jan./Feb. (2001): 14–25.

11 Protection

11.1 INTRODUCTION

Protection relays have gone through major changes, from electromechanical to static and now to microprocessor-based multifunction programmable units. Although the static relays provided similar characteristics as the electromechanical units, they were more sensitive to external transients and needed to be applied with more consideration to interference and electrical noise. Digital relays also need careful implementation, but the new standards for electromagnetic interference (EMC) have made them suitable for use in most industrial environments.

The development of multifunction programmable relays has changed the protection philosophy for many industrial power systems. Many features, including metering, event logging, and additional protective elements, are available in the same unit, and, because the main cost is in the hardware and common input filtering, these additional features can be provided at a very reasonable cost. Such multifunction relays can now provide more coverage with fewer protection packages. For example, a transformer differential relay can include primary and secondary overcurrent elements; an overcurrent unit can include breaker fail, directional overcurrent, thermal models, and negative-sequence elements; and various voltage elements can be provided in most multifunction relays.

However, as noted in chapter 3, one has to be careful in their selection, because a large number of different grades of relays are on the market, and devices appropriate to the installation environment must be selected. This usually means that, for most large-scale industrial applications, the relays must be of utility grade and be of the type accepted by power utility companies.

A power distribution system consisting of a series of components (transformers, motors, cables, circuit breakers, etc.) is shown in fig. 11.1; the primary duty of the protective devices is to detect flashovers on and faults within these devices. Protective devices (automatic circuit breakers, protection relays, fuses, etc.) are also provided to safeguard the downstream equipment, such as motors, feeder cables, transformers, etc., from damage when carrying the short-term currents caused by such short circuits or by overloads within the power system. In this respect, each power equipment or element has two ratings, the thermal rating and the mechanical withstand, as defined below:

$$\text{Thermal (heating)} = K \times I^2 \times t \text{ (ampere squared} \times \text{time)}$$

$$\text{Mechanical} = K \times I_p \text{ (peak current)}$$

where K is a constant, and I is the current in or passing through the device.

The main task for the designer is to know these ratings or to determine the minimum rating that will be needed to match or exceed the system requirements.

In industrial systems (and in commercial and domestic environments), the electrical safety codes (NEC, CSA, etc.) define the minimum or maximum protection limits for each equipment or device. It is the responsibility of the designer to select the equipment rating or protective-device setting that will provide adequate protection. In a practical sense, it is also necessary not to overprotect, i.e., one must recognize that a user must have the ability to operate a given piece of equipment up to its rating limits without causing unnecessary protective tripping.

This chapter outlines the limitations of equipment rating and provides guidelines to assist the designer in selecting adequate protection while ensuring a coordinated system.

11.2 PROTECTION AND COORDINATION PRINCIPLES

11.2.1 Protection Schemes and Relay Selection

11.2.1.1 Current Transformer Connections

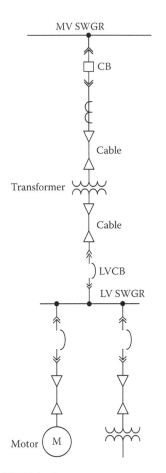

FIGURE 11.1 Power system components

The convention, in North America and in the United Kingdom, is to connect the current transformer (CT) neutral Wye or star point on the side toward the respective protected zone, as shown in fig. 11.2. It is also important to show the CT polarity mark on the drawings, especially where direction of current flow is important, such as protection schemes for transformer or bus differential, directional overcurrent, etc. The instantaneous current flow is into the polarity mark in the primary circuit and away from the mark in the secondary circuit. The method of showing the CT polarity marking and resultant current flow in primary and secondary winding of current transformers is shown in fig. 11.3.

11.2.1.2 Power System Device Function Numbers

The IEEE standard electrical power system device function numbers are defined in IEEE C37.2 [S1]. The device numbers, including typical suffixes used in this chapter, are provided, together with the equivalent IEC symbols and a short description, in

Protection

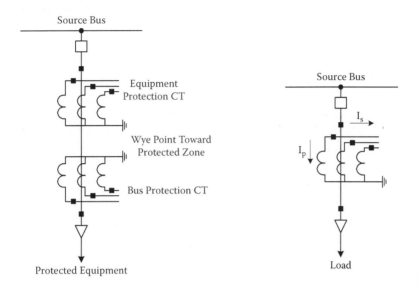

FIGURE 11.2 CT connections

FIGURE 11.3 CT polarity marking

table 11.1. The IEC numbers are enclosed within a circle, and the IEC symbols are enclosed within a rectangle.

11.2.1.3 Selection Criteria

The suggested criteria for the selection of protection relays or schemes are:

- Proven experience for a minimum period of three years
- Relays built to IEEE switchgear standards or equivalent IEC standards
- Relays tested to IEEE C37.90, C37.90.1, C37.90.2 or to IEC 60255
- Where possible, microprocessor-based relays with features such as self-checking, self-diagnostics, RMS sensing, metering, event logging, wave capture, disturbance recording, and with a communications port for data recovery and setting
- Relays that perform satisfactorily in the presence of harmonics and electromagnetic (EMI) and radio frequency (RFI) interferences
- Redundancy requirements for generators, large motors, and transformers
- Availability of application literature and technical support

11.2.2 Necessary Steps to Carry Out Relay-Setting and Time–Current Coordination

The principles of relay-setting coordination have been introduced in chapter 3, including the illustration of a typical software-generated relay-coordination curve. In preparation for such a study, the following needs to be done:

TABLE 11.1
Power System Device Function Numbers

IEEE Device No.	Protection Function
2	Time-delay relay
12	Overspeed
13	Synchronous speed check or centrifugal switch
14	Underspeed device
21	Impedance relay
23, 23Q	Temperature-control device, Oil temperature monitor
24	V/Hz overfluxing relay
26, 26W	Apparatus overheating device, Winding temperature monitor
27	Undervoltage — instantaneous or inverse
32	Directional power relay
38	Bearing temperature monitor
39	Vibration monitor
40	Loss of field protection
41	Field contactor or circuit breaker
46	Phase-balance or negative-sequence relay
47	Phase-sequence undervoltage
48	Incomplete-sequence relay
49	Thermal relay operated by current or winding temperature
50, 50G	Instantaneous OC relay phase, Instantaneous OC relay, ground
51, 51G	Time OC relay, phase, Time OC relay, ground
51N	Residual connected TOC
52	Power circuit breaker
53	Excitation check relay for synchronous motors
55	Power factor relay
56	Field application relay
59, 59GN	Overvoltage relay, Stator ground relay (HR)
60	Voltage balance relay
63	Liquid or gas pressure relay
64	Ground-fault detection from machine to ground
71	Liquid or gas level relay
78	Out-of-step relay
81	Frequency relay
86	Lockout relay
87, 87N	Differential-protection relay, phase, Differential-protection relay, ground
94	Tripping or trip-free relay

- Prepare one (single)-line diagram. Record the data related to the relays, circuit breakers, fuses, current and voltage transformers, and operating equipment.
- Add type and rating of all protective devices (instrument transformers, relays, fuses, etc.).
- Record the rating of major equipment, including transformers, rotating machines, feeders, etc., that influence short-circuit flow.
- Perform a series of short-circuit studies to determine the maximum and minimum fault-current levels.
- Ascertain the maximum load currents that will exist under normal and emergency operating conditions. The minimum and maximum short-circuit currents determine the lower and upper boundaries of current selectivity within which the circuit protective devices must operate, while the maximum load current defines the conditions under which the current-operated protections must not operate.
- Review the applicable electrical code and equipment standards.
- Obtain protective-device time–current characteristics curves and equipment-capability or thermal-limit curves.
- Define the time- and magnitude-setting margins to be used in the study.

Chapter 3, section 3.5, introduces the data needed and the time intervals required for computing the relay settings and for drawing the appropriate time–current coordination curves. Most commercially available coordination software packages contain a built-in database that provides much of the base data and relay characteristics. However, such data should always be checked to ensure that it is applicable to the particular study being undertaken.

11.3 TRANSFORMER PROTECTION

This subject is well covered in IEEE C37.91-2000, "IEEE Guide for Protective Relay Applications to Power Transformers" [S16], and in Chapter 11 of IEEE 242-2001 [S8].

11.3.1 Protective Devices and Features

A typical protection scheme, commonly used in industrial power systems for a medium- or high-voltage oil-immersed power transformer, is shown in fig. 11.4. Descriptions of the protection elements and devices are provided in the following subsections:

11.3.1.1 Devices Provided with the Oil-Immersed Transformers

- *Gas detector relay (63P)*: This device, generally known as a "Boucholz relay," is used with conservator-type transformers. It is mounted on the pipe connecting the highest part of the transformer tank and the conservator. It operates in two stages and has two sets of contacts. The first stage detects the slow accumulation of gas generated from an incipient or minor fault, such as damaged winding or core insulation, loose connections, etc. The second stage operates on the sudden rise in oil pressure in the main tank caused by faults of a major nature, such as internal flashovers, short-

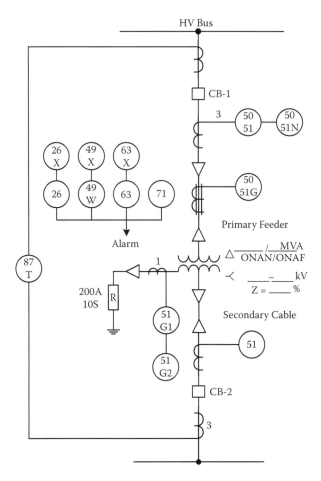

FIGURE 11.4 Transformer protection with primary and secondary breaker

circuited turns, etc. The first stage is used for alarm, and the second stage is used for trip.
- *Sudden gas-pressure relay (63SP)*: This device is used with sealed-tank transformers. It is mounted in the space above the oil and operates on the rise of pressure due to internal faults. The device is wired to trip and disconnect the transformer from the power source.
- *Winding overtemperature (49)*: This is a temperature-measuring device submerged in the transformer oil but also surrounded by a heater that is connected to the secondary of a current transformer in the winding. This is calibrated to measure the winding hot-spot temperature and has two to four adjustable temperature set points. These set-point contacts are wired to start the pumps or fans of a forced-cooled transformer, give an alarm, and trip the primary circuit breaker at the appropriate temperature level.

- *Oil temperature (23Q)*: This device measures the oil temperature and has two, three, or four adjustable set points; it functions in the same manner as the winding-temperature device 49W.

11.3.1.2 Primary Overcurrent Protection

The high-voltage side of a Delta-Wye step-down transformer is typically equipped with:

- *Phase overcurrent, instantaneous (50)*: This element or device protects the feeder and the transformer against short circuits, including faults at the bushing terminals. This element is set at above the maximum asymmetrical secondary through-fault current so that it will not operate for faults on the LV connections.
- *Phase overcurrent, inverse time (51)*: This element protects the feeder and the transformer against overcurrent, and is set at about 2.0 P.U. of the transformer full-load current at base rating. A very inverse or extremely inverse characteristics curve with a time dial or time multiplier, such that the operating characteristic falls below the transformer through-fault current limit and above the operating characteristic of the downstream device, is used.
- *Ground overcurrent, time (51G)*: This element is used in resistance-grounded systems, has the same function as device 51, and is set at 5–10% of the maximum ground-fault current to provide protection against line-to-ground faults. A very inverse or extremely inverse or definite time characteristic can be used.
- *Ground overcurrent, instantaneous (50G)*: The element has the same function as device 51G, except it is set at higher currents.
- *Ground overcurrent, time and instantaneous (50N and 51N)*: These elements are used in solid or effectively grounded systems, and the relay is residually connected to the current transformers. It is set to coordinate with the ground elements on the circuit(s) feeding the high-voltage bus.

11.3.1.3 Secondary Overcurrent Protection

The LV side of a resistance-grounded transformer is typically provided with:

- *Phase overcurrent, instantaneous (50G)*: This element is not normally used unless provided as part of an interlocked LV bus protection.
- *Phase overcurrent, time (51)*: This element is set to protect the transformer secondary and the bus against LV faults and short time overloads, and is usually set at about 1.5 P.U. of the rated secondary current.
- *Ground overcurrent, time (51G1 and 51G2)*: In this illustration, this relay is provided with two stages, usually provided with the same current setting, typically 20–50% of the grounding-resistor rating. The two stages are provided with different time settings, the first, time coordinated with the downstream ground relays and connected to trip the LV breaker, and the second, set to match the short time rating of the resistor and connected to trip the

high-side circuit breaker. In some cases, the first stage is replaced by an LV residually connected ground relay (51N) on the circuit breaker CTs.

11.3.1.4 Differential Protection

This device is generally recommended for transformers rated 7.5 MVA and above, and provides fast protection against faults within the protected zone. In this respect, chapter 11 of the IEEE Buff Book [S8] has recommended differential protection for transformers rated 5.0 MVA and above, and IEEE C37.91 [S16] has recommended this for transformers rated 10.0 MVA and above. Harmonic-restraint features are required to prevent the relay operation from the magnetizing inrush current that occurs upon energizing the transformer. The current transformer ratios and connections, plus those of any interposing CTs, and the relay taps are selected to match the phasing and currents in the primary and secondary windings, although for the digital relays, such settings can be set within the protective relay itself.

11.3.1.5 Time Settings

The inverse-time relays should use very inverse or extremely inverse curves, with the time multipliers set to coordinate with the upstream and downstream devices and the transformer damage curves, as discussed in the next section for fuse-protected transformers. The coordination curve shown in fig 3.14 in chapter 3 illustrates a typical situation, although in that case, the LV devices are static circuit breakers with long and short time characteristics. However, the same principles apply with single or multiple characteristic devices. In some cases, the use of high-side extremely inverse and low-side very inverse characteristics can provide a better margin between the two at high currents. Ground relays typically use definite or standard inverse curves.

11.3.1.6 Typical Settings

Typical settings of protective devices and elements for power transformers are provided in table 11.2.

11.3.2 Transformer Protection with a Primary Fuse

A typical transformer-protection scheme with primary fuse, commonly used with effectively grounded power systems, is shown in fig. 11.5.

11.3.2.1 Minimum Fuse Rating

It is extremely difficult to find a fuse time–current characteristics curve that will coordinate with the transformer damage or short-circuit withstand (SCW) curve. These curves are discussed in more detail in section 11.3.5. However, the fuse curve must be below or to the left of the mechanical limit curve.

Most fuse suppliers provide recommendations for the sizes and types of transformers their fuses can protect. However, the guidelines are:

- *Power fuse*: 140% of the transformer self-cooled rating

TABLE 11.2
Typical Settings of Protective Devices for Power Transformers

IEEE Device Net	Function	CT Primary	Typical Setting
A: Devices Located at the Transformer			
63P	Gas detector/pressure	NA	Factory set
63SP	Sudden pressure	NA	Factory set
49W	Winding temperature:	NA	By manufacturer
	- ONAN (two contacts)		100°C alarms
			125°C trip
	- ONAN/ONAF (three contacts)		80°C fan
			110°C alarm
			125°C trip
	- ONAN/ONAF/ONAF (four contacts)		80°C fan No. 1
			85°C fan No. 2
			110°C alarm
			125°C trip
23Q	Liquid (oil) temperature:		By manufacturer
	- ONAN (one contact)		95°C alarm
	- ONAN/ONAF (two contacts)		65°C fan
			95°C alarm
	- ONAN/ONAF/ONAF (three contacts)		65°C fan No. 1
			70°C fan No. 2
			95°C alarm
B: Primary Overcurrent Protection			
50	Phase overcurrent, instantaneous	200% of full-load current	1.3–1.5 PU through-fault current
51	Phase overcurrent, time		2–4 PU rated current ONAN
51G	Ground overcurrent, time	50/5 or 100/5	5–10% of max. ground current
50G	Ground overcurrent, instantaneous		50% of max. ground current or off if coordination is required (Wye-grounded primary)
50/51N	Alternate for solidly grounded system	Residually connected	20% of phase settings
C: Secondary Overcurrent Protection			
51	Phase overcurrent, time	200% of rated current ONAN	150–200% rated current (ONAN)
51G	Ground overcurrent, time	50% of resistor rating	10% of the resistor current TD to coordinate with feeder breaker

—continued

TABLE 11.2 (Continued)
Typical Settings of Protective Devices for Power Transformers

IEEE Device Net	Function	CT Primary	Typical Setting
	D: Differential Phase – Differential Current		
87T	Biased differential with harmonic restraint	As for LV and LV phase relay	Set tap to correct CT ratio; slope: 15%, with off-circuit taps 25%, with load tap changing 40–60%, for special applications
	E: Lockout		
86	Trip and lockout	…	No setting; must trip primary and secondary breaker to isolate fault

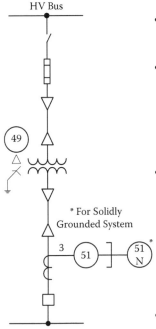

FIGURE 11.5 Transformer protection with primary fuse

- *Current-limiting fuse*: 150% of the transformer self-cooled rating
- *Fuse time–current*: below the transformer short-circuit withstand curve (SCW) and above the magnetizing inrush current
- *Magnetizing inrush current*: points for coordination are 10 to 12 P.U. at 0.1 s and 25 P.U. at 0.01 s

11.3.2.2 Limitation of Fuse for Transformer Protection

- Industrial power systems are resistance grounded, and the limited ground-fault current will not melt the fuse and clear the fault within a short time. A fused load-interrupter switch equipped with a shunt trip and a ground-fault relay, and capable of interrupting a ground-fault current, can provide protection against line-to-ground, phase-to-phase, and three phase faults.
- A phase-to-phase fault on the secondary of a Delta-Wye transformer will melt only one fuse on the primary, thus causing single phasing on the secondary system. This condition requires a switch fuse that will trip if one fuse blows, and if voltage or current are unbalanced, providing protection for downstream three-phase motors.

11.3.3 Protection of Low-Voltage Unit Substation Transformers

A typical low-voltage (LV) unit substation connected to a medium-voltage switchgear is shown in fig. 11.6. The following guidelines can be used for the setting of protective devices.

11.3.3.1 Secondary Circuit Breaker with Static Trip

These devices have separate adjustments for three areas of the characteristics curve for phase faults and a separate adjustment for ground faults as follows:

- *Long time*: This device protects the transformer against overload. The current pickup is set at 110–115% of the transformer full-load current at maximum rating, and the time-delay band is set to coordinate with the largest feeder breaker settings.
- *Short time*: This device protects the transformer against phase overcurrent or short circuit. The current pickup and time delay are set to coordinate with the largest feeder breaker.
- *Instantaneous*: Set to "off" or above the asymmetrical through-fault current to avoid coordination problems with the feeder breakers.

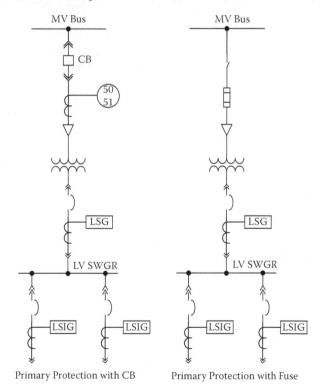

FIGURE 11.6 Protection of LV unit substation

- *Ground fault*: This device is residually connected to the phase sensors and is set at 20% of the long time setting. The time delay is set to coordinate with the feeder breaker settings. This device provides protection against line-to-ground fault in a solidly grounded system and against cross-country faults with phase-to-ground faults in different phases of two feeders.

A typical coordination curve for a transformer and the downstream feeders is shown in fig. 3.14 in chapter 3.

11.3.3.2 Primary Protection with Circuit Breaker

These settings are similar to those noted for power transformers in section 11.3.1.2:

- *Phase overcurrent, instantaneous (50)*: This element or device protects the feeder and the transformer against short circuits, including faults at the terminals. This element is set at 1.4 to 1.6 P.U. symmetrical secondary through-fault current so as not to operate on LV faults.
- *Phase overcurrent, time (51)*: This element protects the feeder and the transformer against overcurrent and is set at about 125–150% of the transformer full-load current at base rating. A very inverse or extremely inverse characteristics curve with a time dial or multiplier that meets the following criteria is used:
    1. The time–current curve lies below the transformer SCW (damage) curve and above the magnetizing inrush current point of 10–12 P.U. of transformer full-load current at natural cooled rating.
    2. The time–current curve coordinates with the largest feeder breaker setting or downstream protective device.
- *Ground overcurrent, time (51G)*: This element is used in resistance-grounded systems, has the same function as (51), and is set at 5–10% of the maximum ground-fault current to provide protection against line-to-ground faults. A very inverse or extremely inverse or definite time characteristic can be used.
- *Ground overcurrent, instantaneous (50G)*: This element has the same function as (51G), except it is set at higher currents.
- *Ground overcurrent, time and instantaneous (50N and 51N)*: These elements are used in solid or effectively grounded systems, and the relay is residually connected to the current transformers.

11.3.3.3 Primary Protection with Fuse

This protection has the same application requirements and sizing recommendations as defined in section 11.3.2.

11.3.4 ELECTRICAL CODE RULES FOR TRANSFORMER PROTECTION

A summary of the current National Electrical Code rules [S14] and Canadian Electrical Code rules [S15], concerning transformer and transformer feeder protection, is shown in figs. 11.7 and 11.8, respectively. These codes specify the minimum feeder size and maximum fuse rating or relay setting for power transformers. The code rules also permit higher protection settings for transformers with impedances above 7.5%. Although electrical codes permit higher settings for the protection of power transformers, it is the designer's responsibility to ensure that the equipment is protected. The primary protection time–current curve must lie below the transformer and feeder short-circuit withstands and thermal-damage curves.

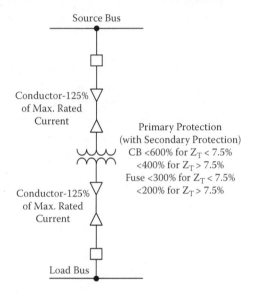

FIGURE 11.7 Canadian Electrical Code rules for transformer protection

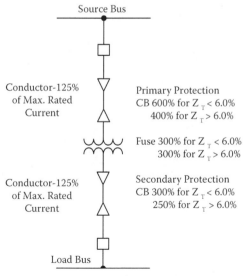

FIGURE 11.8 National Electrical Code rules for transformer protection

11.3.5 Transformer Through-Fault Current Damage Curves

11.3.5.1 Category and kVA Range

Transformer through-fault current damage curves are covered in IEEE C57.12.59-2001 [S19] for dry-type transformers and in IEEE C57.109-1993 [S23] for oil-immersed transformers. The oil-immersed transformers are grouped in four categories and the dry type in two categories, depending on the transformer power rating. The category and kVA range for three-phase transformers and the relevant figure numbers in the respective IEEE standards are provided in table 11.3.

The short-circuit withstand (SCW) curves for each category show the mechanical and thermal limits within which the transformers have to be designed to avoid damage. For category I and category IV, a single curve represents both mechanical and thermal limits. The standards permit the user (at the user's discretion) to apply thermal-limit curves for category II and category III transformers where faults are infrequent. However, good engineering practice is to use a single mechanical thermal-limit curve for all four categories and limit the through-fault current withstand to 2 s, irrespective of the transformer impedance.

Through-fault current-limit curves for two-winding dry-type category II transformers and oil-immersed category II transformers are shown in figs. 11.9 and 11.10, respectively.

11.3.5.2 Short-Circuit Impedance

Short-circuit impedance to calculate the through-fault current is the sum of the primary system impedance and transformer impedance at base rating ($Z_{SC} = Z_S + Z_T$). The primary system fault level to determine the through-fault current capability (SCW) for oil-immersed transformers is given in table 11.4. Determine Z_S from the following system fault levels:

- *Oil-immersed transformers*: IEEE C57.12.109-1993, table 4, as per table 11.4
- *Dry-type transformers*: use 1500 MVA for system voltage <15 kV and 3500 MVA for system voltage of >15 kV and <34.5 kV.

TABLE 11.3
Category for SCW and kVA Ranges

Category	kVA Range (three phase)	Fig. No. and IEEE Standard
I	15–500	Fig. 1, C57.12.59
II	501–5,000	Fig. 1, C57.12.109
		Fig. 2, C57.12.59
III	5,001–30,000	Fig. 2, C57.12.109
		Fig. 3, C57.12.109
IV	above 30,000	Fig. 4, C57.12.109

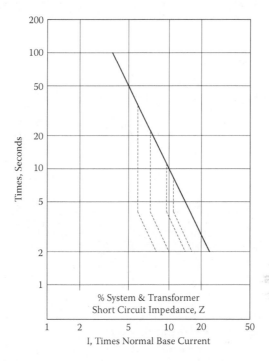

FIGURE 11.9 Through-fault current limit curves for category II dry transformer

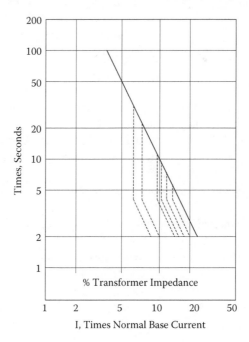

FIGURE 11.10 Through-fault current curves for category II oil-immersed transformer

TABLE 11.4
Primary-System Fault Capacity for Transformer SCW

Max. System Voltage, kV	Voltage Fault Capacity	
	kA, RMS	MVA
Below 48.3		4300
48.3	54	4300
72.5	82	9800
145	160	38200
169	100	27900
245	126	50200
362	84	50200

11.3.5.3 Example

Two-winding oil-immersed transformer rated 90/120 MVA, ONAN/ONAF, 230–34.5 kV, Z = 10%:

- Primary system fault capacity (from table 11.4): 50,200 MVA
- PU system impedance at transformer base, Z_S = 90/50,200 = 0.0018
- PU transformer impedance rating, Z_T = 0.1
- Secondary through-fault current = 90/(0.1 + 0.0018) = 884.1 MVA
- Secondary through-fault current = 884.1/ $\sqrt{3}$ × 34.5 = 14.8 kA, rms symmetrical currentv

11.3.6 Differential Protection of Transformers

11.3.6.1 Phase Differential Scheme

A differential-protection scheme compares the sum of the current entering the protection zone to the sum of the current leaving the protected zone. Overcurrent relays are not now used for differential protection, as CT saturation or magnetizing inrush current may cause the relay to operate. Percentage differential relays with harmonic restraint are used for transformer protection schemes. These relays operate on a percentage ratio of input current to through current, known as the "slope" of the relay. For example, a relay with a 15% slope setting operates when the difference between the incoming and outgoing current exceeds 15% and is higher than the relay minimum pickup.

A typical transformer phase-differential scheme using a percentage differential relay is shown in fig. 11.11. To compensate for the phase shift in the transformer windings, the CT connection to the differential relay needs to be connected in Wye for the transformer Delta winding and in Delta for the transformer Wye winding. However, with microprocessor-based differential relays, the vector shift and ratios can be corrected internally. The main CTs can then be connected in Wye and the relay programmed to correct the transformer winding connections. Some electro-

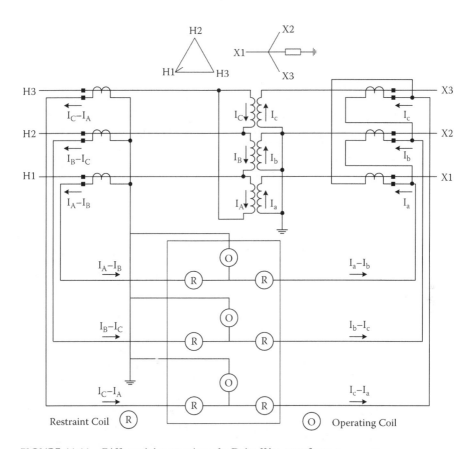

FIGURE 11.11 Differential protection of a Delta-Wye transformer

magnetic and static designs also use external interposing CTs to correct the vector group and ratio, particularly when a ground differential (REF) is also used.

A drawing showing the CT connection, polarity marking, and direction of current flow in primary and secondary circuits needs to be prepared to ensure that the phasing and wiring is correct. Many of the concerns to ensure that the differential protection is correctly connected and will operate correctly can be taken care of if some simple rules and guidelines based on the applicable standards are developed. It is also prudent to on-load test such schemes during commissioning.

11.3.6.2 Ground Differential Scheme

The percentage differential relays may not detect the line-to-ground faults in a Wye winding if the neutral is resistance grounded and the ground-fault current is below the relay pickup level. Figure 11.12 illustrates the application of a ground-differential scheme for a Delta-Wye transformer, where the Wye winding neutral is resistance grounded. Zero-sequence current arrows show an external ground fault for which the ground-fault relay will not operate. The relay scheme will operate correctly for any ground fault within the protected zone.

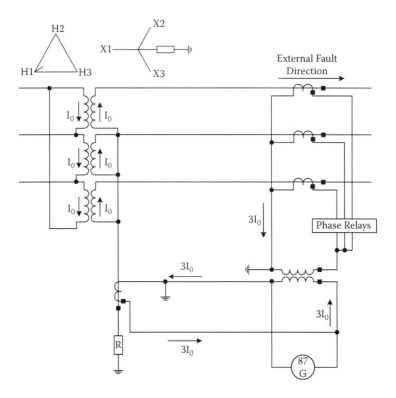

FIGURE 11.12 Ground-differential scheme of a Delta-Wye transformer

11.3.6.3 Guidelines for Phase Differential

For differential protection, the following application and setting guidelines are appropriate:

- CT ratio and connection:
  1. Select ratio or tap of a multiratio CT so that the secondary current corresponding to the current transformer at maximum kVA rating does not exceed the CT secondary rating (5 A in North America).
  2. The combination of the selected CT ratio and relay tap shall provide a match of the relay current within the permissible limit.
  3. Select the CT accuracy voltage that will not saturate at the secondary through-fault current or, at times, the relay rated tap current.
  4. Current transformer connections to compensate for three-phase shift:

Power Transformer Connection	Current Transformer Connection
Delta-Wye	Wye-Delta
Wye-Delta	Delta-Wye
Delta-Delta	Wye-Wye
Wye-Wye	Delta-Delta

Protection

5. For programmable differential relays, connect the CT on both sides in Wye and program the power transformer winding connections.
6. Ground the CT at one point only.

- Matching of relay currents to relay taps:
    1. Connect the relay to receive "in" and "out" currents that are in phase for a balanced load condition.
    2. Select the relay taps close to the current ratios for a balanced load condition.
    3. Calculated the mismatch using the formula:

$$\% \text{ mismatch (error)} = \frac{\left[\frac{I_L}{I_H} - \frac{T_L}{T_H}\right]}{S} \times 100$$

where

I_L, I_H = relay input currents for low (L) and high (H) windings, respectively
T_L, T_H = relay tap settings for low (L) and high (H) voltage side, respectively
S = smelter of the two terms (I_L/I_H or T_L/T_H)

The percent mismatch shall not exceed 5%. The error plus the variation due to ±10% LTC should not exceed 15%. For three-winding transformers, the percent mismatch (error) should be checked for all combinations of currents or taps. In programmable differential schemes, the relay automatically selects the tap to keep the percent mismatch under 5%.

- Percentage slope setting:
    1. Minimum slope setting equals the maximum range of manual or load tap changer (LTC) taps and the net unbalance during an external fault.
    2. The percentage recommended minimum slope setting should typically be 15% for transformers with ±5% off-circuit taps and 25% for transformers with ±10% LTC.

11.3.7 Protection of Grounding Transformers

Zig-zag grounding transformers are often used to provide a ground source on the Delta side of a transformer, and fig. 11.13 illustrates the protection of such a zig-zag grounding transformer and the current flow for an external ground fault. Overcurrent relays connected to Delta-connected CTs are used for this protection. For applications where the transformers are switched, harmonic-restrained overcurrent relays must be used to prevent inadvertent tripping upon energizing the transformer.

Protection of a Wye-Delta grounding transformer is shown in fig. 11.14. The protection scheme and the application is the same as that used for a zig-zag grounding transformer. It is also possible to integrate the protection of the grounding bank into the ground differential protection of the main transformer, allowing the main transformer protection to protect both transformers.

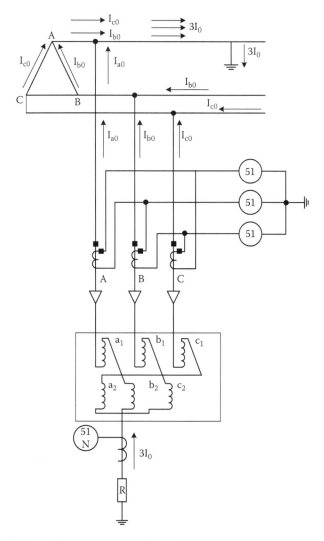

FIGURE 11.13 Protection zig-zag grounding transformer

11.3.8 Overexcitation Protection

When a transformer is subjected to a voltage above its design rating or a frequency below its design level, the core can saturate and produce local heating and the risk of damage. Such overexcitation or overfluxing of a transformer can occur when the ratio of P.U. voltage to P.U. frequency (V/Hz) at the transformer secondary terminals exceeds 1.05 P.U. at full load, 0.8 power factor, and 1.1 P.U. at no load. This condition can occur in generating stations where V/Hz at the generator terminals exceeds 1.05 P.U. due to the malfunction of the voltage regulator or when the unit is running in isolated mode and the frequency falls. The generator step-up and unit auxiliary

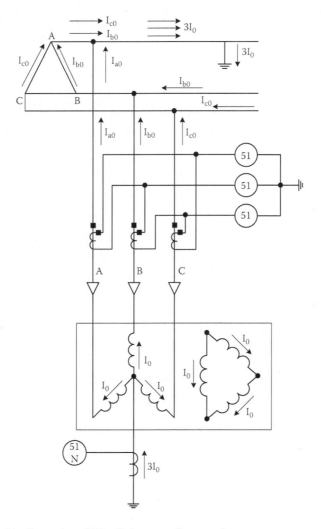

FIGURE 11.14 Protection of Wye-Delta grounding transformer

transformers connected to the generator terminals or bus must be designed for a voltage-to-frequency ratio of 1.1 P.U. or higher at full load.

The V/Hz relay, device 24, is included in the generator protection scheme. The relay pickup and time–voltage curve must lie below the generator and transformer time–voltage curves at rated frequency.

Overexcitation can also occur to transformers on the end of a long line connected to a generating plant. During a load rejection, the transformer may have a significantly higher V/Hz than that at the generator facility as a result of the Ferranti effect.

11.3.9 Surge Protection

Surge protection is required for transformers at locations where they are subject to transient overvoltage produced by lightning and for switching surges from power factor capacitor-bank switching, and other system disturbances. Review the transformer location and system interconnection to determine if an insulation coordination check is required. The following guidelines can be used for the application of surge arrestors:

- A surge arrestor is required if the transformer primary or secondary is connected to overhead line conductors that are exposed to lightning. Mount the surge arrestors as close to the transformer terminals as possible, preferably 1 m or the minimum distance determined from the insulation coordination check.
- A surge arrestor is highly recommended at the primary terminals of dry-type transformers even if they have insulation level compatible with the oil-immersed transformers.

11.4 MOTOR PROTECTION

This subject is well covered in IEEE C37.96-2000, "Guide for AC Motor Protection" [S26]. The driven load and motor characteristics must be taken into consideration for the selection and setting of protective devices, and these aspects are covered in more detail in chapter 9.

11.4.1 DATA REQUIRED FOR RELAY SELECTION AND SETTING

The following subsections provide the information that is required to carry out the selection and setting of the motor-protection schemes.

11.4.1.1 Data from Motor Manufacturer

1. Complete motor data, including the following:
   - The rating expressed as hp or kW, kV, rpm, service factor
   - The power factor at full load and at starting
   - The insulation class acceptable temperature rise
   - The locked-rotor to full-load ratio I_{LR}/I_{FL} for induction motors
   - The direct axis subtransient reactance X_d'' (saturated) for synchronous motors
2. Stator and rotor thermal-limit curves at locked rotor and during acceleration, starting at ambient temperature and at rated full-load temperature (hot and cold)
3. Negative-sequence current capability, I_2 and $I_2^2 \times t = K$
4. Stator and rotor cool-down time when running at full load and stopped
5. Frequency of starting or number of starts
6. Motor performance curves during acceleration or starting, including:
   - Time–acceleration current curves with the driven equipment at 80%, 90%, and 100% voltage
   - Speed–torque curves at 80%, 90%, and 100% voltage
   - Speed–PF curves during acceleration

Protection

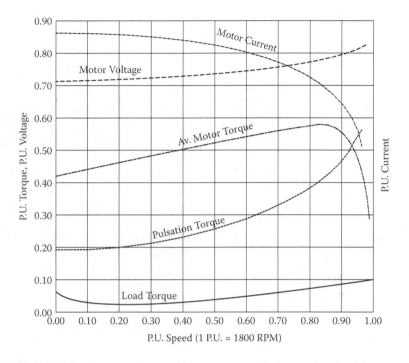

FIGURE 11.15 Synchronous-motor performance curves during acceleration (plot generated during motor design)

The above performance curves are generated during the motor design stage. A typical plot from a motor manufacturer is shown in fig. 11.15.

11.4.1.2 Data from Mechanical Equipment Vendor

1. Load speed–torque curve during acceleration converted to motor speed
2. Load inertia converted to motor speed

11.4.1.3 Data from Power Utility Company

1. Minimum and maximum fault levels and X/R ratio or system fault impedance (R + JX)
2. Auto reclosing practice (It is preferable that there be no autoreclosing on the lines feeding the plant, but if there is, the dead time needs to be known. All motors must be tripped or disconnected using an underfrequency element or other "islanding" detection protection during the dead time if the power utility company has an autoreclose practice on the in-feeding line.)
3. Percent voltage unbalance
4. Permissible voltage flicker curve

11.4.2 Motor-Protection Schemes

The following guidelines are recommended for motor-protection scheme choice and when choosing a protective device or relay setting.

11.4.2.1 Selection of Current Transformers

- Refer to chapter 6, section 6.2, "Current Transformers," for general selection criteria.
- Select the highest CT ratio that can be used within the relay, setting the range to reduce the secondary current on faults and hence the risk of CT saturation. A CT primary current rating of 2.0 P.U. of motor full-load current is adequate for a majority of motor-protection relays. Microprocessor-based relays have a wide pickup range and can be set even when a higher CT ratio is used.
- Specify dual-accuracy CTs, e.g., 1200/5-C400/0.3B1.8 and 600/5-C200/0.3B0.9, to get good accuracy for fault and load-current measurement.
- For fused starters, the CT must not saturate at 2.0 P.U. motor locked-rotor current or for the asymmetrical starting current.
- For a switchgear-type controller, the CT must not saturate at the maximum asymmetrical fault current, or the calculated excitation voltage, at maximum rms symmetrical fault current, must be less than 50% of the CT knee-point voltage.

11.4.2.2 Relay Characteristics and Modeling

To prevent damage to a motor during start and normal operation, the motor must not overheat. The motor therefore needs to operate below the motor thermal or heating curve, which is composed of the overload and locked-rotor or acceleration curves. The overload curve is closer to a long time inverse characteristic, whereas at the locked rotor it is almost an $I^2 \times t$ type of curve. The majority of the protection schemes available in the market use only one curve to cover both regions and use a model responsive to the motor current as measured by the relay. The protection function needs to be based on a thermal model and provide full coverage for overload conditions and during acceleration.

11.4.2.3 Motor Acceleration Torque

For a motor to start successfully the difference between the motor torque and the load torque, the net acceleration torque, must always be positive. At the same time, the current taken by the motor, for the time taken to start, must remain below the motor acceleration thermal-limit curves. This acceleration torque is proportional to the motor voltage, and it is necessary to ensure that the motor can bring the load up to the rated speed, at the available voltage, in time to avoid motor (rotor) damage or the protective relay from tripping. Therefore, there must be an adequate margin between the motor-capability curves and the load–torque curves over the range of normal starting voltages. Figures 11.16 and 11.17 illustrate the motor acceleration and protection problem, respectively.

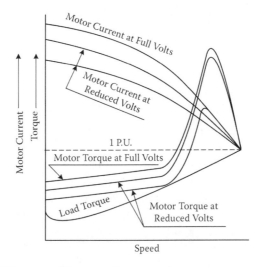

FIGURE 11.16 Motor speed: torque and current

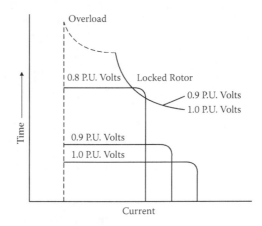

FIGURE 11.17 Motor time–current curve

11.4.2.4 Protection Redundancy

In the past, motor protection comprised a number of proven relays, each with individual features, in separate cases. Modern microprocessor-based relays combine most of these protective features into one single package (case). It is normal practice in generator protection for the protective features to be distributed in two separate relays and divided into two groups, A and B. Each has separate tripping capability with some degree of overlapping features. Motor-protective features should also be divided into a minimum of two physical groups. For the protection of large AC motors, the overtemperature element(s), connected to the stator winding resistance-temperature detector(s) (RTDs), and the flux-balance differential relays should be kept out from the main protection scheme, or two independent protection schemes should be provided.

11.4.2.5 The Use of RTD Bias into Motor Current

Some protection schemes provide a bias from the stator RTD into the motor current measurement. The relay operating time–current characteristic curve is raised or lowered to compensate for higher or lower stator winding temperature. This permits overloading of the motor if it is located in a location with a cooler ambient temperature. The thermal lag in the RTD is too slow to provide protection during starting. Moreover, for large induction motors and all synchronous motors, rotor heating is the limiting factor. The rotor damage curve is usually based on mechanical considerations and very little or no shift occurs with the change in initial temperature. RTD bias can also cause a problem for the motor feeder in cases where the motor is located in a cooler ambient temperature. The bias will permit an overload, depending on the ambient temperature, but the feeder is normally sized for only up to 125% of the rated motor full-load current and could therefore be overloaded.

11.4.2.6 Motors Driving High-Inertia or High-Torque Load

The acceleration time of motors driving a high-inertia or high-torque load is close to, and can exceed, the permissible stall time. This makes it difficult to use current measurement alone to detect the difference between a normal start and a delayed start that could damage the motor. The alternative is to provide a speed-monitoring switch or an impedance relay for locked-rotor protection in such applications.

11.4.2.7 NEMA E2 Motor Controllers

Motor controllers or starters, such as those designed to NEMA E2, consist of load-break vacuum contactors and series fuses, with the latter providing the fault protection. The fuse rating needs to be based on the guidelines provided in chapter 9, section 9.8.1. Although fused starters are available for squirrel cage induction motors rated up to 5000 hp at 4.16 kV, the large fuse current rating required for motors rated above 1500 hp creates a coordination problem with adjacent protections. It is necessary, therefore, to draw a time–current coordination curve during the initial phase of the project and determine the maximum motor size above which a switchgear-type controller is required to achieve coordination. For contactor-type starters, it is essential that the fuse protects the interrupting duty of the contactor, in that the fuse must always rupture before the contactor attempts to open for all current conditions in excess of the contactor's interrupting rating.

11.4.2.8 Electrical Code Requirements

It is always necessary to ensure that the protective-device settings are in compliance with the electrical codes having jurisdiction over the installation. A summary of the U.S. National Electrical Code (NEC) [S14] and Canadian Electrical Code (CEC) [S15] rules, pertinent to motor protection, is provided in section 11.4.5.

11.4.3 Recommended Protection

Motor protection can be divided into four categories, as listed in the following subsections.

11.4.3.1 Protection against Abnormal Power Supply Conditions

The following power supply conditions can damage the motor when they deviate significantly from the standards the motor was designed to operate under:

- Undervoltage, detected by undervoltage protective device 27
- Overvoltage, detected by overvoltage protective device 59
- Voltage or current unbalance, detected by unbalance or negative-sequence protective device 60 or 46
- Underfrequency, detected by frequency-protective device 81

11.4.3.2 Protection against Faults in the Motor or Motor Feeder

Ground or phase faults in the motor circuit can be detected by the following:

- Ground fault, detected by ground overcurrent of differential-protective devices 50G, 51G, or 87M
- Short circuit detected by phase overcurrent or differential-protective devices 50, 51, or 87M or, for fuse-protected contactors, by the series fuse

11.4.3.3 Protection against Abnormal Conditions Caused by the Driven Load

The following unusual operating conditions can be detected by the following:

- Locked rotor or fail to accelerate, detected by a current relay and timer, device 51S
- Overload, detected by thermal protection, device 49, 49W
- Bearing overtemperature, detected by a direct temperature measurement, device 38
- Vibration, detected by vibration detectors, device 39

11.4.3.4 Protection against Abnormal Conditions Caused by Environment

Unusual environmental conditions affecting the motor can be detected by the following:

- High ambient temperature, detected by a thermal measurement, device 49W
- Blocked ventilation or cooling, detected by a thermal measurement, device 49W
- Lightning and switching surge, detected by surge arrestors and controlled by surge capacitors

11.4.3.5 Recommended Protection Schemes

The recommended protection schemes for medium-voltage motors are shown in the following figures:

- The typical protection features of a MV induction motor controlled by fused starters. In this scheme, shown in fig. 11.18, the fuse provides the short-circuit protection for the contactor and motor feeder. It is necessary to program the 50 element to "off," if it is included in the relay, to prevent the contactor being instructed to open on short circuits. The contactor drops out at about 60% of its voltage rating and prevents the motor from restarting. Some relay models include device 27; program these to open the contactor at 80% voltage with a time delay of about 2 s.
- Typical protective features for an MV induction motor controlled by a circuit breaker. Figure 11.19 shows the protective features for a large motor, usually larger than 1500 hp, and controlled by a circuit breaker. Short-circuit protection is provided by the device 50 operating the circuit breaker. Two options are shown for the differential, one a core-balance scheme covering the motor alone, and the other an overall differential, covering the whole circuit. Protective features 87M, 49W, and 38 have been kept separate from the main protection scheme. For very large and important motors, it is preferable to provide two protection schemes to achieve overlapping and component redundancy.

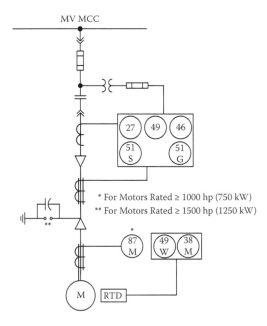

FIGURE 11.18 Typical protection features for MV induction motor (≤1500 hp) controlled by fused starters

Protection

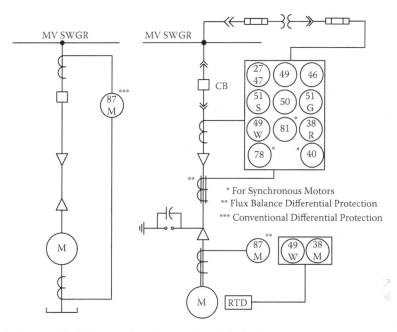

FIGURE 11.19 Typical protection features for MV induction motor (>1500 hp) controlled by circuit breaker

- Typical protective features for MV motor and captive (unit) transformer. This scheme, shown in fig. 11.20, is very similar to the protection scheme show in fig. 11.19. The motor controller, operating into a circuit breaker, is provided on the transformer primary side so that the protective devices protect the motor and transformer, and a differential relay (87T) is added for the transformer.

11.4.4 Protective Devices and Suggested Settings

As an example of a typical set of device settings for a medium-sized motor, a list of the protection typically applied and the recommended settings are summarized in table 11.5 and discussed in more detail in the following subsections.

11.4.4.1 Undervoltage Protection (Device 27)

Motors are built to perform satisfactorily with a terminal voltage variation of ±10% of the rated voltage. The inherent undervoltage dropout characteristics of NEMA controllers or magnetic-type starters do not protect the motor against undervoltage. The dropout voltage of 55–65% is too low, and a sustained operation at voltages between this value and 90% will result in motor damage. Moreover, undervoltage protection of short duration (100 to 250 ms) may be required to prevent the motor from restarting upon autoreclosing of utility lines.

Two schemes, one with an instantaneous undervoltage relay with a timer and the other with an inverse-characteristic relay similar to an induction disc type, are used:

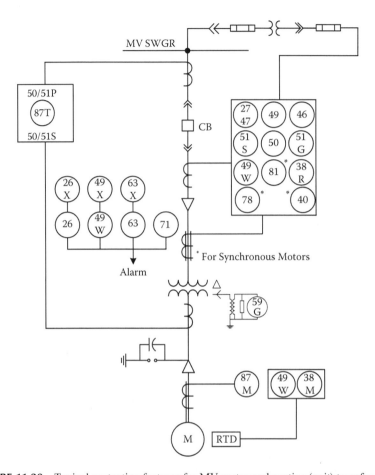

FIGURE 11.20 Typical protection features for MV motor and captive (unit) transformer

- *Instantaneous undervoltage*: The relay is normally set to drop out at about 80% of the nominal system voltage, and the timer is set to operate in about 2–3 s, as seen in fig. 11.21.
- *Inverse-time undervoltage*: The relay is set at 0.8 PU motor rated voltage and a delay of 1.25–2 s to give a reduction from normal to approximately zero volts. The relay automatically gives a longer time for intermediate voltage values, as seen in fig. 11.22.

For applications where voltage drop during starting is high, set the dropout to a lower value or activate the relay by a timer after completion of the starting period.

11.4.4.2 Unbalanced Voltage/Current (Device 46)

NEMA MG1 has established a curve for derating the motor output according to the voltage unbalance in percent. This approach results in a compromise, as it assumes

TABLE 11.5
Motor Protection: Typical Settings

IEEE Device No.	Function/Purpose	CT/VT	Typical Setting
27	Under voltage -Instantaneous timer - Inverse time	...	UV dropout at 80% of nominal system voltage Timer: 2–3 s Set 80% of system voltage with the TD set to give a delay of 1–2 s for a voltage drop from normal to zero
46	Negative-sequence or current unbalance	Phase CT 4.2.1 (b)	Negative-sequence setting in percent of motor rated full-load current - 8% alarm - 12–15% trip
81	Underfrequency	...	58.5-Hz trip
49	Overload	Phase CT	105% for motors with SF_6-1.0 115% for motors with SF_6-1.15
49W	Winding overtemperature	NA	130°C alarm 135°C trip
51S	Locked rotor or fail to accelerate	Phase CT	10–15% more than slowest normal start time
50	Phase overcurrent or short circuit	Phase CT	1.5 to 2.0 PU locked-rotor current
51G	Ground fault	50/5 or 100/5 C30	2–5% of the neutral resistor current rating; time delay of three to six cycles
87M	Differential (flux balance) Differential (overall)	100/5 C50	Primary pickup of 5–10 A; refer to manufacturer brochure, but on the order of 20% of CT rating

a typical response for all motors. Individual motor capability in terms of negative-sequence current gives a much more specific means of protecting a particular motor by the application of a negative-sequence current relay. This is the method used for generators. Negative-sequence current can be estimated by dividing the voltage unbalance by the negative-sequence locked-rotor impedance of a motor. For a system voltage unbalance of 4% and motor locked-rotor impedance of 0.16 PU, the negative-sequence current will be approximately 0.04/0.16 = 25%.

In general, negative-sequence current relays with an inverse-time characteristic are used. Their operating time is inversely proportional to the square of the negative-sequence component of the three-phase current ($I_2^2 \times t = k$). A negative-sequence pickup setting of 8–12% of motor full-load current for alarm and 12–15% of motor full-load current for trip is suggested. However, this must be verified with the motor design data.

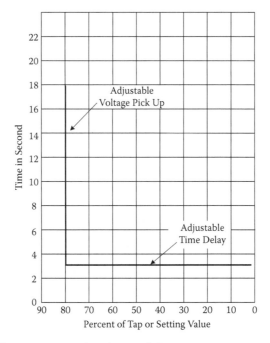

FIGURE 11.21   Instantaneous undervoltage and timer

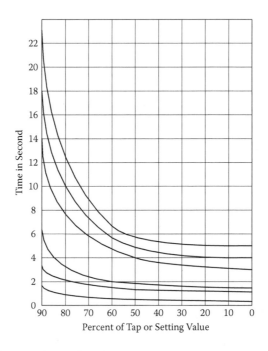

FIGURE 11.22   Inverse-time undervoltage

Protection

11.4.4.3 Underfrequency (Device 81)

Underfrequency protection is required against the following conditions:

- When autoreclosing is provided by the utility company on the transmission line feeding the plant, or there is another means of reenergizing the motor supply, synchronous motors and large induction motors must be disconnected from the supply bus before the line is reenergized to avoid unsynchronized restart. An underfrequency relay set at 58.5 Hz for a 60 Hz system with a time delay of 100 ms to trip is generally adequate to protect the machines. One underfrequency relay with an auxiliary tripping relay is adequate where several motors are connected to the same bus, or alternatively, the frequency relay can trip the main supply from the utility line(s).
- Upon loss of utility power supply to the plant, large synchronous motors driving low-inertia loads, such as refiners, act as generators and can keep the system back energized for a considerable duration, but the frequency will start to fall. Because the majority of the plant main transformers have Delta winding on the HV side, the HV system will become ungrounded and can cause overvoltages on the in-feeding circuit if the main circuit breaker is not opened. The undervoltage relay will be slow to act, and an underfrequency relay is required. A setting of 58.5 Hz on a 60 Hz system with a time delay of 100 ms is generally adequate for such applications.

11.4.4.4 Overload Protection (Device 49)

The majority of overload relays are simple relays that are responsive to motor current alone and hence do not provide a direct overload protection. This type of relay is usually set at 105% of full-load current for motors with a service factor (SF) of 1.0 and at 115% of full-load current for motors with a SF of 1.15. Some relays have an $I^2 \times t$ characteristic that may cause premature trip upon short time overload. It is not recommended to use the RTD bias feature to shift the overload curve. For large induction motors and all synchronous motors, rotor heating is the limiting factor; there is very little or no shift in the damage curve with a change in initial stator temperature.

11.4.4.5 Winding Overtemperature (Device 49W)

Resistance-temperature detectors (RTD), consisting of three elements of 100-ohm platinum, embedded in the stator winding, are utilized to measure the winding temperature. This is the most direct protection for overtemperature while the motor is running, as the RTD responds both to motor overload and to blocked ventilation. For motors with a Class 155 (F) insulation system and an 80°C temperature rise, a setting of 135°C for alarm and 140°C for trip is suggested.

11.4.4.6 Locked Rotor or Fail to Accelerate (Device 51S)

- *Motor-driving load with low or normal load inertia*: For a normal application, there is a substantial margin between the motor current–time acceleration curve and the motor-acceleration thermal-limit curve. After drawing the motor-acceleration thermal-limit and motor current–time acceleration curves on log-log paper, or plotting the same using appropriate software, select the relay-characteristic curve that will provide complete coverage for a successful acceleration to rated speed. Do not use the RTD bias option, as the shift in the overload curve may prevent one start from hot as permitted by the motor design standards.

- *Motor-driving high-inertia load*: Time overcurrent relays generally do not provide protection against locked rotor or failure to accelerate to the rated speed for motors driving high inertia loads or where acceleration time is close to the motor thermal damage curve. The following methods can be used for such applications:

  1. *Application of an impedance relay (21)*: Motor impedance increases as it accelerates from standstill to the rated speed. For example, locked-rotor impedance increases gradually from subtransient to transient and then to the rated impedance. An impedance relay, device 21, has a mho characteristic set to close its output contact for the locked-rotor condition and to reset once the motor speed has started to increase. The contact will therefore open when the measured motor impedance falls outside the circle. The contact of the mho relay (21) is wired in series with a timer (62) or a time-overcurrent relay (51), as shown in fig. 11.23, and the output is used to trip the motor if it fails to accelerate within the designated time.

  2. *Application of a speed switch*: In this scheme, the contact of a speed switch (14) is wired in series with a timer to trip the motor if it fails to accelerate within the designated time. For large and critical motors, such as large refiners in paper mills, the speed switch needs to measure the rate of acceleration to ensure that the motor will reach the rated speed within the designed time.

  3. *Application of time and instantaneous overcurrent relays (50 and 51)*: In this scheme, two instantaneous relays (50) and one time overcurrent relay (51), with an inverse or very inverse characteristic, are used. The first device 50, a high dropout (HDO) unit, is set at 1.15 P.U. motor-rated current and drops out at 0.85 to 0.90 P.U. of the set current with no delay.

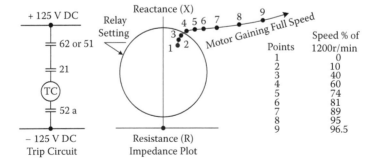

FIGURE 11.23 Application of device 21 for the protection motor driving high-inertia load

The second device 50, which is set at about 2.0 P.U. rated motor current, or above the asymmetrical starting current, provides short-circuit protection. For starting, a timing relay with a time delay of 100–250 ms prevents the first device 50 (HDO) from tripping due to asymmetrical starting current and provides short-circuit protection after the set time delay. The device 51 is set at 1.05–1.15 P.U. motor rated current with the time multiplier set so that the time–current curve lies below the motor thermal limit and above the time–current acceleration curve. A typical time–current plot and DC trip circuit are shown in fig. 11.24.

11.4.4.7 Phase Overcurrent or Short Circuit (Device 50)

For motors controlled by contactor-type starters, a current-limiting fuse provides protection against short-circuit current. Guidelines for the selection of fuse rating are given in chapter 9. For motors controlled by circuit breakers, a three-phase instantaneous overcurrent element is required. Usually a setting of 1.6 P.U. to 2.0 P.U. of locked-rotor current is provided. A minimum setting of 1.6 P.U. is provided to override the asymmetrical current at the start, except in the case of more sophisticated relays that are immune to the asymmetrical component that can be set lower.

11.4.4.8 Ground Fault (Device 51G)

Medium-voltage power systems in industrial plants are generally resistance grounded to minimize the damage to the motor windings from internal line-to-ground faults. The system neutral resistor is generally rated at 100 to 400 A. The ground-fault-sensing CT and relay-setting combination must be sensitive enough to provide a minimum coverage of 95% of the motor winding.

A zero-sequence CT, with a ratio of 50/5 A or 100/5 A and an accuracy of C30 and a ground-fault relay set to provide a primary pickup of 2–5% of the resistor rating, is adequate to provide such protection sensitivity. A time delay of three to six cycles is required to prevent nuisance tripping.

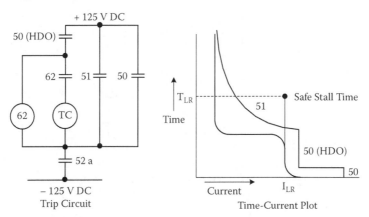

FIGURE 11.24 Application of device 50/51 for the protection of motor driving high-inertia load

11.4.4.9 Differential (Device 87M)

Differential protection is recommended for motors rated 1000 hp and above. The flux-balance differential scheme is more economical compared with a conventional differential-protection scheme and offers reasonable protection for the motor itself. The flux-balance differential-protection scheme, shown in fig. 11.25, utilizes three current transformers, ratio 100/5 A with a C50 accuracy, and a three-phase instantaneous overcurrent relay. The relay is usually set to provide a primary pickup of 5–10 A with a time delay of 50–100 ms. For the sake of redundancy and increased reliability, differential protection should be kept separate from the integrated motor protection.

An overall differential scheme, shown in fig. 11.26, utilizes three CTs at the starter and three at the neutral end of the motor winding. This scheme provides protection against phase-to-phase and three-phase faults for both the motor winding and the associated cable. It should be noted that the second CT ground connection shown should only be used if the CT inputs to the relay are fully isolated in the relay.

11.4.4.10 Loss of Field Protection for Synchronous Motors, Device 40

Upon loss of field, a motor draws power reactor VARs from the system. For medium-size brushless machines, a relay that measured VARs into the motor is applied. The relay is usually set to trip or disconnect the motor when the VAR flow into the motor exceeds 10% of the motor rated kVA. For large machines, an impedance relay similar to the generator loss-of-field protection is used.

11.4.4.11 Out-of-Step Protection (for Synchronous Motors), Device 55 or 78

For medium-size machines, a power factor relay device 55 is used. Loss of synchronism causes the motor to pull out of step with the power system, resulting in high current pulsations. The relay with an adjustable power factor angle is set to operate watts flow out of and VARs flow into the motor.

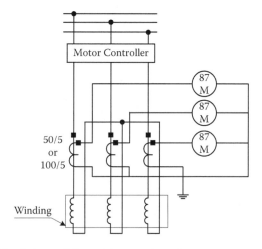

FIGURE 11.25 Flux-balance differential-protection scheme

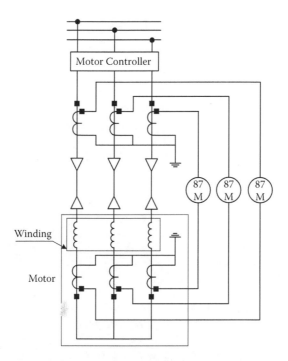

FIGURE 11.26 Overall differential-protection scheme

11.4.4.12 Rotor Protection and Monitoring for Synchronous Motors

A system for the protection and monitoring of a synchronous motor consists of a transmitter mounted on the rotor, a receiver module mounted on the machine, a serial data cable, and a PLC for data processing mounted in the excitation cubicle, as shown in fig. 11.27. The system provides thermal protection on starts, temperature monitoring and protection of the amortisseur and field windings, and rotor ground-fault monitoring and protection.

11.4.4.13 Slip-Ring Flashover Protection for Wound Rotor Induction Motors

The semiconducting dust from the carbon brush wear accumulates in the slip-ring housing and creates a path from the rotor winding to the ground. The rotor winding or secondary voltage is higher during start, which can cause flashover from the slip ring to ground or across the slip rings. The arc flash can create fire in the accumulated

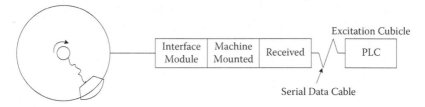

FIGURE 11.27 Rotor protection and monitoring of a brushless synchronous motor

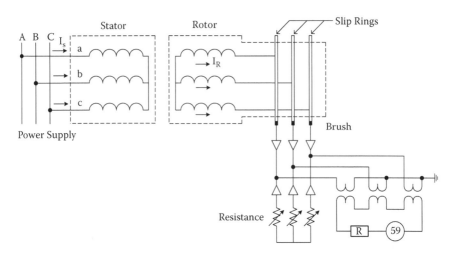

FIGURE 11.28 Slip-ring flashover ground protection

carbon dust and damage the slip rings and the rotor winding terminal. The protection scheme to detect the ground and prevent fire is shown in fig. 11.28.

11.4.4.14 Out-of-Phase Reenergizing Protection

An inrush current up to 2.5 times normal locked-rotor current can be produced in a motor when it is reenergized soon after a power interruption. This magnitude depends on the saturation of motor magnetic paths, the system impedance, and resulting voltage and phase angle between the motor back EMF and the supply voltage at the instant of reenergizing. One of the following methods can be used to avoid out-of-phase reenergization:

- *Fast-transfer scheme*: This scheme is designed to reenergize the motor before the residual voltage angle has diverged significantly from that of the supply bus by minimizing the time that normal and alternative sources are both open (less than 10 cycles). Phase-angle relays should be applied to block the transfer if the angle between the sources were such that excessive current would flow into the motor if the transfer was completed.
- *Parallel-transfer scheme*: In this case, both the normal and the alternative sources are closed for a short time during the transfer. Switchgear short-circuit rating may be exceeded during this transfer, and the logic needs to ensure that the transfer will complete and not leave both sources connected.
- *Residual voltage-transformer scheme*: In this scheme, the residual voltage (back EMF) of the motor bus is monitored after the normal source is lost. The relay then prevents transfer to the alternative source until the motor bus voltage drops to a predetermined value. All the connected motors are then reaccelerated. Special voltage relays, which are not affected by the decaying frequency, are used for such an application. In such a scheme, the

Protection

contactors of the essential motors must be latched, and the second source must be strong enough to accelerate those motors.

11.4.4.15 Surge Protection

Surge protection comprising three metal oxide surge arrestors and three surge capacitors in series with a damping resistor is installed in the motor line terminal box. The commonly used equipment rating is provided in table 11.6.

11.4.4.16 Variable-Frequency Drive Protection

A single-line diagram for a typical medium-voltage variable-drive-frequency (VDF) system and the recommended protection is shown in fig. 11.29. The protection scheme comprises:

- Feeder and drive-isolation transformer protection (provided in the supply switchgear; setting considerations are covered in section 11.3, "Transformer Protection"):
  1. Device 50/51: feeder and transformer overcurrent protection
  2. Device 51G: ground overcurrent protection
  3. Device 49: transformer-winding overtemperature
  4. Device 87: transformer differential protection
- Drive-controller protection:
  1. Device 27 or 47 and 60: Phase-sequence undervoltage and phase-voltage unbalance;
  2. Device 51 and 49: Converter overcurrent and overtemperature;
  3. Device 51 and 49: Inverter overcurrent and overtemperature.
- Motor protection:
  1. Device 59 and 81O or device 24 and 50/51: overvoltage and overfrequency or V/Hz, and overcurrent protection
  2. Device 87M and 49: motor flux-balance differential and overtemperature protection; drive-controller protective devices covered under 2 and 3 (above) are provided by the drive vendor and are located in the drive controller.

TABLE 11.6
Commonly Used Surge-Arrestor and Surge-Capacitor Rating

Motor-Rated Voltage, kV	Surge Capacitor		Surge Arrestor	Damping Resistor ohm
	µf	BIL kV_p	Rated kV	
2.4	0.5	60	3.2/2.7	...
4.16	0.5	60	4.5	25–100
7.2	0.5	75	7.5	25–100
13.8	0.25	95	15	25–100

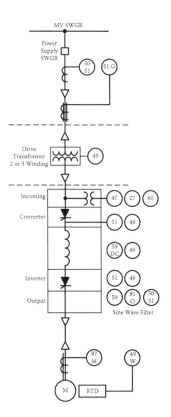

FIGURE 11.29 Medium Voltage V.F.D. Protection

11.4.5 Electrical Code Requirements

11.4.5.1 Canadian Electrical Code, CSA C22.1-2002

Figure 11.30 shows Canadian Electrical Code rules pertinent to individual motor and branch circuit protection as well as feeder protection for a group of motors. Stringent code requirements sometimes impose additional relays that are not necessary.

The Canadian rules specify that only motor overload and overcurrent (short circuit) protection is required. Other protection requirements, such as against stall or locked rotor (51S), phase current unbalance (46), ground fault (51G), fail to accelerate, etc., are not covered. It is the responsibility of the application and protection engineer to ensure that a complete motor-protection scheme is provided against all abnormal conditions covered in this chapter and in IEEE recommendations for industrial protection [S8] and in the IEEE motor-protection guide [S26]. However, Rules 28-204 (1) and (2), related to protection of feeders dedicated for a group of motors, create coordination problems with downstream protective devices. Rule 28-204 (2) permits an increase in settings where two or more motors are required to start simultaneously. However, the current setting must not exceed 300% of the feeder ampacity.

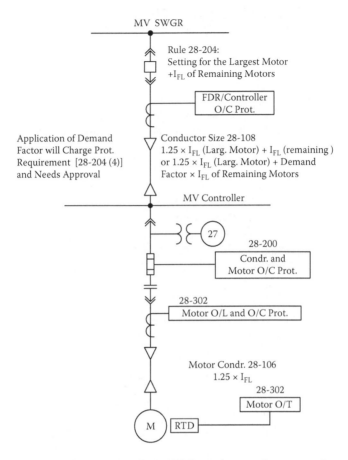

FIGURE 11.30 Canadian Electrical Code (CEC) requirement for motor and motor-feeder protection

Before this rule was introduced, the feeder overcurrent protection was set to coordinate with the largest motor protection as long as the feeder was protected against three-phase short circuits. One of the following two solutions could be used to achieve a coordinated system while complying with the code requirements:

- Select the feeder size based on Rule 28-108. Select the overcurrent relay setting that will coordinate with the largest motor protection while protecting the feeder against three-phase short circuits. Plot time–current coordination curves for all devices in the circuit including feeder short-circuit withstand and request inspection authority for an approval.
- Provide an additional single-phase overcurrent relay with long time characteristics and set it at 125% of the feeder ampacity. One single-phase element is sufficient for a three-phase balanced load and designates it as a device 51 C. The main three-phase overcurrent relay can be set to coordinate with the largest motor and protect the feeder against short circuits.

11.4.5.2 National Electrical Code, NEC-2002

National Electric Code (NEC) requirements pertinent to the motor, motor conductor, and motor protection are covered in the following rules and shown in fig. 11.31:

- Part II — Motor Circuit Conductors:
  1. 430.22 (A) for Single Motor Rated 600 V or less
- Part III — Motor and Branch-Circuit Overload Protection:
  1. 430.32 Continuous Duty Motors
  2. More than 1 hp
     A separate overload device that is responsive to motor current and selected to trip at not more than 115–125% of the motor nameplate full-load current rating Thermal Protection (RTD)
- Part IV — Motor Branch Circuit, Short Circuit, and Ground-Fault Protection
  1. 430.52 for Rating or Setting for Individual Circuit
- Part V — Motor Feeder Short-Circuit and Ground-Fault Protection
  1. 430.62 Rating and Setting

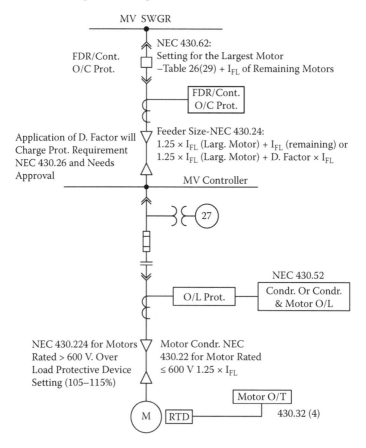

FIGURE 11.31 National Electrical Code (NEC) requirement for motor protection

11.4.6 Typical Coordination Curves

Relay setting and coordination curves are provided to illustrate the methodology used for the setting of relays and plotting of time–current coordination curves. Computer software has been used for plotting the following typical curve.

11.4.6.1 5000 hp Compressor Time–Current Coordination Curve

Figure 11.32 illustrates the settings and coordination curves for a 5000 hp compressor motor. Continuity of operation is essential for the process, and for this reason, a well-proven scheme and component was selected to eliminate unwanted trips.

Salient features of the design are as follows:

- *Captive transformer*: This transformer is provided to limit the voltage drop on a 13.8 kV bus and to provide high-resistance grounding for the motor. Extra bracing was provided for impact loading due to frequent starting.
- *Motor-protection relays*: Westinghouse COM provides complete coverage. Westinghouse COM and GE IAC66M used to be extremely reliable protection relays at that time. For critical applications such as nuclear and large thermal stations, current-responsive thermal overload elements are avoided.
- *Temperature relay*: A separate temperature relay operated from the winding RTDs was used for thermal protection.
- *Time–current coordination curve*: Motor acceleration, motor thermal damage, captive transformer, thermal damage and relay curves are shown.

11.4.6.2 1750 hp Time–Current Coordination Curve

Figure 11.33 illustrates the settings and coordination for a 1750 hp motor and feeder supplying power to the motor control center. In this case, a GE Lodtrak was used for motor protection.

The main feeder overcurrent relay, a GE IAC-53, was set to protect the feeder against short circuits and coordinate with the 18R (390 A) motor fuse. This relay is set at 2400 A, primary, to achieve the objective. However, this high setting violates the electrical code requirements.

The maximum setting of feeder overcurrent relay as allowed by electrical code rules corresponds to the sum of 390 A (largest fuse) plus 640 A (full-load current of remaining motors). To maintain the coordination and meet the code requirements, another relay (51C), an IAC-66 with long time characteristics set at 960 A, was added. This extra relay provides a long-time trip for low-level faults and high overload conditions. However, as can be seen, the other elements already provide significant margin over the cable-damage curve. As noted above, although these figures illustrate the use of electromechanical relays, the curves used are still available in modern digital relays, and these coordination curves would still be valid for the application of digital relays.

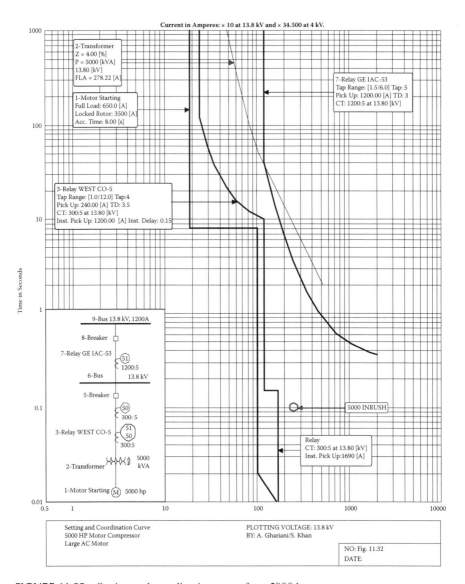

FIGURE 11.32 Setting and coordination curve for a 5000 hp compressor motor

11.5 GENERATOR PROTECTION

This subject is well covered in the two IEEE generator protection guides [S29, S30] and in the IEEE tutorial on the same subject [S37]. A review of chapter 14 in this book, "Electrical Aspects of Power Generation," and chapter 8 of Walter Elmore's book [2] is also suggested before carrying out the selection and setting of generator-protection relays.

Protection 297

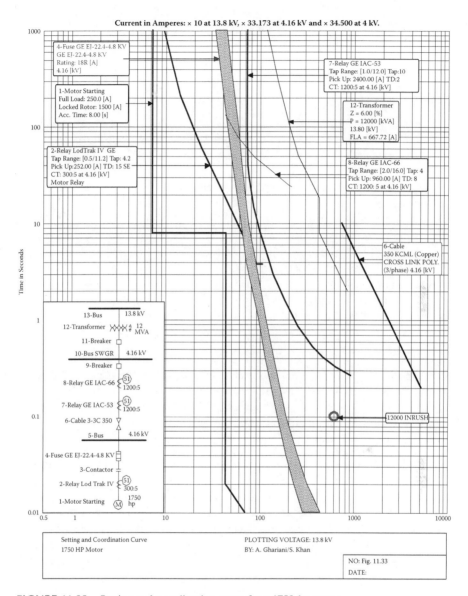

FIGURE 11.33 Setting and coordination curve for a 1750-hp motor

11.5.1 Data Required for Relay Selection and Settings

The following information is required to carry out the selection and setting of generator protection schemes:

- Normal characteristics:
  1. Standards, ANSI, IEEE, IEC

2. Power factor, speed (rpm), overspeed
  3. The rating voltage, frequency
  4. The insulation class (stator and rotor) and temperature rise
  5. Winding connection
  6. Cooling method
  7. Short-circuit ratio
- Reactance:
  1. Direct axis subtransient (X_d'') — nonsaturated/saturated
  2. Direct axis transient (X_d') — nonsaturated/saturated
  3. Synchronous (X_d)
  4. Negative sequence — nonsaturated/saturated
- Resistance:
  1. Stator and rotor
- Time constants:
  1. T'_{d0} — transient, open circuit
  2. T'_d — transient, short circuit
- Excitation system:
  1. Static or brushless
- Reactive power capability curve

11.5.2 Recommended Protection Schemes

A brief description of recommended protection schemes for process plant applications is provided. These are shown in the following figures.

11.5.2.1 High-Resistance Grounded Generator

The protection scheme for a distribution-transformer-grounded, high-resistance (HR)-grounded, 24 MW brushless generator is shown in fig. 11.34. Redundant protective devices, synchronizing panel instrument transformers (CT/VT), and other devices are shown. An out-of-step of protection relay 78 is shown as an integral part of one of the multifunction relays. This feature can be provided separately. The scheme illustrates the use of multifunction digital relays by enclosing the various elements used in a box. Any or all of these devices could be provided separately. Some users will also turn on complementary functions in the two relays and may use an impedance element (21) in place of the voltage-controlled overcurrent (51 V) in one relay.

11.5.2.2 Unit-Connected Generator

Figure 11.35 shows the protection schemes for the same generator connected to a 120 kV bus through a step-up transformer. In this scheme, the generator is synchronized to the system through a HV (120 kV) circuit breaker, and with the exception of adding an overall differential protection (87), the generator protection would be the same as shown in fig. 11.35.

A generator circuit breaker is used for systems where station auxiliary power is taken from the system through the same transformer by tapping the auxiliary transformer onto the connections between the generator and the step-up transformer.

Protection

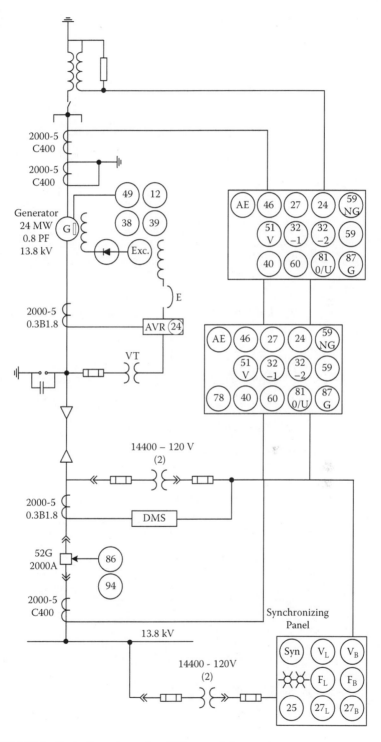

FIGURE 11.34 Protection scheme for a HR-grounded generator

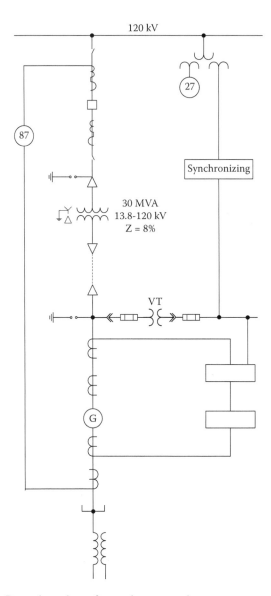

FIGURE 11.35 Protection scheme for a unit-connected generator

In this case, the protection becomes a combination of that shown in the two figures with, usually, an overlapped protection zone for the station service transformer. This scheme is more common with the combined-cycle power plants and hydroelectric generators connected to EHV systems.

11.5.2.3 Resistance-Grounded Generator Connected to LV-Grounded System

The single-line diagram of the same generator, grounded via a directly connected resistor connected in parallel with the low-resistance grounded system, is shown in fig. 11.36. This drawing shows the protective features, instrument transformer (CT/VT), synchronizing, etc., and serves as a guide in designing such a system. A single multifunction generator protection is shown with separate overall differential (87) and out-of-step (78) relays. In this case, the voltage-operated stator ground relay (59NG), fed from the secondary of the grounding transformer, is replaced by current-operated overcurrent (51NG) elements connected to CTs in the neutral connection. Most multifunction relays are provided with both of these options built into the relay, and only the appropriate input connections and program logic need to be changed. Redundant relaying could also be provided, as shown in figure 11.34.

11.5.3 PROTECTIVE DEVICES AND SUGGESTED SETTINGS

This section covers the detailed features and setting principles of the following protective devices as used for generator protection.

11.5.3.1 Accidental or Inadvertent Energizing

An interlock must always be provided to prevent human error in the closing of the generator breaker when the turbine is not running or is coasting to stop. The generator behaves like an induction motor and can be damaged within a few seconds when it is energized while it is off line, on turning gear (thermal units), or coasting to stop. A relatively new protection is called inadvertent energizing (AE) or (50/27), which is a sensitive element that must be armed and operational when the machine is not running. Several schemes are available as a part of the integrated generator-protection package in multifunction relays, and these are described in the manufacturer's descriptions and in the IEEE tutorial [S37]:

- Frequency-supervised overcurrent-relay scheme
- Voltage-supervised overcurrent-relay scheme
- Directional overcurrent-relay scheme
- Auxiliary contact-enabled overcurrent-relay scheme

A typical voltage-supervised current-relay-scheme logic is shown in fig. 11.37, where "VT" indicates a VT supervision scheme.

11.5.3.2 Volts per Hertz or Overfluxing Protection (Device 24)

This feature protects the generator and transformer against high flux density or overfluxing, leading to core saturation, if full voltage is applied while the frequency is reduced. This condition may occur during overloaded conditions in isolated operation or malfunction of the voltage regulator. It can also occur if the exciter is left operational when the machine is being shut down. This feature must be provided in the generator-protection package even though the V/Hz control and

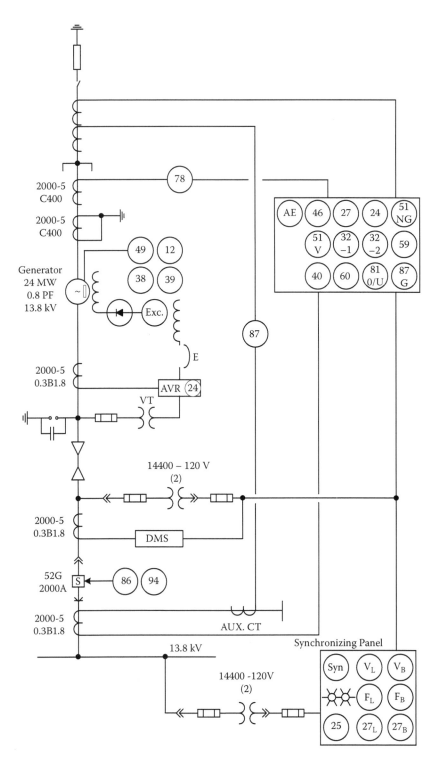

FIGURE 11.36 Resistance-grounded generator to LV-connected system

Protection

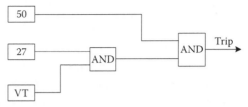

FIGURE 11.37 Inadvertent emerging (AF) using voltage-supervised current relay

protection feature is also included in the voltage regulator. V/Hz-limit curves are available from the generator and transformer manufacturers. An inverse-time V/Hz relay is set so that the relay operating curve lies below the generator and transformer V/Hz-limit curves. The set points or relay operating curve shall be below the generator and transformer limit curves. A typical plot is shown in fig. 11.38.

11.5.3.3 Reverse Power or Antimotoring (Device 32)

Motoring of a generator occurs when, for some reason, the energy supply to the prime mover is cut off while the generator is still on the line. A primary indication of motoring is the flow of real power into the generator.

The estimated power required to motor the idling prime mover is:

- Diesel engine, about 25%
- Hydro unit, 0.2–2% when the turbine blades (runner) are above the tail-race level
- Hydro unit, >2% when the turbine blades are under the tail-race level
- Steam turbine (condensing), 0.5–3%
- Steam turbine (noncondensing), 3% or more
- Gas turbine, 5%

Reverse-power relays are supplied with time delay (up to 30 s) to prevent operation during power swings caused by system disturbances or immediately after

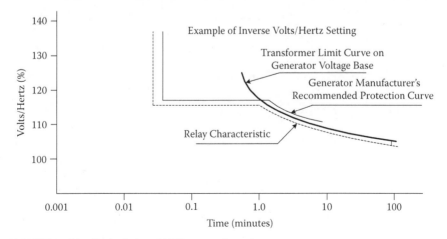

FIGURE 11.38 Typical plot of V/Hz protection relay

synchronizing. These elements are set at about 50–70% of the motoring power, as provided by the manufacturer and measured during commissioning. Two reverse-power elements (32-1 and 32-2), each with an integrating timer, are required. The timer for 32-1 is set at about 3 s to protect against loss of power input (steam, etc.) to the prime mover and is used for sequential tripping. The timer for 32-2 is set at about 10–30 s to override the power swings.

A reverse or low forward-power element is also used in the control circuit for sequential tripping of steam turbine generators. This control feature prevents the steam unit from possible overspeeding upon simultaneous tripping of the circuit breaker and the steam valve, and is discussed further in section 11.5.5.1.

11.5.3.4 Loss of Field/Excitation (Device 40)

Excitation can be lost due to field open circuit, field short circuit, regulation system failure, or loss of power supply to the excitation system. For most generators, the unit will overspeed and operate as an induction generator. It will supply power to the system and take significant reactive power from the system. The stator current can be 2.0 PU, and a high level of current is induced in the rotor.

For small and less important machines, a single-zone offset mho relay characteristic is used to detect this condition by measuring the apparent impedance of the machine as seen from the power system. The relay is set with its mho circle offset by $X_d'/2$ and having a diameter equal to X_d. The timer is set at 0.5–0.6-s delay to prevent tripping from power swings.

For large machines, two-zone offset mho characteristics are still used, as shown in fig. 11.39, with an offset of $X_d'/2$:

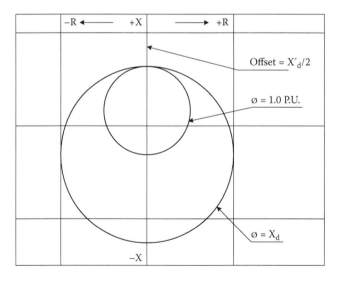

FIGURE 11.39 Two-zone offset mho relay characteristic for loss of field protection

- *Zone 1*: Diameter set at 1.0 PU of machine rating with instantaneous tripping; this relay detects loss of field from full load down to 30% load
- *Zone 2*: Diameter = X_d, time delay of 0.5–0.6 s.

The relay senses the variation of impedance as viewed from the generator terminal. The particular relay manufacturer's literature should always be checked for detailed setting calculations.

11.5.3.5 Negative-Sequence Current (Device 46)

This feature protects the generator against sustained unbalance or negative-sequence currents that induce triplen harmonic currents into the rotor. The ability of a generator to withstand the unbalanced current in terms of negative-sequence current is specified in the relevant ANSI generator standards [S39, S40]. These levels of negative sequence are summarized as follows:

Type of Generator	Permissible I_2	Permissible $I_2^2 \times t$
Salient pole with connected amortisseur winding	10%	40
Salient pole with nonconnected amortisseur winding	5%	40
Cylindrical rotor, indirectly cooled	10%	30
Cylindrical rotor, directly cooled up to 960 MVA	8%	10

The protection relay characteristic is set to match the $I_2^2 \times t$ withstand of the generator. The relay is set to give an alarm at 3–5% of the rated stator current, which permits the operator to reduce the load and avoid tripping. The trip pickup is set at the continuous negative-sequence current capability of the machine. The characteristics of a typical negative-sequence time-overcurrent relay are shown in fig. 11.40.

11.5.3.6 Generator Thermal Overload Protection (Device 49)

The generator output can be increased for a short time up to the permissible short-time overload capability given in ANSI C50.13 [S40]. Resistance-temperature detectors embedded in the stator winding are used to give an alarm at a preset temperature and eventual trip. The alarm stage is provided to give the operator an opportunity to reduce the load in an orderly manner. Some generator-protection relays also include thermal replicas that utilize stator current to approximate the heating effect in the generator.

11.5.3.7 Winding Overtemperature Protection (Device 49W)

Resistance-temperature detectors (RTD) embedded in the stator are used for monitoring the winding temperature of indirectly cooled generators. For directly cooled generators, the stator-bar-coolant discharge temperature is used along with the embedded RTD to monitor the winding temperature.

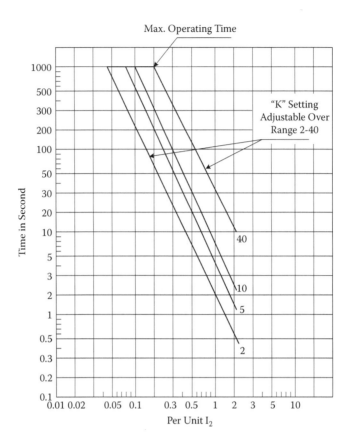

FIGURE 11.40 Characteristic of a negative-sequence time-overcurrent relay (From IEEE Cat. No. 95 TP 102, IEEE Tutorial on the Protection of Synchronous Generators. With permission.)

11.5.3.8 Voltage-Restrained Overcurrent Protection (Device 51V)

This is the final backup feature to trip the generator if faults on the generator or its connections are not cleared by other system protections. The relay becomes more sensitive when the voltage is reduced during a fault, and hence the protection needs to be blocked for voltage-transformer fuse blown conditions. This is usually from a device 60 balance relay, which compares the protection VT output with that of the exciter VT. Device 51V can be replaced by an impedance element, device 21, if coordination is needed with high-voltage distance relays or if both elements are used in conjunction to get more backup coverage.

Time settings for the overcurrent element (51V) are set long enough to permit other protections to clear with a current pickup set at approximately 2 PU with 100% volts. The impedance element (21) is usually set to reach just through the generator transformer to the high-side bus.

11.5.3.9 Overvoltage Protection (Device 59)

This device protects against overvoltage that might occur if there were a malfunction of the voltage regulator or a severe external system problem. The usual setting for turbogenerators will be 110% of nominal volts and a time delay of 15–20 s.

11.5.3.10 Out-of-Step Protection (Device 78)

When a generator loses synchronism, the resulting peak currents and off-frequency operations cause winding stresses, pulsating torques, and possible mechanical resonance. When generators slip, there is a point where they are 180° out of phase with the system, and this causes a point of zero voltage to occur between the generation and the system. This is known as the system center. Out-of-step protection is applied when, during a loss of synchronism, the electrical center is in the step-up transformer or in the generator zone. Loss of excitation relay (40) may provide some degree of protection but cannot be relied upon to detect the loss of synchronism.

This protection is highly recommended even if the system-stability study shows that the loss of synchronism can be detected by this protection. System parameters change with time, and a stability study is costly and time consuming, so the settings need to have a margin for such changes. The protection can be wired to trip only the generator circuit breaker, provided that the unit has full-load breaking capabilities, allowing the machine to run at speed-no-load and be available to be resynchronized quickly.

11.5.3.11 Frequency — Under or Over (Device 81 U/O)

Steam turbine capability for abnormal frequency operation has more restrictions than the generator itself. Underfrequency on a turbine generator is more critical than overfrequency. In isolated operations, overloading of generators will cause the system frequency to decay, and the unit may be subjected to prolonged operation at reduced frequency. Because the damage to the turbine is cumulative, the turbine needs to be operated for 50% of the design lifetime within a specified frequency band. Combustion or gas turbine generators have greater capability than steam units for underfrequency operations. There is no restriction on hydrogenerators, where such protections are usually used to detect total system collapse. Multiple set points for frequency and timers are required for steam turbine generators.

Turbine manufacturers provide limits in the form of permissible operating time within a specified frequency band. A typical operating limitation during abnormal frequency is shown in fig. 11.41 from ANSI C37.106 [S31]. The setting should not be less than 57.5 Hz, even if the turbine limit is lower, as the associated motors cannot be operated at less than 57 Hz and the rated voltage.

11.5.3.12 Differential Phase (Device 87G)

Generator internal faults must be cleared as quickly as possible to minimize the costly damage to the insulation, winding, and core. Differential relaying will detect three-phase, phase-to-phase, double phase-to-ground, and phase-to-ground faults, depending upon the generator grounding method. Two schemes, variable-slope-percentage differential and high-impedance differential, are used for differential

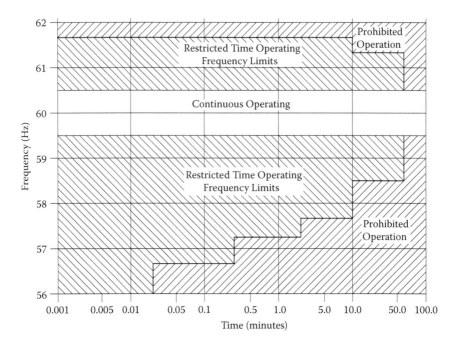

FIGURE 11.41 Typical steam-turbine limitations during abnormal frequency operation

protection. The variable-slope-percentage differential scheme is more widely used, and its characteristics result in a relay that is very sensitive to internal faults and insensitive to CT error current during severe external faults.

11.5.3.13 Differential, Ground (Device 87GN)

Because differential relays (87G) are not as sensitive to internal ground faults, a separate ground-differential element is required to provide a low-current pickup and minimize iron-core damage. A directional overcurrent relay connected in a differential form, shown in fig. 11.42, provides an excellent protection against internal ground faults for low-resistance-grounded machines. For high-resistance-grounded machines, the stator ground protection provides this sensitive protection requirement. This differentially connected scheme provides excellent security against misoperation for external faults and provides very sensitive detection of internal ground faults.

The relay is connected to receive differential current in its operating-coil circuit and generator-neutral ($3\ I_0$) current in its polarizing circuit. The differential comparison is biased to ensure that a positive restraint exists for an external fault, even though the current transformers, R_{CN} and R_{CL}, have substantially different performance characteristics. Figure 11.42 illustrates an example for a 20 MVA machine. Ratios for neutral, line, and auxiliary CT are indicated. A GE-type TCC51 relay with a range of 0.1–0.8 A and set at 0.2 A (8 A primary) would be a typical electromechanical relay that can be used for a 20 MVA machine.

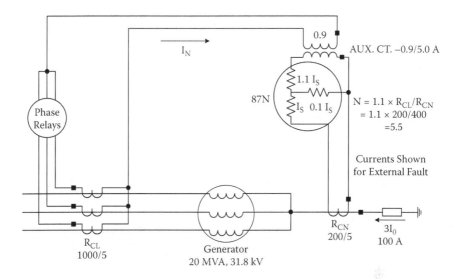

FIGURE 11.42 Generator ground differential using a direction relay

11.5.3.14 Stator Ground-Fault Protection

The IEEE guide for generator ground protection [S29] provides a detailed explanation of the different schemes used for ground-fault protection and the applicable setting considerations. The more common applications are summarized below:

- *Low-resistance-grounded machine*: A time-overcurrent relay with inverse or very inverse characteristics is used for this application. The setting should be about 5% of the resistor continuous current rating. The relay operating curve must be coordinated with the voltage-transformer primary fuse to prevent relay operation on ground faults at the voltage-transformer secondary circuits.
- *High-resistance-grounded machine*: A voltage relay immune to the third harmonic voltage is connected across the grounding resistor that is connected on the secondary side of the grounding transformer. The setting is about 2.5–5% of the voltage developed across the resistor on line-to-ground faults and usually covers some 90–95% of the winding. The pickup and time-delay setting must be coordinated with the VT primary fuse.

For machines directly connected to a separately grounded load bus, an additional zero-sequence CT and a current relay with a pickup set at about 10–20% above the generator ground current, and a short time delay (51G), is used and set to operate on the ground-fault in-feed from the load bus. This setting offers a fast trip on machine internal ground faults and prevents tripping upon an external ground fault. A typical scheme is shown in fig. 11.43.

If a stable setting for the voltage-operated relay cannot be achieved when operating in parallel with the load bus, that element may need to be blocked until the generator circuit breaker opens.

- *100% ground-fault protection*: The following schemes are generally used for the detection of ground fault in 100% of the stator winding:
  1. Third-harmonic neutral undervoltage
  2. Third-harmonic terminal residual undervoltage
  3. Third-harmonic comparator between the neutral and terminal voltages
  4. Neutral or residual voltage injection at a nonharmonic AC voltage with tuned undervoltage detection

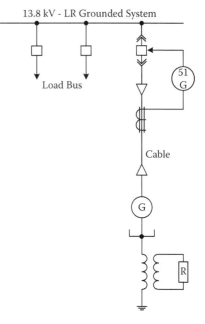

FIGURE 11.43 Ground-fault protection for HR-grounded machine

The first three of the above schemes will work only with machines that produce sufficient third-harmonic voltage to have a significant change in measured third harmonic from normal conditions to a fault close to the neutral. The schemes work in conjunction with a standard stator ground relay, as the schemes only work for failures near the neutral. All schemes are usually only applicable on unit-transformer-connected generators.

11.5.3.15 Field Ground Protection

The field circuit of a generator is an ungrounded DC system. A single ground fault will not generally affect the operation of a generator, nor will it produce any immediate damaging effect. However, the probability of the second ground fault occurring is greater after the first ground fault has occurred. The second ground fault will produce unbalanced fluxes that result in machine vibration and damage. The IEEE tutorial on generator protection [S37] provides more detail on the different schemes for static and brushless machines. For a brushless machine, the ground detection for the exciter and field winding will normally be included as a part of the generator package. A typical field ground-detection scheme for a brushless generator is shown in fig. 11.44.

11.5.3.16 Generator Breaker Fail Protection (Device 52BF)

In a breaker-fail protection scheme, the protection relays detect the fault, provide a trip signal, and initiate a breaker-fail timer. If the breaker does not trip within the specified time, the breaker-fail timer will trip the necessary breakers to isolate the generator from the system. The usual method of detecting whether the circuit

Protection

breaker has opened is to measure the current through the circuit breaker. However, for generator breakers, where the trip can come from mechanical failures that do not cause fault current, circuit breaker status is also added into the current check logic. Two breaker-failure schemes reproduced from the IEEE Guide C37.102 [S30] are shown in figs. 11.45 and 11.46, respectively.

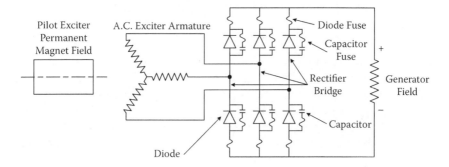

FIGURE 11.44 Field ground detection for a brushless generator

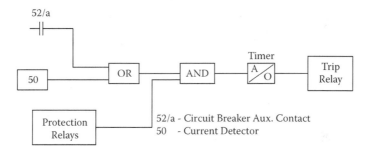

FIGURE 11.45 Logic diagram of a generator breaker-failure scheme

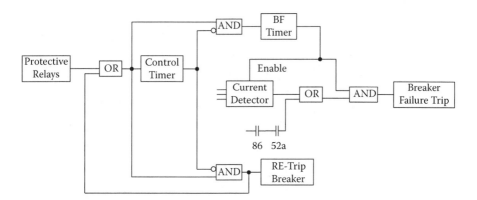

FIGURE 11.46 Logic diagram of generator breaker-failure scheme

11.5.4 Diesel Generator Connected to a Low-Voltage Bus

Diesel generators rated 500 kW or less are used for emergency and standby power in industrial plants. These are usually provided with a built-in control device that combines control and protection appropriate to the size of the generator. The generator neutral can be high- or low-resistance grounded, depending on the load being served. A recommended protection scheme is shown in fig. 11.47.

11.5.5 Generator Tripping Schemes

Generator-tripping schemes must be developed with care and in consultation with plant operators. Where possible, the generator trip relays shall provide redundancy in trip paths and trip functions. A tripping logic adapted from the IEEE generator protection guide [S30] and from the associated IEEE tutorial [S37] is shown in table 11.7. As noted earlier, devices 51GN and 87GN are appropriate to low- and medium-impedance grounded generators while 59GN (stator ground fault) is applicable to high-resistance-grounded machines.

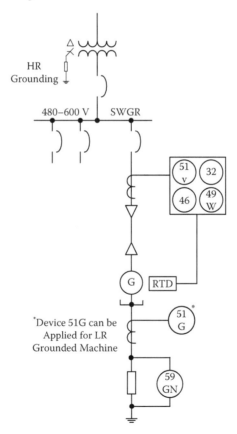

FIGURE 11.47 Protection scheme for diesel generator (≤500 kW) connected to a low-voltage bus (≤600 V)

Protection

TABLE 11.7
Suggested Trip Logic

Device	Generator Breaker Trip	Field Breaker, Trip-Excitation Off	Prime-Mover Trip	Alarm Only
AE	X	X	X	...
24	X	X	X	...
32	X	X	X	...
40	X	X	X	...
46	X	...	...	...
49	...	...	...	X
51GN	X	X	X	...
51V	X	...	...	...
59	...	...	...	X
59GN	X	X	X	...
51G	X	X	X	...
78	X	...	...	...
81	X	...	...	...
87G	X	X	X	...
87GN	X	X	X	...

11.5.5.1 Sequential Tripping

Sequential tripping is the preferred control scheme used for delayed tripping of steam turbine generators. This scheme is implemented to prevent possible overspeed upon simultaneous tripping of a generator circuit breaker and a steam inlet valve. The contact of a reverse or low forward power device 32 is included in the control logic, which trips the turbine inlet valves first, followed by the generator circuit breaker and field circuit breakers. Because a sequential tripping prevents overspeed, this is the preferred tripping mode for the boiler; a backup protection is required to ensure tripping of the generator main and field breakers in case the turbine inlet valve limit switch or reverse-power relay fails. This backup protection is provided by a separate reverse-power relay. A similar scheme is used for mechanical faults on hydroelectric machines, but in that case, all electrical trips trip the main circuit breaker immediately.

11.6 FEEDER PROTECTION

The application considerations of power and control cables are covered in chapter 10. Feeder protection is required at the source or sending end of radially operated feeders, with the exception of taps, where the protective device is permitted at the receiving end and loop feeds, which require protection at both ends. The minimum feeder ampacity and maximum protective device setting is determined by the electrical code rules. A higher relay setting or fuse rating is permitted in cases where the initial current is higher upon energizing, such as on motor starting, or to allow for transformer-magnetizing current.

The protective-device setting or fuse rating must protect the feeder against:

- Overcurrent that exceeds the feeder ampacity determined for the installation
- Thermal and mechanical damage caused by short-circuit current; the short-circuit withstand capability can be determined using the formulae in Chapter 10 or from the curves published by the cable manufacturers

The following steps are suggested:

- Select the feeder size based on the established ampacity and electrical code rules, such as 125% of the motor rated full-load current for motor feeders
- Determine the maximum through-fault current; increase the short-circuit current by a correction factor (K_0), which depends upon the system X/R ratio and fault duration
- Determine the feeder short-circuit withstand capability using the adjusted fault current determined in step 2
- Draw the time–current plot of the cable damage; the protective-device-setting time–current curve must lie below or on the left side of the feeder short-circuit withstand curve

11.7 CAPACITOR PROTECTION

The application of capacitors in industrial power systems is covered in chapter 15. The capacitor protection consists of:

- Fuses for individual capacitor unit protection
- Circuit breakers with fault- and condition-monitoring relaying or circuit breaker for capacitor-bank protection

11.7.1 FUSE FOR INDIVIDUAL CAPACITOR UNIT PROTECTION

The following criteria are applied for the selection of capacitor fuses for individual units and for externally fused capacitors used in capacitor banks. The internal fuses for internally fused units used in capacitor banks follow the same basic criteria, but in those cases, the fuse characteristics are applied by the manufacturer:

- *Voltage rating*: Must be larger than the capacitor unit voltage rating.
- *Continuous current rating*: The fuse must carry at least 165% of capacitor current for grounded banks and 150% for ungrounded banks. This includes allowance for harmonics, capacitor unit tolerance, and overvoltage. The fuse for an individual unit in a capacitor bank must withstand the energy contributed to the failed unit by other capacitors in the same phase group.

- *Short circuit (interrupting)*: Must be greater than the short-circuit current that will flow when the capacitor unit is shorted.
- *Time–current characteristics*:
  1. The fuse must clear the overcurrent due to a failed unit, preferably in 30 s or less or 300 s maximum.
  2. The time–current curve must lie below or to the left of the case (can) rupture curve.

11.7.2 Relaying for Capacitor-Bank Protection

Relaying for capacitor-bank protection includes overcurrent (for fault protection), overvoltage, system problem detection, and current or voltage unbalance, depending on bank configuration, for monitoring the condition of the capacitor units. The protection scheme for a typical 12.6 MVAR (2 × 6.3 MVAR connected in double Wye) capacitor bank with external fuses and a series detuning reactor is show in fig. 11.48.

11.7.2.1 Overcurrent Relay for Capacitor-Bank Protection

A time-overcurrent relay, device 51, with an inverse or very inverse characteristic, is used for capacitor-bank fault protection. The current pickup is set at about 150–200%

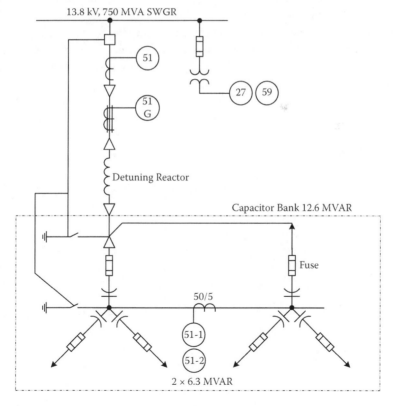

FIGURE 11.48 Protection scheme for capacitor unit and bank

of the bank current rating, and the time dial is adjusted to override the maximum inrush current upon energizing or switching. The relay time–current curve must also coordinate with the capacitor-unit fuse. The same bank protection would be used for internally or nonfused banks, but without the latter coordination requirement.

11.7.2.2 Under- and Overvoltage Protection

The bus-connected overvoltage relay is a time-delayed relay and is set to trip the bank if the system voltage exceeds the total voltage rating of the bank and before the unit voltage withstand time is exceeded. The undervoltage is provided to trip the bank for the loss of system voltage and is time-delayed to allow for parallel fault clearance and other transient voltage excursions.

11.7.2.3 Current- or Voltage-Unbalance Protection

Current- or voltage-unbalance relays are used to detect the loss of capacitor units within a bank and protect the remaining units against overvoltage. The relays must be set above the inherent unbalance that is caused by the capacitor tolerance, system voltage unbalance, and harmonic current or voltage. The tolerance of a capacitor unit is 0–15%, with the average being about 4%. However, the units can be purchased with a manufacturing tolerance of 0–4% for special conditions.

The schemes to detect voltage or current unbalance between the neutrals of two ungrounded banks are shown in figs. 11.49 and 11.50, respectively. Both schemes are insensitive to system voltage unbalance and third-harmonic current or voltage. This scheme is applicable to double Wye-configured banks.

A similar scheme can be applied on a phase-by-phase basis when multiples of four capacitor units are connected in "H" configuration with the CT (or VT) connected in the connection between the midpoints of the two pairs of capacitors that make up the "H." This variant can be used for a single Wye- or for Delta-connected banks.

More details of the application of protection to the various configurations of capacitor banks can be found in the CEA report reviewing the protection of capacitor units [16].

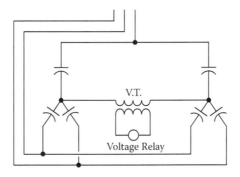

FIGURE 11.49 Voltage unbalance

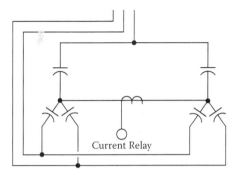

FIGURE 11.50 Current unbalance

11.8 REACTOR PROTECTION

In industrial power systems, reactors are used for the following applications:

- Current-limiting reactors to limit or reduce short-circuit current
- Detuning reactors in series with shunt capacitors to shift the resonant frequency away from the characteristic harmonic frequencies
- Harmonic filters

The devices used for the protection of feeders, capacitors, or filters usually provide protection for the reactor in addition to the other equipment. The exception is for interbus series reactors and large shunt reactors, which may be provided with their own differential protection, particularly if an oil-filled design is used. In this case, the conventional protections applied to an oil-filled transformer also apply.

11.9 BUS PROTECTION

The possibility of a bus fault is high in metal-enclosed switchgear with noninsulated bus-work and open bus installations, such as outdoor substations. High-speed bus protection is required to limit the damage on the equipment, to maintain system stability, and to maintain stability for as much of the load as possible.

Bus protection also has a role in minimizing the risk to staff working in the vicinity of such switching equipment. In the majority of cases, bus protection works on the closed-zone, current-balance principle and can therefore detect faults quickly.

11.9.1 CURRENT-TRANSFORMER SATURATION

In a bus differential-protection scheme, the vector sum of the current entering and leaving the bus must be same unless there is a fault within the protective zone. During an external fault, the current through the CT on the feeder where the external fault is located is the sum of all the currents entering the bus. This high level of current can result in that "out-feeding" CT becoming saturated and hence produces little or no output, thus risking destabilization of the differential zone. The following methods are generally used to avoid or minimize CT saturation:

- Select CT with a high ratio and accuracy voltage that will not saturate at maximum asymmetrical fault current.
- Apply a high-impedance differential scheme with a resonant circuit, such as an ABB/Westinghouse KAB relay system. This scheme limits the sensitivity to CT saturation in the method of setting the device.
- Apply a restraint differential scheme with moderately high impedance to limit sensitivity to CT saturation.
- Apply a linear coupler (LC) system to eliminate the iron core in the CT.

11.9.2 Bus-Differential Schemes

11.9.2.1 Nondirectional Overcurrent Relays

This scheme uses the principle of CT summation but uses a conventional overcurrent relay as the measuring device. It is applied where bus construction minimizes the possibility of bus faults. This scheme is not recommended for the following reasons:

- The relay is set less sensitive to minimize the possibility of incorrect operation
- The relay is limited to locations where the system fault current and system X/R ratio is small

For radial systems with no fault source on the feeders, a bus-protection scheme using interlocked instantaneous overcurrent relays can be used. This scheme will be slower than the conventional bus-differential schemes, as it must allow time for the outgoing relays to operate, but it will be faster than the time-overcurrent relay used for bus protection. Figure 11.51 shows such a scheme. Digital versions of this scheme are also available with high-speed interrelay communication, which results in a faster and, due to the enhanced monitoring possible, a more reliable scheme.

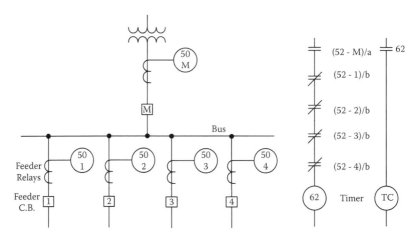

FIGURE 11.51 Bus-differential circuit for a radial system

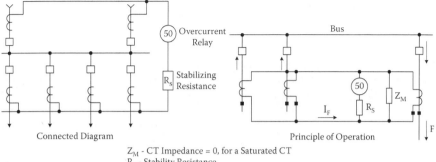

Z_M - CT Impedance = 0, for a Saturated CT
R_S - Stability Resistance
R_L - Lead Resistor
R_C - R_{50} + R_S
R_{CT} - CT Winding Resistance + Internal Lead
Difference Current $I_C = I_F \times [R_{CT}/(R_C + R_{CT})]$

FIGURE 11.52 Bus-differential scheme using overcurrent and a stabilizing resistor

11.9.2.2 Stabilizing Resistor in Series with the Overcurrent Relay

The sensitivity of an overcurrent differential scheme can be improved by connecting a resistor in series with each overcurrent relay, as seen in fig. 11.52. These resistors are called "stabilizing resistors." If an external fault causes the CT on the faulted feeder to saturate completely, the CT excitation reactance will approach zero. The error current I_d that flows through the overcurrent unit can be computed using the equation:

$$I_d = \frac{I_F \times (2 \times R_L + R_{CT})}{(2 \times R_L + R_{CT} + R_d)}$$

where
  R_L = lead resistance from junction point to the most distant CT (one-way lead length for phase fault and two-way for ground faults)
  R_{CT} = DC resistance of the CT secondary winding, including the internal leads to the terminals
  R_d = resistance in the differential path
  V_{CT} = CT accuracy class voltage

To reduce the error current I_d in the differential circuit and improve the sensitivity of the scheme, increase the value of R_d. The limitation of R_d is determined by the overvoltage to the CT circuit and the maximum available internal fault current. R_d is limited to

$$R_d = \frac{V_{CT}}{4 \times I_{MIN \; pickup}}$$

where the multiplier 4 includes a safety factor of 2.

11.9.2.3 High-Impedance Voltage Differential Scheme

High-speed voltage-actuated relays are designed to circumvent the effect of CT saturation during external faults. This scheme discriminates between external and internal faults by the relative magnitude of the voltage that appears across the differential junction points during those faults. This arrangement tends to force any false differential current through the CT rather than the relay operating coil. In this scheme, all current transformers should have the same ratio and knee point on the selected tap. The external connections and DC trip circuit of a high-impedance bus-differential scheme is shown in fig. 11.53.

11.9.2.4 Moderately High-Impedance Bus-Differential Scheme

This scheme combines the features of high-impedance and percentage differential characteristics in one, which provides a reliable operation for internal faults and a secure restraint for external faults. This scheme can accommodate different CT ratios, and other relay burden may be included in the same CT circuit. The relay is connected to the system with a special auxiliary CT, required for each restraint circuit when a 5-A rated CT is used. One of such schemes available in the market is the ABB type REB 100.

11.9.2.5 Linear Coupler Bus-Differential Scheme

This scheme uses air-core mutual reactors, known as linear couplers, in place of the conventional CT with an iron core. The linear couplers have negligible DC response, so only steady-state conditions need to be considered. This is a voltage-differential scheme in which the linear couplers are connected in series. For further reading and understanding of this scheme, refer to chapter 11 of Walter Elmore's book [2].

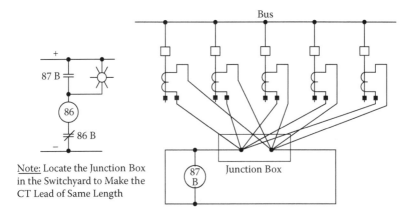

FIGURE 11.53 External connection of a high-impedance bus-differential scheme

REFERENCES

Standards
General

S1. IEEE C37.2-1996, IEEE Standard Electrical Power System Device Function Numbers, 1996.

S2. IEEE C37.90-1989 (R 1994), IEEE Standard for Relays and Relay Systems Associated with Electric Power Apparatus, 1994.

S3. IEEE C37.90.1-2002, IEEE Standard Surge Withstand Capability (SWC) Tests for Relays and Relay Systems Associated with Electric Power Apparatus, 2002.

S4. IEEE C37.90.2-1995, IEEE Standard Withstand Capability of Relay Systems to Radiated Electromagnetic Interference from Transceivers, 1995.

S5. IEEE C37.90.3-2001, IEEE Standard Electrostatic Discharge Tests for Protective Relays, 2001.

S6. IEEE C37.110-1996, IEEE Guide for the Application of Current Transformers Used for Protective Relaying Purposes, 1996.

S7. IEEE C37.112-1996, IEEE Standard Inverse-Time Characteristic Equations for Overcurrent Relays, 1996.

S8. IEEE 242-2001, Recommended Practice for Protection and Coordination of Industrial Power Systems, 2001.

S9. IEEE C62.21-2003, IEEE Guide for the Application of Surge Voltage Protective Equipment on AC Rotating Machinery 1000 V and Greater, 2003.

S10. IEEE C62.22-1997, IEEE Guide for the Application of Metal-Oxide Surge Arresters for Alternating Current Systems, 1997.

S11. IEC 60255-22-2 Ed. 2.0 (1996-09), Electrical Relays, Part 22: Electrical Disturbance Tests for Measuring Relays and Protection Equipment, Part 2: Electrostatic Discharge Tests, 1996.

S12. IEC 61000-4-2 Ed. 1.0 (1995-01), Electromagnetic Compatibility (EMC), Part 4: Testing and Measurement Techniques, Section 2: Electrostatic Discharge Immunity Tests, Basic EMC Publications, 1995.

S13. IEC 60255-22-4 (1992-03), Electrical Relays, Part 22: Electrical Disturbance Tests for Measuring Relays and Protection Equipment, Section 4: Fast Transient Disturbance Tests, 1992.

S14. ANSI/NFPA 70-2005, National Electrical Code (NEC), 2005.

S15. Canadian Standard CSA C22.1-2002, Canadian Electrical Code, 2002.

Transformer Protection

S16. IEEE C37.91-2000, IEEE Guide for Protective Relay Applications to Power Transformers, 2000.

S17. IEEE C37.108-2002, IEEE Guide for the Protection of Network Transformers, 2002.

S18. IEEE C57.12.44-2000, IEEE Standard Requirements for Secondary Network Protectors, 2000.

S19. IEEE C57.12.59-2001, Guide for Dry-Type Transformer Through-Fault Current Duration, 2001.

S20. IEEE C57.91-1995, IEEE Guide for Loading Mineral-Oil Immersed Power Transformers, 1995.

S21. IEEE C57.92-2000, Guide for Loading Oil-Immersed Distribution and Power Transformers up to and Including 100 MVA with 55°C or 65°C Winding Rise, 2000.

S22. IEEE C57.96-1999, IEEE Guide for Loading Dry-Type Distribution and Power Transformers, 1999.

S23. IEEE C57.109-1993, Guide for Transformer Through-Fault Current Duration, 1993.

S24. IEEE C57.115-1991, IEEE Guide for Loading Mineral-Oil-Immersed Power Transformers Rated in Excess of 100 MVA, 1991.

S25. IEC 76-5, Power Transformers Ability to Withstand Short Circuit and Amendment No. 1.

Motor Protection
S26. IEEE C37.96-2000, IEEE Guide for AC Motor Protection, 2000.
S27. IEEE 620-1996, IEEE Guide for Presentation of Thermal Limit Curves for Squirrel Cage Induction Machines, 1996.

Generator Protection
S28. IEEE C37.95-2002, IEEE Guide for Protective Relaying of Utility-Consumer Interconnection, 2002.
S29. IEEE C37.101-1993, IEEE Guide for Generator Ground Protection, 1993.
S30. IEEE C37.102-1995, IEEE Guide for AC Generator Protection, 1995.
S31. IEEE C37.106-1987 (R 1992), IEEE Guide for Abnormal-Frequency Protection for Power Generating Plants, 1992.
S32. IEEE 122-1991, IEEE Recommended Practice for Functional and Performance Characteristics of Control Systems for Steam Turbine Generator Units, 1991.
S33. IEEE 487-1992, IEEE Recommended Practice for the Protection of Wire-Line Communication Facilities Serving Electric Power Stations, 1992.
S34. IEEE 502-1985 (R 1992), IEEE Guide for Protection, Interlocking and Control of Fossil-Fuel Unit Connected Steam Stations, 1992.
S35. IEEE 1046-1991, IEEE Guide for Distributed Digital Control and Monitoring for Power Plants, 1991.
S36. IEEE 1129-1992, IEEE Recommended Practice for Monitoring and Instrumentation of Turbine Generators, 1992.
S37. IEEE Cat. No. 95 TP 102, IEEE Tutorial on the Protection of Synchronous Generators.
S38. ANSI C50.10-1990, American National Standard for General Requirements of Synchronous Machines, 1990.
S39. ANSI C50.12-1982, American National Standard Requirement for Salient-Pole Synchronous Generators and Generator/Motors for Hydraulic Turbine Applications, 1982.
S40. ANSI C50.13-1989, American National Standard Requirement for Cylindrical Rotor Synchronous Generators, 1989.
S41. ANSI C50.14-1977, American National Standard Requirement for Gas Turbine Driven Cylindrical Rotor Synchronous Generators, 1977.

Capacitor Protection
S42. IEEE 18-2002, IEEE Standard for Shunt Power Capacitors, 2002.
S43. IEEE 1036-1992, IEEE Guide for Application of Shunt Power Capacitors, 1992.
S44. IEEE C37.99-2000, IEEE Guide for the Protection of Shunt Capacitor Banks, 2000.

Bus Protection
S45. IEEE C37.97-1997, IEEE Guide for Protective Relay Applications to Power System Buses, 1997.

BIBLIOGRAPHY

General
1. ALSTOM-T&D, *Network Protection and Automation Guide*, Lavallois Perret, France: ALSTOM-T&D (now AREVA T&D), 2002.
2. Walter A. Elmore, ed., *Protective Relaying, Theory and Applications*, New York: Marcel Dekker.
3. Westinghouse Electric Corp., *Applied Protective Relaying*, Westinghouse Electric Corp.
4. C. R. Mason, *The Art and Science of Protective Relaying*, New York: Wiley, 1956.

5. D. V. Fawcett, "How to Select Overcurrent Relay Characteristics," *AIEE* May (1963).

Transformer Protection

6. R. W. Brown and S. Khan, "Planning and Protection Implications of Reduced Fault Withstand Time of Power Transformers," *IEEE Trans. Power Delivery* 1 (4): 60–67 (1986).

Motor Protection

7. R. L. Nailen, "What Is Terminal Box Fault Capacity," *Electrical Apparatus Mag.* Mar. (1988).
8. S. Khan, R. W. Brown, and Z. Naamani, "Protection Philosophy and Relay Selection for Large AC Motors," *CPPA* (1991).
9. R. H. Kaufmann, "Surge-Voltage Protection of Motors as Applied in Industrial Power Systems," Schenectady, NY: General Electric.
10. E. M. Hunter and N. E. Dillow, "Surge Protection: AC Rotating Machines," publication GER-294, Schenectady, NY: General Electric.

Generator Protection

11. H. K. Clark and J. W. Feltes, "Industrial and Cogeneration Protection Problems Requiring Simulations," *IEEE Trans. IAS* 25 (4): (1989).
12. M. Shan Griffith, "Modern AC Generator Control Systems: Some Plain and Painless Facts," *IEEE Trans. Ind. Applic.* 1A-14 (6): 481–491 (1976).
13. C. H. Griffin, "Relay Selection of Generator Station Service Transformer," *IEEE Trans. Power Appar. System* PAS-101 (8): 2780–2789 (1981).
14. J. Berdy, M. L. Crenshaw, and M. Temoshok, "Protection of Large Steam Turbine Generator during Abnormal Operating Conditions," CIGRE: International Conference on Large High Tension Electric Systems, paper 11-05.

Capacitor Protection

15. D. F. Miller, *IEEE Trans. IAS* 12 (5): 444–459 (1976).
16. SNC-Shawinigan Inc., "Capacitor Can Protection Review," CEA report no. 388 T 893, Montreal: SNC-Shawinigan, 1995.

Bus Protection

17. H. J. Li, "Station Bus Protection," chap. 11, ABB/Westinghouse Applied Protective Relaying.
18. U.S. Department of Agriculture, RUS Bulletin 1724-30, chap. 12, Washington, DC: USDA.

12 High-Voltage Substation Design Considerations

12.1 INTRODUCTION

The previous chapters have discussed the design of the individual components in an electrical system that comprises an electrical network. Chapters 2 and 3 additionally cover the system configuration and design of the networks for industrial networks and cover some of the design aspects relative to the reliability and configuration of the interface and internal substations within an industrial system. Chapters 11 and 14 discuss, respectively, the transformer configurations that affect the protection and the interconnection of generators into industrial systems and at industrial interface networks. This chapter deals with some of the specific items that need to be considered in the configuration and location of both indoor, cabin-type metal-enclosed switchgear and outdoor air-insulated substations, noting that some air-insulated open switchgear is also located inside buildings, with the appropriate screening and access protection.

The majority of industrial distribution stations are cable-connected, metal-clad, or metal-enclosed switchgear where the basic design details are chosen by the supplier to a basic specification provided by the purchaser, and hence only the location, space allocation, voltage, basic insulation level (BIL), and current rating need to be specified, along with the basic environmental factors to be considered in the design. However, many more factors need to be considered for air-insulated designs, as most such stations are purposely designed for the application, location, space available, and more specific environmental considerations, such as pollution levels and altitude. For some applications, particularly when a site is severely restricted, or where there are high levels of pollution, even the highest-voltage stations need to be located indoors and, in such cases, gas-insulated switchgear (GIS), SF_6-insulated equipment, including bus connections, may need to be considered.

12.2 NECESSARY INFORMATION FOR THE DESIGN OF SUBSTATIONS

12.2.1 CONSIDERATIONS FOR THE LOCATION OF THE SUBSTATIONS

One of the important factors in the design of a substation is its location. Special consideration needs to be given to the location to achieve the optimal compromise between the line entry to the site and the load centers in the plant (see fig. 12.1).

The main factors that must be considered in the selection of the site for the construction of a substation (see fig. 12.2) are:

FIGURE 12.1 Typical outdoor high-level industrial substation

FIGURE 12.2 Typical outdoor low-level A-frame high-voltage transformer station

High-Voltage Substation Design Considerations

- Present and future location of the load centers
- Present and future location of on-site generation
- Availability of access for the incoming transmission or subtransmission lines and the distribution and communication systems circuits within the plant
- Alternative uses of the considered area
- Location of any existing transmission or distribution lines
- Facilities for the transport of heavy equipment to the site
- Environmental impacts of the site related to its appearance, noise, or electrical interference or its impact on other facilities
- Conditions of the ground, its facility of drainage, and its load bearing capability
- Cost of excavations and earthwork
- Atmospheric pollution, whether industrial or saline contamination
- Necessary space for present use and future extensions
- Governmental or municipal restrictions at the site
- General topographical considerations, including seismic levels and the risk of flooding and the proximity to marshy areas
- Security requirements, including any heightened risk of vandalism, theft, or sabotage
- Total cost, including the provision of distribution circuits to the plant and the required control and communication facilities

12.2.2 Basic Information Required

Before beginning the design of a substation, it is necessary to obtain information on which the design will be based. One set of information required will concern the system to which the substation will be connected, such as the levels of active and reactive power available, the voltage regulation, the system short-circuit levels, the lightning risk, etc. Other information includes the climatic data of the site, the seismic risk levels, the altitude, and the environmental requirements that must be taken into account in the design of the facilities. It should also include any other data required to obtain approval from the various responsible governmental departments. It is also necessary to make the topographical survey of the site and to obtain the characteristics of the ground, such as its bearing strength and electrical ground resistivity.

12.3 STANDARDS AND DESIGN PRINCIPLES FOR SUBSTATION DESIGN

12.3.1 System of Units

North America is moving slowly toward the adoption of metric rather than imperial measurements. While international standards have become more widely used in more countries, there has been a significant increase in the usage of the metric SI (System International) system of measurements in most countries, and most electrical equipment is currently manufactured to metric prime dimensions.

12.3.2 Applicable Standards

In North America, there is a comprehensive set of national standards produced by the IEEE and ANSI for the United States and CSA standards for Canada. Countries that do not have a good domestic base of local standards use the standards of their main trading parties, but are tending to have the design of their electrical systems and substations based on the IEC standards. In addition, any electrical substation, including the switchgear installed within an industrial plant, should be designed, manufactured, built, tested, and commissioned in accordance with the local laws and regulations.

The applicable standards from the following organizations are typical of those used for normal substation work, and a selection of the most appropriate documents is listed at the end of this chapter:

- North American industry and regulatory standards: ANSI, IEEE, NEC, CEC, CSA, etc.
- International Electrotechnical Commission standards (IEC)
- Various electrical manufacturers' associations: NEMA (United States), EEMAC (Canada), BEAMA (United Kingdom), etc.
- American Society for Testing and Materials (ASTM)
- American Society of Civil Engineers (ASCE)
- U.S. Department of the Interior, Bureau of Reclamation
- International Conference of Building Officials
- U.S. Department of the Interior, Water and Power Resources Services
- American Concrete Institute
- American Institute of Steel Construction, Inc.
- American Welding Society (AWS)

In addition to these standards, references are provided at the end of this chapter to the standards and reference documents that are appropriate to the design of each of the electrical components that make up industrial main interface switchyards and the associated distribution substations.

12.3.3 Standard Transformer Connections and System Phasing

System phasing and transformer vector group choice is an important part of substation design, as it defines the angle difference between systems of different voltages. In many cases, transformers that interconnect the high- and extra-high-voltage networks often use Wye-Auto or Wye/Delta/Wye connections, which keep the vectors of these higher-voltage networks in phase. Such transformer winding configurations fully link the high-voltage (HV) and low-voltage (LV) systems' zero-sequence networks, and hence each system contributes to the ground-fault currents on the other voltage levels and does not allow grounding policy to be separately defined. This practice often leads to ground-fault levels in these networks that will tend to be above the three-phase fault level. This can improve protection performance in some cases, but could require higher equipment fault ratings. Chapters 11 and 14 discuss these aspects in more detail.

High-Voltage Substation Design Considerations

As discussed in detail in Chapter 14, it is normal practice to connect generation into the system via individual or group Delta-Wye transformers with the high-side Wye solidly grounded. One reason for this is to ensure that no part of the network can be fed by an ungrounded generation source, unless the system is specifically designed to operate ungrounded or ineffectively grounded. The provision to the Delta on the generation side also ensures that triplen harmonics are blocked and that ground faults on the main system do not reflect directly into the generator(s). The generator grounding system can then be specifically rated to limit damage and achieve fast stator ground-fault protection.

Within any one industrial plant, it is necessary to keep the vector relationship of the various distribution voltages the same, particularly in the same area, to enable standby connections to be operated between the various circuits without phasing problems. This standard vector arrangement also allows transformers to be relocated in an emergency and simplifies the holding of strategic spares.

At the lower levels of high voltage and for the transformers that connect the subtransmission and distribution levels, including the transformers used within utility or industrial distribution systems, the most commonly used transformer vector group in both North America and internationally is a Wye-Delta or Delta-Wye connection, where the secondary network operates with a 30-degree lag with respect to the primary side. These are defined as Yd11 or Dy1, respectively.

It should be noted that it is also possible to purchase transformers with reconnectable Delta windings so that the transformer can be connected to have either a 30-degree lead or lag vector group to allow for locations where the high-side connection does not physically match the A-B-C to H_1-H_2-H_3 standard primary phase connection, with the secondary phasing appropriately connected, while still maintaining the desired system vector shift (see fig. 12.3).

The Delta-Wye connection is also often used to provide a low-voltage-side ground source without the need for a grounding transformer, although grounding resistors are often used to limit network ground-fault levels. These transformers are somewhat more expensive than the Y-D units, as their high-side winding is fully insulated, but are commonly used where there is no LV generation or remote back-feed into the HV system. On the other hand, substations with high-side Delta-connected transformers can suffer from higher ground-fault voltage rise problems, as there is no local high-voltage ground-fault return path in the station, and any ground-fault current caused by a high-voltage ground fault in the station must flow to remote grounded neutral points.

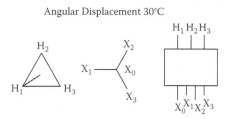

FIGURE 12.3 Delta-Wye transformer vector diagram

12.4 SUBSTATION ELECTRICAL CONFIGURATION

12.4.1 Switching Configuration

One of the first design factors to be considered before a substation is designed in detail is the switching configuration to be used. This sets the size and switching complexity to be used and should be based on the number of circuits to be controlled and the resultant reliability to be achieved.

The main factors that need to be decided and balanced, as noted in a CEA Working Group report [1], are:

- Service, security, and reliability
- Operational flexibility
- Simplicity of operation
- Short-circuit limitation (system sectioning)
- Protective relaying complexity
- Maintainability of equipment
- Extensibility
- Standardization
- Cost

Similarly, service reliability, in terms of substation switchability, can be divided into four categories as follows, based on the consequences of a circuit loss:

- No single circuit interruption causes the loss (outage) of any major equipment
- A circuit interruption causes only a short interruption of supply to a major equipment (minutes)
- A circuit interruption causes a long supply interruption to a major equipment (hours)
- No standby — loss of a circuit causes a loss of power feed until the circuit is repaired. A supply loss also occurs when maintaining the circuit.

To consider these factors correctly, the overall system service reliability objectives relative to costs, as discussed in the 1979 CIGRE Electra paper [9], need to be defined together with a consideration of the inherent reliability built into the power system and the specific requirements of the industrial process. Where a single station controls a major part of the network, or where the distribution network itself does not consist of many alternative routes, the station design will dominate the overall system reliability.

Another major factor that affects reliability is the effect of maintainability. In this respect, any items of equipment that cannot be taken out of service for maintenance, adjustment, or cleaning will, in time, become a hazard to the system. This is particularly true of bus-bar disconnects and bus insulators, which require total station shutdown to maintain.

A further consideration should be given to the decrease in maintenance needs for certain equipment as their design improves. Of particular note is the SF_6 circuit breaker, which now requires little maintenance except for inspection and monitoring, allowing it

High-Voltage Substation Design Considerations

to be inspected when the respective controlled circuit is taken out of service. Therefore, this may not justify special equipment (bypass) for circuit breaker access.

12.4.2 Commonly Used Substation Configurations

The CEA 1980 report on commonly used single-line diagram configurations [1] covers the general arrangements in use at transmission voltages in North America, while the 1987 *IEEE Journal* article on transformer substations [5] discusses the possible configurations for distribution transformer stations which, although based on U.K. distribution practice at the time, provides the basic design criteria to be considered in such station types. The CIGRE 1993 technical brochure on outdoor substation design [6] covers those options more commonly found in international practice.

Substation configurations have been developed to provide the flexibility and reliability appropriate to the duty of the particular station, and the most common practices found for the three major substation types are as follows:

1. Generation stations:
   a. Large generators:
      - Double bus, single breaker
      - Double bus, double breaker
      - Combination double/single, with double breakers for the generator circuits and a single breaker for outgoing lines or transformers
      - Double bus, breaker and a half.
   b. Small generators:
      - Single bus
      - Various forms of line tap
2. Network switching stations:
   a. Breaker and a half
   b. Double bus, usually single breaker
   c. Single bus, single breaker
   d. Ring bus
3. Major transformer stations:
   a. High-voltage side
      - Ring bus
      - Single bus, double section
      - Double bus
      - Breaker and a half
   b. Low-voltage side
      - Single bus, double section
      - Double bus
4. Minor transformer stations:
   a. High-voltage side:
      - Single breaker tap
      - Three-breaker line loop
      - Disconnect only tap
   b. Low-voltage side:
      - Single bus, single or double section

In the configurations above, ring and 1½ breaker, including double breaker and double bus, all use two shared circuit breakers to control each circuit, and providing that the ring is closed or the two buses are tied, each circuit breaker provides an alternative connection. In such an arrangement, any circuit breaker can be removed from service without interrupting power flow through the station (see fig. 12.4, 12.5).

Single- and double-bus layouts have a dedicated circuit breaker per circuit; hence, to access that circuit breaker would require the outage of the main circuit. In the case where such a maintenance outage would degrade the service reliability of the station, there are three basic practices for dealing with such an outage (see fig. 12.6, 12.7):

- Breaker bypass via a disconnect directly to the station bus
- Breaker bypass with the disconnect replaced by reconnectable bus-work
- Bypass to a separate transfer bus

The transfer bus can be used with either a disconnect or bus-work connection, and the transfer bus is then connected to the main bus, or buses, via a tie breaker that then takes over the tripping function of the bypassed breaker. The most complex example of this practice is the double bus and transfer bus complete with dual-use transfer/bus tie breaker, as found in Sweden. However, fully implemented versions of this arrangement are rare.

The more common alternatives are double bus with bypass disconnect connections to either the main bus, via the bus selector disconnects, or by a single connection directly to one (reserve) bus only. The single- and transfer-bus arrangement is also common for intermediate-sized stations as per fig. 12.8. Direct bypass to the main bus is only applicable to small, lower-voltage stations, where it is acceptable to trip all bus infeeds for a fault on the circuit controlled by the bypassed breaker.

All the bypass arrangements add complexity to both the protection and the operating and interlocking arrangements. On the other hand, developments in circuit breaker design have significantly reduced their required maintenance time, so the need for bypass connections has diminished. However, the provision of temporarily reconnectable bus-work is preferable where layout and site conditions allow (see fig. 12.9, 12.10).

In all large double- or single-bus configurations, including breaker-and-a-half stations, it is common to divide the buses with section breakers, typically when a station has more than four 1½ switch bays (12 breakers) or eight bays for single- and double-breaker arrangements. Similarly, large ring-bus configurations are normally broken into multiple interconnected rings when the number of breakers in the ring exceeds six to eight.

The bus arrangement chosen for any particular station needs to be matched to

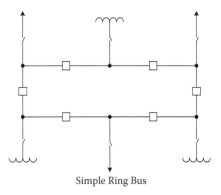

Simple Ring Bus

FIGURE 12.4 Ring-bus arrangement

High-Voltage Substation Design Considerations

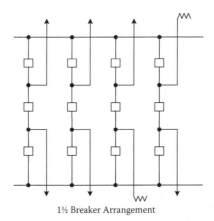

FIGURE 12.5 1½-breaker configuration

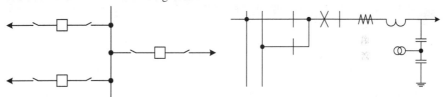

FIGURE 12.6 Basic single-bus configuration

FIGURE 12.7 Double-bus arrangement

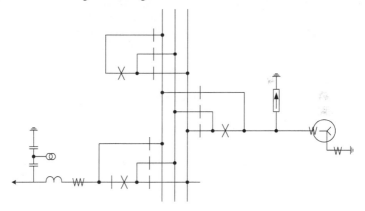

FIGURE 12.8 Full ABC double-bus configuration with selectable bus coupling

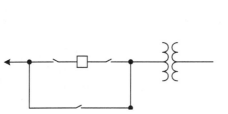

FIGURE 12.9 Basic bypass arrangement

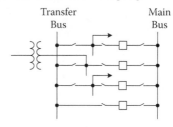

FIGURE 12.10 Breaker bypass with transfer bus

the reliability objectives with an optimum balance of complexity, outage risk, and maintenance or repair time. For these reasons, a number of special arrangements that balance the reliability features of the earlier described arrangements, with the switching flexibility provided by these various arrangements, have been developed. One such configuration, defined as a 4 × 6 network arrangement, is discussed in a 1989 IEEE paper [4].

12.4.3 Recommended Configurations

Most industrial substations fall into the category of "minor transformer stations," depending on the network configuration, while for larger users, such as aluminum smelters, the main substations can fall into the "major network" category. In this latter case, single-breaker–double-bus configurations are common, even in North America, where 1½-switch configurations are more common for power utility network switching stations.

For medium-power users with multiple transformers or with a network-connected cogeneration unit, a ring-bus configuration would give the appropriate combination of features, providing reasonably high reliability for relatively low cost as compared with a single-bus configuration. The ring bus has superior reliability than the single-bus arrangement and, because it uses the same number of breakers, has a similar cost. The two major problems with ring-bus arrangements are protection complexity and greater restriction in expandability, although if laid out in two lines of breakers with bus crossovers at each end, it can be expanded at each end. Any ring-bus configuration is limited in maximum size before the risks of cross outages become unacceptable.

In terms of expansibility, if it is known that future expansion is likely to require a full 1½-switch configuration, a station can be laid out as a two-bay 1½ switch, a configuration that can be initially utilized as four to six breaker rings using the bus connections as circuit bays. It is also possible to lay out a single bus to be expandable to add a second bus and to lay out a single-tap or -loop station to be ultimately expandable to a single or double bus. For small stations, where the site access restrictions will always limit their development to the maximum of two incoming circuits and two transformers or where there is little prospect of load growth, then a specific transformer station layout is justified.

For transformer stations, the simplest configuration is the simple single-tap transformer, with the choice of a fault thrower, direct-transfer trip, high-side fuse, or circuit breaker for transformer protection. For the first three options, the provision of an auto-opening, motorized disconnect can permanently isolate a faulted transformer in the absence of a transformer circuit breaker.

The first two of these options will result in a line trip for a high-side fault. Although line autoreclose can be used to restore the line quickly, for industrial transformer stations, the fuse and breaker options, which avoid the need for a LV line trip for transformer faults, are normally required. However, inside an industrial network, such an arrangement, which protects the transformer feeder and transformer together, is often used.

High-Voltage Substation Design Considerations

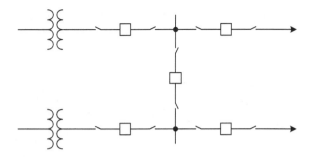

FIGURE 12.11 Typical transformer station

At high voltages, the fuse solution is only practical for very small tap stations, and hence the breaker layout is used more often. For more important stations where the reliability objectives mandate alternative line feeds or two transformers, a full-split single bus, where line breakers and line protection are provided, becomes a suitable option. If the transmission line is looped into an existing circuit, then the three-breaker partial ring, which omits the two line breakers, is the most suitable, with the tie breaker directly connecting the lines and providing a thorough line circuit even if a transformer breaker trips. For a double-circuit feed, the alternative arrangement that omits the transformer circuit breakers would be appropriate (see fig. 12.11).

For transformer feeders, different options apply. As there is normally no thorough line feed, the circuit is dedicated to feed a particular transformer station. In such cases, overreach distance protection and fault throwers can provide an economic alternative to providing high-side breakers if the fault level is adequate to operate the remote end protection and if the effect on the system of a solid ground fault is acceptable. The alternative is for a direct-transfer trip where communication links exist. However, if neither of the options is practical, then the provision of a transformer high-side breaker gives the required facilities and the simplest option for protection and operational safety. A typical arrangement is shown in fig. 2.1 in Chapter 2. Some other transformer tap configurations suitable for the distribution transformers within an industrial plant are also discussed in that chapter.

12.5 SECONDARY SYSTEM ASPECTS OF SUBSTATION DESIGN

12.5.1 SUBSTATION PROTECTION

For the ring-bus arrangements, the current transformers for the individual circuits are normally located on each circuit breaker, and hence the sections of ring bus and the circuit breakers are covered in the transformer, feeder, or line protection zones. At the higher system voltages, live tank circuit breakers can be used. In this case, it is normal practice only to provide current transformers on one side of the breaker. In this case, the section of bus between the current transformer (CT) and the circuit breaker will be detected in one zone, but will not be cleared by the tripping of that circuit breaker. For those conditions, circuit breaker fail protection will detect this as a "failed breaker" and will trip the appropriate "next in line" circuit. Although faults in this area will be cleared more slowly (≈300 ms), all equipment is adequately protected. In North America it is more common to use dead-tank circuit breakers that have current transformers on both sides, and in such a case full CT overlap can be obtained and both sides will trip simultaneously.

For single-bus stations, with either a bypass or transfer bus, the line and transformer current transformers are located on the line entry or circuit breaker feeding the circuit of transformer protection, and hence they do not directly provide protection for the bus. The remote backup protections may give acceptable performance for small radial systems, but as the number of circuits increases or the system becomes interconnected, direct protection of each bus becomes advantageous. Chapter 11 provides more details of the options available.

For bypass and transfer-bus designs, it is important that the circuit current transformers be located on the circuit side of the bypass or transfer disconnect, so that the main-protection CT circuits remain in service even when the breaker is bypassed. It is then only necessary to transfer the trip circuits to the transfer-bus breakers or the other breakers on the bus. By adding the relevant trip logic using disconnect auxiliary switches, this transfer can be made automatic.

The high cost of transformers justifies fast, reliable protection for all sizes, although additional protection is justified above about 5 MVA. The simplest form of protection for transformers below about 2 MVA and at 69 kV and below is the fuse. Although fuses are available for 115 kV, their low fusing current can make them unreliable. For transformers up to this size that are equipped with circuit breakers, phase-instantaneous and phase-ground inverse-time overcurrent relays (50/51) can provide good protection if the fault level is large enough to get satisfactory settings. For transformers rated at 5 MVA and above that are important to system operation, or where high sensitivity is necessary (generator transformers), differential protection is recommended in addition to high- and low-side overcurrent relays. Buchholz protection should be standard for all conservator-type transformers together with winding-temperature trip for transformers of more than 5 MVA or just a temperature trip for smaller units.

12.5.2 LOCAL AND REMOTE CONTROL AND METERING AND INDICATION REQUIREMENTS

With the provision of modern substation control systems, also known as substation automation or distributed control systems (DCS), the need for full hard-wired metering and control panels at each substation has diminished. The main need is for sufficient standby control and indication to operate the station in an emergency if there is a breakdown in the control system or in the communication links/local area network (LAN) to the plant control center. It is therefore more appropriate to use separate relay panels rather than locating the relays in a duplex arrangement with "conventional" control panels. The introduction of digital protective systems, and the earlier static devices, made it more appropriate for the relay panels to use a fixed or swing-rack arrangement, although steel panel construction could also be used (see fig. 12.12, 12.13).

High-Voltage Substation Design Considerations

FIGURE 12.12 Swing-rack-type relay panel

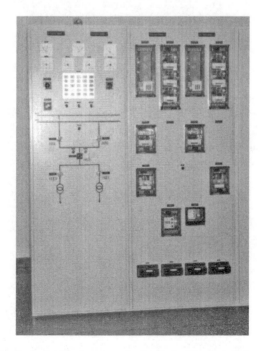

FIGURE 12.13 Steel panel with electromechanical relays and control panel

It may still be appropriate to retain some form of local standby control with a circuit breaker control switch and a digital measurement unit that can provide a local display for MW, volts, and amps while providing a local area network (LAN) interface to the control system. These could be located on the appropriate circuit relay panel. For switchgear cubicles, the minimum of a single-phase ammeter with ammeter switch would be recommended if a measuring unit is not justified. The provision of a local control switch and local/remote control-transfer switch would usually be provided, noting that colored light-emitting diode (LED) long-life indicators are tending to replace the earlier incandescent-lamp status indicators.

12.5.3 Relay Building Location

The most suitable location for a relay building is in the center of any outdoor switchyard, but this location is not always practical. Where two outdoor stations are involved, the best location is between the two switchyards and between the interbus transformers. For sites where indoor switchgear is used for the low-voltage-side switchgear, the best choice is often a suitable location for the associated protection relays. However, where packaged switchgear having its own metal-enclosed structure is used, it is sometimes more appropriate to locate the protection within the plant buildings, particularly if a main electrical room is suitably located.

The main criterion for conventional CT and VT cabling and 110–125 V DC circuits is to limit the cable runs to all equipment to less than 400 m, although with appropriate design features, somewhat longer lengths are possible. For most conventionally sized stations up to 220 kV, a single control relay equipment location is adequate. However, for large stations, particularly where the altitude or process (e.g., aluminum smelters) requires enhanced bay spacing, there will be some cases where it will be necessary to put in a second relay room to limit cable lengths. The alternative option of having individual relay buildings per circuit or per bay has also been used, particularly at the 300–500 kV voltage level.

The earlier designs of control and communication systems, which used low signal level cables that required screening and other precautions for interference reduction, made the use of distributed relay buildings less attractive, but the availability of fiber-optic cables for such control links now makes distributed locations more attractive in large sites.

12.5.4 Utility Telecommunication Requirements

Efficient utility network operations require reliable communications, and to avoid the reliability and cost impact of leasing such circuits, utilities prefer to operate their own independent systems. Although some utilities invested in microwave networks, the use of power-line carrier has been the telecommunications system of choice for SCADA (supervisory control and data acquisition) data, speech, and protection signaling in utility systems, especially in less populated regions. These systems are now being supplemented and, for some networks, replaced by fiber-optic ground wire (OPGW)-based communications. The power-line carrier systems inject radio frequency signals (usually in the 150–500 kHz range) and require line traps and capacitor voltage transformers (CVT) connected directly to the utility incoming

High-Voltage Substation Design Considerations

lines to inject those signals onto the line. It is also necessary to have the line ground switch on the station side of the line trap so that the communications system can still be functional with the line out of service and grounded. Although such equipment would usually be supplied by the connecting utility, the station layout needs to allow for the provision of capacitor voltage transformers, line traps, and impedance-matching units on the line entries when these technologies are used.

The line traps are needed to prevent the radio frequency (rf) signals used by the power-line carrier on the line from being shunted by the substation equipment. These are effectively rf filters insulated for the system voltage with a current rating that should match the line rating, rather than the station ampacity, with a fault rating to match the station and system short-circuit design levels. Line traps can be located on the associated CVT or suspended from the incoming line terminal structure. For the larger ratings, typical of extremely high voltage (EHV) systems, self-supported insulator structures may be used, but these would be rare for industrial interface stations.

For fiber-optic-type interconnections, the fibers for OPGW are contained within the line ground (shield) wires, although underslung self-supporting insulated fiber (ADSS) cables may also be used. In such cases, only junction boxes at the line terminal structure base, plus suitable routes for the interconnecting fiber-optic cables to reach the telecommunications terminal equipment, would be required. In both cases, space in the relay/control equipment area for the associated communication panels would be necessary, in addition to the utility's line relay panels, including the provision of power supplies and cable routing.

12.5.5 Current and Voltage Transformers Requirements

Current and voltage transformers are the primary measurement devices for the protection and control and tariff measuring systems. These need to be located to provide the required zones of protection, to provide the best protection for the respective network components, and to provide the required inputs for the measuring systems. Good protection zoning is required to provide good fault location so that any faulted primary component can be isolated quickly and the appropriate level of power supply retained.

Some of these design aspects have been covered earlier in this section, and for single circuit breaker configurations and for transformers, the CT location on the transformer or the associated circuit breaker is appropriate. However, in ring-bus and 1½-switch configurations, the current transformers are usually located on the ring (bay) circuit breakers, and their secondary outputs are added to produce the measuring zone required. CT matching is then essential. Depending on the class of the individual cores and the system operating voltage, the maximum number of cores that can be accommodated in a standard separate CT stack varies from four to six, depending on system voltage, while bushing-type current transformers on transformers and dead-tank circuit breakers will usually have fewer than four cores. This needs to be considered when the protection zones and CT locations are chosen.

Specific discussions of these aspects and the rating and performance of such measuring systems are discussed further in chapter 6 and their application in chapter 11, "Protection."

12.6 HIGH-ALTITUDE CONSIDERATIONS

12.6.1 Effect of Altitude on Insulation Levels

The electrical strength of air insulation is inversely proportional to the air density, which varies with altitude and temperature. The appropriate derating factors are given in ANSI/IEEE C37.30 [S1] or IEC 60071-2 (originally IEC 71 A) [S2], which require that the equipment voltage be proportionally higher than the actual system operating voltage at altitudes greater than 1000 m. Although transmission or distribution line insulation can be more finely adjusted, most primary equipment is designed for standard base voltages which, for both North American and international use, are specified in IEC 60038 [S4].

Based on these standard values, the standard maximum system operating voltage that can be safely used up to 1000 m for sea-level-rated 115 kV equipment is 123 kV and for the 69 kV systems is 72 kV. To obtain equivalent sea-level design and test figures, both the maximum operating voltage and the BIL must be adjusted to obtain equivalent performance for the actual system conditions when operating at high altitude.

For example, for a system operating at 115 kV, the next equipment voltage level generally available is 145 kV, and equipment rated at this voltage level would cover a typical altitude range of 1000 m up to about 2700 m. For higher altitudes, the next standard voltage level above 145 kV is 245 kV. There is more flexibility in cap and pin insulators to adjust the insulator length to intermediate voltage ratings, but for all factory-assembled air-insulated components, it is more economical to utilize the next available standard operating level and associated BIL-class equipment.

The above enhancement of insulation level applies for all air-insulated equipment operating at high altitude, and a similar enlargement of both phase-to-phase and phase-to-ground clearance is necessary to allow for this decrease in air-insulation capability.

The increase in size can lead to significant cost increase, not only for the switchyard equipment, but also for the site preparation, particularly in those areas where flat land is difficult to obtain. For these reasons, it has been found worthwhile to use GIS, SF_6 gas insulated, switchyards at high elevations. In these cases, the internal insulation of the substation is identical to that of the standard design for 0–1000 m. The only special designs are for the gas-to-air bushings and the line entry equipment. The use of GIS gas-to-oil interface bushings for the transformers also eliminates the need for special oversized transformer tanks to accommodate the high external BIL oil-air bushings.

12.6.2 Effect of Altitude on Conductor Size

The substation bus and line entry conductors need to match the line conductors as far as practicable in terms of the radio noise (RI) levels and other electrical performance parameters. Similarly, the current rating of the line bay must either match or exceed the line current rating.

At high altitudes, RI and corona performance become an important consideration in conductor selection. The low air density at high altitudes also affects the

High-Voltage Substation Design Considerations

heat convection and, consequently, the temperature rise of the conductor. For more details of these effects, the derating factors are detailed in ANSI/IEEE C37.30 [S1].

The radio interference levels in the transmission line, and hence on the substation entry, need to be controlled both to avoid noise interference to other radio users and also to avoid interference with the utility's power-line carrier and radio systems. It is similarly necessary to avoid noise interference in the industrial plant's electronic and other communication systems. Work done by IEEE committees [10, 11] resulted in an interpretation of this regulation in terms of acceptable signal-to-noise ratios. Unfortunately, the RI level does not uniquely set the design criteria.

The noise generated by the conductor falls off as a function of lateral distance from the line or conductor, and hence factors based on interference alone would vary with the distance from inhabited areas. For the transmission line itself, the criteria could be relaxed in view of the general remoteness of line routes at very high altitude. It should be noted, however, that a further increase in the RI would result in rapid escalation of corona losses under normal operating conditions and would adversely affect the operation of the power-line carrier and maintenance radio systems because of the high noise generation of the phase conductors.

The methods of calculation for transmission lines are recommended in the EPRI *Transmission Line Reference Book* [12] and in CIGRE publications [13].

Relative to surface gradients, the lower surface gradient with the larger conductors does not result in a similar reduction in the RI. This phenomenon is also demonstrated in the IEEE guide [10], which shows that the bigger the conductor, the lower is the permissible surface gradient.

Relative to RI, the corona loss is basically a foul-weather phenomenon. Under dry conditions and with the conductors selected for an acceptable RI performance, corona losses are very small, usually on the order of up to 3 kW/km, only 1 or 2% of $I^2 * R$ losses. They are in the same order of magnitude as corona losses generated by the conductor and insulator fittings.

In rain, fog, or particularly for ice on conductors, there will be a substantial increase in corona losses, which can be ten times or more of the fair-weather losses. However, in most locations, the number of hours for which such conditions exist do not justify that this should be taken into account in selecting the size of the conductor.

12.6.3 Definition of Altitude Zones for Substation Design

For each voltage level, there are internationally agreed design levels for maximum continuous operating voltage (MCOV), power frequency withstand, and BIL (LWIL). Above 300 kV MCOV, there are similar figures for switching surge levels (SWIL). IEC 60038 [S4] and IEC 60071-1, 2, and 3 [S2] provide base levels and the factors for use for altitude correction. As there are specified insulation classes for equipment, it is appropriate to use the next equipment design insulation level when the altitude exceeds the 1000 m standard by a reasonable amount. As air-insulated systems have a finite risk of flashover related to their design and operating voltages, operating with a standard voltage level above the standard altitude increases the risk of flashover, and hence once the altitude is above 1000 m (in practice, above between 1100 and 1200 m), the higher voltage class should be used. Similarly, the next step in voltage class will be at an altitude when the altitude derating factor makes the risk of using that particular voltage class unacceptably high.

12.7 SUBSTATION DESIGN CONSIDERATIONS

12.7.1 LAYOUT AND CONDUCTOR SPACING

When designing outdoor air-insulated substations, sufficient spacing must be provided to limit the risk of flashover to grounded structures and between phases to acceptable levels. There must also be consideration for safe access for personnel and vehicles and other equipment access systems that will provide the ground-clearance requirements. In general, the minimum level of ground clearance where personnel are allowed is between 2300 and 2400 mm from ground level to the bottom of any support insulator, and that support insulator must be long enough to meet the phase-to-grounded-structure clearance requirements. These minimum ground clearances are based on no or minimal snow levels, and in areas where snow accumulation can be significant, higher clearances must be allowed.

Typical values for the phase and ground clearances at standard altitudes are provided in the CIGRE guidelines [6], and standardized values are given in the relevant ANSI/IEEE standard [S5] and IEC standard [S7]. It should be noted that these dimensions have a built-in safety margin, as they are relevant to what are basically site-assembled and untested systems. It is noticeable that three-phase equipment assemblies, such as circuit breakers and disconnects, often use closer phase spacing, as the units have been factory tested and their actual insulation withstand levels proven.

It is therefore not uncommon to have to reduce the specified bus spacing when connecting to such factory-assembled equipment, noting that the transition needs to be carefully designed to keep the largest possible spacing during the transition. The other factors that need to be considered include industrial plants, where local codes such as the NEC [S8] or CEC [S9] apply; ground clearances may need to be larger than the clearances given in the associated standards. This also applies to the ground clearances that would be obtained with a full BIL-tested disconnect or circuit breaker installed with its bushing base at the normally accepted 2300 to 2400 mm above grade. In such cases, the standard support structures may need to be extended to provide higher ground clearance.

This difference can also be found with utility substations that follow the standard phase-to-ground requirements rather than the safety code levels. This is because most utilities are exempt from the industrial safety codes on the basis that utility staff have had additional training for working in switchyards. It should be noted that code requirements for overhead lines that cross an industrial area can also require larger phase-to-ground clearances than would apply to utility-owned lines outside the site fence.

12.7.2 GENERAL CONSIDERATIONS FOR BUS DESIGN

The overall design of the station bus must be such as to be able to withstand the forces impressed on the bus by the combination of wind, wind plus ice (if appropriate), and short-circuit forces while continuing to meet the design clearances noted in

High-Voltage Substation Design Considerations 343

section 12.7.1. These forces will cause a strain bus to move and cause a rigid bus to transfer forces directly to a bending movement on the structures.

The calculated forces need to be used to calculate the swing angle and hence the air clearance of the strain bus and the strengths of the support structures. Similarly, the dead load must be checked to confirm the strain bus sag and the strain insulator strengths and the appropriate maximum span between support insulators for a rigid bus.

It is important that a realistic value be taken for the system fault level, as the substation must be able to withstand any fault level likely to occur over the life of the station. Another important factor is the realistic fault time, which will depend on the protection philosophy at the various voltage levels. If the circuit breaker standard (1 or 3 s) short-time rating is used for the design, then the design will be consistent for the whole station.

Two options exist for designing the bus-work for substations. These are rigid or strain bus and, for any station, they may be used exclusively or in combination relative to the site size, visual impact, and height restrictions. There are also two basic structural configurations: high profile, box girder structure and low profile, individual structure layouts. In the latter case and when rigid bus is used, it is possible to use A-frame structures mounted on one bus to support an upper, crossing bus on the same phase (see fig. 12.4, 12.5).

For these various options, a number of basic configuration options need to be considered. These are typically:

- **Bus location:** Equipment maintenance and removal should be considered in locating buses and support structures. The bus should also be chosen to allow entrance of construction and maintenance equipment into the substation. When flexible conductors are used for strain bus construction,

FIGURE 12.14 Switchyard using rigid bus-work

FIGURE 12.15 Switchyard using strain bus

they permit significant conductor movement. Consequently, the conductors have to be carefully positioned to prevent contact with other equipment and infringement upon minimum electrical clearances under all loading and climatic conditions;

- **Future expansion:** It is important to plan for future expansion by sizing and positioning buses to facilitate modifications. Strain buses usually require large supporting structures. These structures can limit future expansion if not properly positioned.

The other basic decision is the conductor material to be used, noting that the choices are between aluminum (either in rigid shape or ACSR), all-aluminum alloy conductor, AAAC stranded conductors, or the equivalents in copper. The conductivity of aluminum is 50–60% of copper, and hence larger sizes are needed for aluminum, which can increase wind and ice loading. On the other hand, copper is about twice as heavy as aluminum, so the supports need to be closer. At higher voltages, a larger size is preferable to meet the RI levels, and this favors aluminum, while copper is usually found on older and lower-voltage systems. One disadvantage of aluminum is the special techniques required for welding, and this needs to be considered when site-fabricated bus-work needs to be used.

The bus conductors are selected based on ampacity, physical properties, cost, and, at high voltages, the RI noise levels. Conductor type and size needs to be selected to have sufficient size and capacity to withstand system faults and overloads without damage from overheating. The conductor is therefore selected based on ampacity, physical properties, and cost, and, at high voltages, the ability to meet the station RI requirements.

12.7.3 Rigid Bus Design

The design of a rigid bus system involves a number of special factors, such as:

- **Short-circuit conditions:** During short circuits, large forces can be developed in the bus system. The rigid bus design needs to include consideration of these forces to prevent damage during short-circuit conditions. The bus centerline-to-centerline spacing and the short-circuit current both have effects on these forces.
- **Wind and ice loads:** If not properly considered, wind and ice loads can cause extensive damage to bus conductors and overstress support insulators. It is also necessary to consider local conditions, since they may necessitate the use of more severe loading criteria than standard practice.
- **Insulator strength:** Because the number of different insulator ratings is limited, it is necessary to exercise care in the bus layout so that a practical system is achieved. The strength of the insulators required is based on the total bus loading and, particularly, the short-circuit forces.
- **Conductor deformation:** The sag of the bus conductors needs to be minimal, as a flat horizontal bus looks much neater than one with excessive deformation. The conductor sag is influenced by the conductor weight and section modulus, the span length, and the vertical loading.
- **Aeolian vibration:** Long conductor spans can be damaged by vibrations caused by winds. Excessive conductor sag can add to this problem. Span lengths whose natural frequency is near that set up by a wind that has a high risk of recurrence (prevailing wind conditions) can cause vibration and fatigue.

The details and methods of calculation are provided in a series of IEEE papers [19, 20] and the associated IEEE guide [S12], which deals in detail with the methods of calculation and the data required. However, the following procedure should be used in designing a rigid bus system:

- Select the type of conductor material and the size of the bus conductors based on the continuous current requirements. In higher-voltage systems with longer bus spans, the structural capabilities of the bus conductors and the surface gradient levels may be the factors that determine the conductor and size. Over and above these requirements, the conductors selected must be capable of carrying the required continuous current.
- Calculate the conductor spacing requirements, which will determine the minimum bus conductor centerline-to-centerline spacing.
- Calculate the short-circuit forces for the maximum short-circuit forces the bus has to withstand.
- Determine the total bus conductor loading. Base this on the wind and ice loading for the actual site location. For sites located in the United States, the minimum values are defined in the National Electrical Safety Code [S10]. It is also necessary to consider extreme wind conditions.

Because the spans for rigid bus-work are often fairly long, damaging vibrations may occur. Consequently, a means for controlling the vibrations should be provided. Prefabricated dampers can be attached to the buses, or cables can be installed inside the buses. If cables are used, the cable weight has to be added to the conductor weight for the bus calculations.

The other factor that will need to be considered is the method of clamping. To allow for expansion and contraction with temperature, slip fittings need to be strategically located to prevent stress being applied to support insulators, and in particular to devices such as disconnects, which could otherwise be forced out of alignment.

12.7.4 Strain Bus Design

Some of the specific factors that should be considered in strain bus design are as follows:

- **Wind and ice loading:** Wind and ice can increase conductor sags and tensions appreciably. Local conditions should also be considered, since they may necessitate the use of more severe loading criteria than provided in the standard literature.
- **Insulator strength:** The suspension insulators need to be selected based on the anticipated maximum loading conditions. The maximum loading for porcelain insulators should not exceed 40% of the manufacturer's design ratings.
- **Span length:** The span length influences the conductor sag because the sag increases as the span length increases, if the same tension is maintained. To limit the sag, the tensions can be increased, but this influences structure and insulator strength. Spring tensioning can be used to limit the tension and sag.
- **Sag and tension:** Strain buses are usually positioned above other substation equipment, and conductor breakage could result in equipment damage or outage. To prevent breakage and to minimize support structure size, the conductors need to be installed at conservative levels of tensions. If support structures deflect, the sag may increase, and this also needs to be factored into the design.
- **Temperature**: Temperature variations cause changes in conductor lengths. As conductor temperature increases, the sag increases and the tension decreases.
- **Other loads:** Taps from the conductors to other buses or equipment should be limited in tension to prevent damage to equipment. The taps are usually installed as slack connections.

Short-circuit current aspects of strain bus design are covered in two 1980 IEEE papers [18, 21]. The following procedure should typically be used to design a strain bus system:

- Select the material size and associated characteristics of the bus conductors.

High-Voltage Substation Design Considerations

- Determine the bus conductor centerline-to-centerline spacing for the voltage and altitude. For strain bus, the minimum metal-to-metal, bus centerline-to-centerline, and minimum ground clearances should be at least 50% more than the specified base clearances when rigid conductors are used.
- Select the strength and type of suspension and tension insulators.
- Determine the total bus conductor loading for wind and ice loading for the actual station location. The values defined in the base literature for a particular area should always be considered as minimum values.

Once the main bus data and configurations are decided, then the same studies need to be done to confirm the design of the conductor drops to the individual equipments and then the design of the main buses readjusted based on the loading levels of these drops.

12.8 GROUNDING DESIGN AND LIGHTNING PROTECTION

12.8.1 Introduction

The primary objective of the grounding system of any electrical substation is to provide a safe working environment for all persons within the switchyard and to ensure the safety of any other person who may be in the vicinity of the switchyard, but outside the security or access fence line. The main principles of calculation are given in IEEE 80 and are discussed in detail in chapter 13. The second factor influencing both personnel and equipment operational safety is the shielding of the station from direct lightning strikes and providing the conditions so that the installed arrestors can function effectively.

Station shielding and arrestor effectiveness are also linked to good grounding design. The main points for consideration when deciding the overall station layout, prior to the detailed design of the grounding grid itself, are discussed in the following sections.

12.8.2 Substation Area

The whole operational area of the switchyard must be provided with a reasonably uniform grounding grid over the whole area, and this area must equally be covered with a layer of rock or crushed stone as a means of providing an equipotential surface to control step potentials. Similarly, the area has to be shielded from direct lightning strikes.

The cost of correctly preparing and installing the ground grid is significant, and hence the area to be so protected must be selected based on the initial and future expansion plans for any given station. Where the operational area is initially small but the site is large, or if the switchyard is a small part of a larger site used for other purposes, it may be appropriate to limit the area to be provided with a ground mat. In such a case, the fence line may be outside the ground mat. In such cases, the fence is provided with its own grounding rods along the fence line. It is essential that the fence be kept well separated from the grid so that it is not possible for anyone to bridge, by touch or by step, from the grid to the fence. If the fence has to cross the grid to join, say, the control building, then the section within the touch of any individual (at least 3 m each side of the grid boundary) must be constructed of non-conducting material (wood, concrete, or brick), and therefore barbed wire must not be used to top the transition fence.

In the choice of the protected area, it is also important to ensure that all directly connected cables do not leave the area of the grid. This requires that the relay and control buildings and the station service transformers and switchgears all be within the main grid area. Cable connections that leave the protected site area must be minimized, and these need to be isolated from the grid-potential rise during ground faults, and any ground conductor or sheath must be cut back and insulated at one end. As it is possible, in mountainous areas where ground resistivity is high, to get ground-potential rises of up to 10 kV, such isolation can be difficult and expensive to provide.

The best technique is to eliminate such connections by using radio or optic fiber to bring in any links, such as communication circuits, into a site. However, there are isolation units available for telephone system isolation up to the 5 kV level of ground-potential rise.

The provision of auxiliary supplies directly to other users outside the fence line should similarly be avoided, as the neutral ground link in the substation will be subjected to the station ground-potential rise, which could be a hazard in another area. If such supplies need to be provided, then isolation transformers must be provided, either by a 1:1 interface transformer or by a high-voltage feed with a Delta connection or split ground connections.

The most important provision for the protection of the general public, when the fence is connected to the ground grid, is to provide an equipotential grid outside the fence at such a distance that the fence cannot be touched while standing outside the line of the outer grid. This also needs to be taken into account when the gate opens outward. In areas of high potential rise, it may also be necessary to provide the stone covering out beyond the outer grid line.

12.8.3 GENERAL SURGE PROTECTION

Substation electrical equipment, as described in the chapters covering the particular equipment, are specified to be able to withstand particular levels and characteristics of high-frequency surges (BIL, LWIL, and SWIL), which are typical of the surges that arrive at switchyards due to lightning surges, switching surges, and faults on the system. Any abnormal conditions that cause surges with more severe characteristics than the design rating can result in equipment flashover or insulation failure. To prevent equipment damage or system shutdown, protective devices are used to limit the surge voltages to reasonable levels. When surge protection is applied, it is also possible to reduce the basic insulation level (BIL) of the equipment used and reduce the clearances used in the station design.

As these surge arrestors are relatively expensive and need to be located close to the equipment being protected, the application of these devices is usually a compromise between the costs of the surge arrestors and the degree of protection desired. However, surges caused by a direct lightning stroke onto substation equipment is significantly worse than the typical surges, and arrestor characteristics and location

High-Voltage Substation Design Considerations 349

will not be able to provide protection to critical switchyard equipment. The surge protection provided for substations and substation equipment can be broken into two main parts:

- **Surge protection:** employed to protect the equipment from damaging overvoltages caused by lightning surges, switching surges, and system faults that enter the station from the connecting overhead lines
- **Direct stroke protection:** employed to protect the equipment from direct lightning strokes

12.8.4 Equipment Surge Protection

Surge arrestors are used to protect equipment against high-voltage surges entering the station or produced by switching equipment operations. The arrestors function by discharging surge current to the ground system and then interrupting or limiting that current to prevent the flow of normal power frequency current (follow-through current) to ground. Surge protection can also be produced using rod gaps but, in that case, the follow-through current has to flow, and hence a system trip always follows such a gap flashover. A detailed discussion of insulation coordination, the coordination of the insulation levels of the various pieces of equipment within a switchyard, and the application and selection of surge arrestors can be found in the various parts of IEC 60071 [S2] and the respective IEEE 1313 [S6].

12.8.5 Direct-Stroke Shielding Principles

Because the effects of a direct lightning stroke to any electrical equipment in an unshielded substation can be devastating, it is recommended that some form of direct-stroke protection be provided. Direct-stroke protection normally consists of shielding the substation equipment by using lightning masts, overhead shield wires, or a combination of these devices. The types and arrangements of protective schemes used are based on the size and configuration of the substation equipment and the location of sensitive equipment such as transformers, power cables, and instrument transformers. The concepts are discussed further in two IEEE reference papers [28, 29] (see fig. 12.16, 12.17).

12.8.5.1 Overhead Shield Wires

Overhead shield wires are often used to provide direct stroke protection. The shield wires need to extend over the substation area and can be supported by the circuit pull-off structures, if conveniently located. Because these shield wires are located above substation buses and equipment, any breakage could result in outage of or damage to the equipment below. To minimize possible breakage, the overhead shield wire systems are constructed from high-quality, high-strength materials and should be limited in tension to provide an added safety margin over the calculated loading conditions.

This tension also has to be coordinated with support structure design. Such tensions may need to be lower if the capabilities of the support structures are lim-

FIGURE 12.16 Switchyard protected by overhead shield wires

FIGURE 12.17 Switchyard protected by masts

ited. The sag also has to be considered to ensure adequate clearance from energized equipment, again under the worst loading conditions.

To be complete, the overhead shield wire system should include the protection for the overhead circuits entering or leaving the substation, as a direct conductor strike close to the substation will also exceed the ability of the surge arrestors to adequately protect the substation equipment. In areas not employing transmission

High-Voltage Substation Design Considerations 351

line shielding, the substation shield wire systems should be extended at least 800 m (one-half mile) away from the substation.

Strokes occurring on the circuits beyond the shielding will usually be attenuated enough by the time they reach the substation to be discharged successfully by the surge arrestors without causing equipment damage. For adequate protection, the line overhead shield wire circuits should be directly connected to the substation shield wire system.

12.8.5.2 Masts

Shielding masts can be used for nearly all types of substations to provide protection against direct lightning strokes. They are particularly useful in large substations and those of low-profile design. Shielding masts can be guyed or self-supporting steel poles or lattice-type towers and are usually made of steel. Other materials, such as pre-cast concrete or aluminum, can also be used. In all cases, a good electrical connection from the tip of the mast to the station ground grid is required. In some instances, shielding masts can also be used to provide support for substation lighting equipment.

12.8.6 Lightning Protection Shielding Calculation Methods

There are two widely used methods for designing substation lightning protection:

- Fixed angle
- Rolling sphere

The zone of protection of a shielding system is the volume of space inside which the equipment is considered to be adequately protected by the system. A shielding system that allows statistically no more than 0.1% of the total predicted number of lightning strokes to penetrate the shield and strike protected equipment is considered adequate for most situations:

- **Fixed-angle method:** The fixed-angle design method uses vertical angles to determine the number, position, and height of shielding masts and wires. The zones of protection afforded by a single mast or shield wire are considered to be between 20°C and 60°C, normally accepted as 35°C, from vertical on either or any side. For a single mast, the zone of protection consists of a cone, while for a single shield wire, the zone of protection is a wedge. When two or more masts or shield wires are used, the zones of protection of each should overlap to provide complete coverage. In this case, the protected angle rises to 40°C to 60°C, with the normally accepted objective at 45°C.
- **Rolling-sphere method:** The rolling-sphere method involves rolling an imaginary sphere of a prescribed radius over the substation. The sphere rolls up and over the lightning masts, shield wires, and other grounded metal objects intended for lightning shielding, all of which need to be spaced so as to support the sphere's diameter. A piece of equipment is protected from a direct stroke if all parts of the equipment remain below the curved surface

produced by the locus of the sphere's surface. If any part penetrates this surface, it is not protected.

Both these methods are discussed further in the two IEEE papers [31, 32] and in the IEEE guide, IEEE 998 [S11].

12.9 MEDIUM- AND LOW-VOLTAGE SUBSTATIONS AND SWITCHGEAR

The design process for medium- and low-voltage substations is similar to that detailed for high-voltage outdoor air-insulated stations and, for air-insulated designs, many of the same considerations need to be discussed. However, switchgear and switchboards are factory-assembled units that are designed and built to meet all the required standards and safety codes relevant to the plant location, local industry standards, and local regulations and, as such, these aspects do not need to be designed specifically. In such cases, most of the design considerations for the individual components that make up an industrial distribution substation, such as transformers, switchgear, cables, etc., are defined individually, and the design aspects and their application are discussed in the chapters covering each of these systems. However, all such equipment and systems need to be installed to meet the local standards and regulations, including fire and escape requirements as well as operational clearances, and must be based on ensuring reliable cable connections and operational requirements. There must also be suitable clearances and access to maintain and replace any of the main components.

12.10 SITE TESTING AND COMMISSIONING

12.10.1 SCOPE OF WORK

The commissioning of a new substation or the extension of an existing station needs a program for the site testing of all the primary plant, such as switchgear, bus-work, and instrument transformers included in the new works. The testing and commissioning of the larger sizes of autotransformer, power transformer, and oil-filled reactor should normally be carried out by the manufacturer in conjunction with the other site contractors.

The commissioning and testing should also include all indicating meters, relays, and control, alarm, and communication equipment and similar functions associated with the primary plant, including the necessary interface with existing installations or equipment supplied by others. All station auxiliaries, including such items as pumps, fans, compressors, lights, batteries, chargers, diesel generator, and air conditioners, should be included.

Testing of the energy meters and the certification of their accuracy may require special tests, depending on the legal requirements in any particular region. Testing of the interface with equipment installed by each contractor should be done jointly with the other contractors as required.

Testing of control system and any data transmission to the utility's dispatch center(s) should also be included.

High-Voltage Substation Design Considerations

12.10.2 Objectives

The objectives of the testing and commissioning works, prior to the successful energization of the plant at full voltage and connection to the system, are to:

- Confirm the integrity (correctness) of installation
- Confirm the integrity of insulation, connections, and phasing
- Ensure proof of equipment characteristics
- Review workmanship
- Confirm the correct implementation of the design
- Check equipment ratings
- Make final field checks

Measurements taken during and after equipment energization and on-load checks to meet these objectives should follow equipment connection to the system. Tests should be specifically included to:

- Check the performance of the various manual and automatic control features
- Check settings and operation of protective relays
- Assess the ability of the system to accommodate scheduled or unscheduled connection and disconnection of the plant without undue disturbance
- Initiate loading and performance tests of all components (including verification of losses in some cases)
- Conduct integrated system-performance tests
- Check and measure resistivity of earthing grid and earthing system

12.10.3 Responsibilities

To ensure that the test jurisdiction and transfer of responsibilities is regulated by strict safety and handover procedures, it is necessary to agree on the interface between the contractor and the owner/operator and, where appropriate, the engineer.

To ensure that this is done satisfactorily and prior to the start of commissioning, a detailed commissioning program and testing procedures should be prepared. Similarly, handover procedures consistent with the terms of each contract should also be prepared in advance.

The owner/operator should always retain full jurisdiction over all commissioning activities that might affect the operation of the existing system.

12.10.4 Safety Procedures

It is important that the contractor and the owner share the responsibility for safety procedures. In addition, they should jointly develop a system for a work permit, a tagging system, and the associated safety procedures for all equipment, systems, and areas not covered by the owner's safety procedures. At the same time, the owner/operator should always assume responsibility for the establishment and implementation of tagging, safety, and work-permit procedures for the protection of personnel and equipment as soon as equipment and systems are connected to or can be energized from the existing system.

12.10.5 Training of the Employer's Staff

The commissioning plan should include the owner's staff, either continuously or on a regularly recurring basis, even after the commissioning has been completed. This will allow the owner/operator's staff to become familiar with the operating and maintenance aspects of the new equipment. It is also important that the precautions required in, or possible consequences of, initial energization of equipment also be assessed jointly by all interested parties.

12.10.6 Maintenance Aspects

Many of the tests made during commissioning will need to be repeated in future, either as routine checks or, if a problem develops, during fault investigations. It is therefore important that good records be taken and retained for future reference. Commissioning is also an opportunity to use and check that the operating manuals are clear and cover the equipment that has been supplied in sufficient detail to enable local staff to perform the required routine testing and, where necessary, perform sufficient fault investigation to identify the severity of the problem and whether site repair is feasible.

REFERENCES

Standards

S1. ANSI/IEEE C37.30-1992, IEEE Standard for High-Voltage Switches, 1992.
S2. IEC 60071-1,-2,-3, Application Guide for Insulation Coordination, parts 1–3.
S3. CSA C1-8.3.1-1975, Interference Produced by Corona Effect of Electric Systems, 1975.
S4. IEC 60038, IEC Standard Voltages.
S5. ANSI C37.32-1996, American National Standard, High-Voltage Air Disconnect Switches, Schedule of Preferred Ratings, Construction Guidelines and Specifications, 1996.
S6. IEEE 1313-1993, IEEE Standard for Power Systems — Insulation Coordination, 1993.
S7. IEC 60621-1,-2-1987, Electrical Installations for Outdoor Sites under Heavy Conditions, Part 1 and Part 2, 1987.
S8. NFPA 70, National Electrical Code.
S9. CSA C22.1, Canadian Electrical Code.
S10. ANSI C2, National Electrical Safety Code.
S11. IEEE 998-1996, Guide for the Direct Lightning Stroke Shielding of Substations, 1996.
S12. ANSI/IEEE 605-1998, IEEE Guide for Design of Substation Rigid-Bus Structures, 1998.
S13. NEMA CC1-1975, Electric Power Connectors for Substations, New York: National Electrical Manufacturers Association, 1975.
S14. EEI-NEMA CC3-1973, EEI-NEMA Standards for Connectors for Use between Aluminum or Aluminum-Copper Overhead Conductors, New York: National Electrical Manufacturers Association, 1973.

S15. NEMA HV 1-1973, High Voltage Insulators, New York: National Electrical Manufacturers Association, 1973.

BIBLIOGRAPHY

1. Canadian Electricity Association, CEA Working Group, "Report on Basic Station Single Line Diagrams in Use by Canadian Utilities and Contractors," Canadian Electrical Conference, Mar. 1980.
2. AIEE Committee Report, "Substation One Line Diagrams," *AIEE Trans.* 72 (1953): 747–751.
3. K. R. Knight and L. M. Gordon, "Method of Comparing High Voltage Substation Diagrams," AIEE Conference, paper no. 58-874, June 1958.
4. R. Page, K. Nishikawa, and F. Stromotich, "The 4x6 Network: A New Power Switching Scheme," *IEEE Power Eng. Rev.*, Oct. 1989,
5. C. R. Baylis and D. W. Moore, "The Transformer Feeder Substation," *IEEE Power Eng. J.* Sep. (1987): 269–273.
6. L. Demoulin et al., "General Guidelines for the Design of Outdoor ac Substations," CIGRE Study Committee 23, WG 04 brochure, 1993.
7. E. G. Norell, "Basic Designs for Large High Voltage Substations," *AIEE Trans.* 75 (1956): 1157–1169.
8. G. E. Hertig, "High Voltage and Extra High Voltage Substation Design," AIEE paper 62-246, Jan. 1972.
9. R. J. Cakebread, K. Reichert, and H. G. Schutte, "Reliability in the Design of EHV Substations, Availability as a Function of Component Failure Rate: Repair and Maintenance Times," *CIGRE Electra* 65 (July 1979).
10. IEEE Noise Subcommittee, "Radio Noise Guide for High Voltage Transmission Lines," *IEEE Trans. PAS* 90 (2): (1971).
11. IEEE Noise and Corona Subcommittee, "Review of the Technical Considerations on Limits of Interference from Power Lines and Stations," *IEEE Trans. PAS* 99 (1): (1980).
12. Electric Power Research Institute, *Transmission Line Reference Book, 345 kV and Above*, Palo Alto, CA: EPRI, 1975.
13. CIGRE, "Interference Produced by Corona Effect of Electrical Systems," CIGRE publication, 1974.
14. G. B. Furst, "Electrical Design Aspects for Transmission Lines at Very High Altitudes."
15. T. H. Frick, J. R. Stecart, A. R. Hileman, C. R. Chowgnic, and T. E. McDermott, "Transmission Line Insulation Design at High Altitude," *IEEE Trans. PAS* 103 (12): (1984).
16. N. Nimura, "Influence of Air Density on Flashover Voltages of Air Gaps and Insulators," *IEEE Trans. PAS* 89 (4): (1970).
17. A. Philips, A. F. Rohlfs, L. M. Robertson, and R. L. Thomason, "The Influence of Air Density on the Electrical Strength of Transmission Line Insulation," *IEEE Trans. PAS* 86 (Aug.): (1967).
18. M. B. Award and H. W. Huestil, "Influence of Short-Circuit Currents on HV and EHV Strain Bus Design," *IEEE Trans. PAS* 99 (2): (1980).
19. W. Stefnik, Jr., G. A. Volta, and J. M. Stipcevich, "Short-Circuit Test on a 3-Phase, 230-kV Rigid Bus Assembly," *IEEE Trans. PAS* 98 (3): (1979).
20. J. J. Landin, C. J. Lindqvist, L. R. Bergstrom, and G. R. Cullen, "Mechanical Effects of High Short-Circuit Currents on Substations," *IEEE Trans. PAS* 24 (5): (1975).
21. D. B. Craig and G. L. Ford, "The Response of Strain Bus to Short-Circuit Forces," *IEEE Trans. PAS* 99 (2): (1980).

22. .J. P. Nelson and P. K. Sen, "High-Resistance Grounding of Low-Voltage Systems: A Standard for the Petroleum and Chemical Industry," *IEEE IAS Trans. IAS* 35 (4): 941–948 (1999).
23. Westinghouse Electric Corp., *Electrical Transmission and Distribution Reference Book*, 4th ed., East Pittsburgh, PA: Westinghouse, 1964.
24. Westinghouse Electric Corp., *Electrical Utility Engineering Reference Book*, vol. 3, East Pittsburgh, PA: Westinghouse, 1965.
25. Aluminum Company of America, *Alcoa Aluminum Bus Conductor Handbook*, Pittsburgh, PA: Alcoa, 1957.
26. Aluminum Association, *Aluminum Electrical Conductor Handbook*, New York: Aluminum Association, 1971.
27. Anderson Electric Corp., *Technical Data: A Reference for the Electrical Power Industry*, Leeds, AL: Anderson Electric, 1964.
28. M. A. Sergent, "Monte Carlo Simulations of the Lightning Performance of Overhead Shielding Networks of High Voltage Stations," *IEEE Trans. PAS* 91 (4): (1972).
29. A. M. Mousa, "Shielding of High-Voltage and Extra-High Voltage Substations," *IEEE Trans. PAS* 95 (4): (1976).
30. G. W. Brown and E. R. Whitehead, "Field and Analytical Studies of Transmission Line Shielding: Part II," *IEEE Trans. PAS* May (1969).
31. G. N. Alexandrov, V. Y. Kisvetter, V. M. Rudakova, and A. N. Tushnov, "The AC Flashover of Long Air Gaps and Strings of Insulators," *Elecktrichestwo* (1961): 27–32.
32. T. Harqda, Y. Arhara, and A. Hoshima, "Influence of Humidity on Lightning and Switching Surge Overvoltages," *IEEE Trans. PAS* 92 (3): (1978).

Related Documents
- a. Regulatory Standards
    - CSA C22.1, Canadian Electrical Code
    - ANSI C2, National Electrical Safety Code
    - NFPA 70, National Electrical Code
- b. North American Standards
    - ANSI C37.06, Switchgear-AC High-Voltage Circuit Breakers Rated on a Symmetrical Current Basis: Preferred Ratings and Related Required Capabilities
    - ANSI C84.1, Electrical Power Systems and Equipment-Voltage Ratings (60 Hz)
    - ANSI/IEEE 141, Recommended Practice for Electric Power Distribution for Industrial Plants (IEEE Red Book)
    - ANSI/IEEE 142, Grounding of Industrial and Commercial Power Systems (IEEE Green Book)
    - ANSI/IEEE 242, Recommended Practice for Protection and Coordination of Industrial and Commercial Power Systems (IEEE Buff Book)
    - ANSI/IEEE 399, Recommended Practice for Power System Analysis (IEEE Brown Book)
    - ANSI/IEEE 446, Emergency and Standby Power Systems for Industrial and Commercial Applications (IEEE Orange Book)
    - ANSI/IEEE 493, Recommended Practice for Design of Reliable Industrial and Commercial Power Systems (IEEE Gold Book)
    - ANSI/IEEE 605, IEEE Guide for Design of Substation Rigid-Bus Structures

- ANSI/IEEE C37.010, Application Guide for AC High-Voltage Circuit Breakers Rated on a Symmetrical Current Basis
- ANSI/IEEE 100, Dictionary of Electrical and Electronics Terms
c. National Electrical Manufacturers Association (USA)
    - NEMA 107, Methods of Measurement of Radio Influence Voltage (RIV) of High Voltage Apparatus
d. American Society for Testing and Materials (ASTM)
    - A 36M-84A, Specification for Structural Steel
    - A 53-81A, Pipe, Steel, Black and Hot-Dipped, Zinc-Coated, Welded and Seamless
    - A 120-84, Pipe, Steel, Black and Hot-Dipped Zinc-Coated (Galvanized) Welded and Seamless, for Ordinary Uses
    - A 121-81, Zinc-Coated (Galvanized) Steel Barbed Wire
    - A 123-78, Zinc (Hot-Galvanized) Coatings on Products Fabricated from Rolled, Pressed, and Forged Steel Shapes, Plates, Bars, and Strip
    - A 153-82, Zinc Coating (Hot Dip) on Iron and Steel Hardware
    - A 307-84, Carbon Steel Externally and Internally Threaded Standard Fasteners
    - A 325M-84A, High-Strength Steel Bolts for Structural Joints, Including Suitable Nuts and Plain Hardened Washers
    - A 392-84, Zinc-Coated Steel Chain-Link Fence Fabric
    - A 529-84, Structural Steel with 42 ksi (290 MPa) Minimum Yield Point (½ in. (12.7 mm) Maximum Thickness)
    - D 653-85, Definition of Symbols Relating to Soil Mechanics
    - D 1557-78, Test for Moisture-Density Relations of Soils Using 10 lbs (4.5 kg) Rammer & 18 in. (457 mm)
    - D 4253-83, Drop Test Methods for Maximum Index Density of Soils Using a Vibratory Table
    - D 4254-83, Test Methods for Minimum Index Density of Soils and Calculation of Relative Density
    - F 567-84, Installation of Chain-link Fence
    - F 626-84, Fence Fittings
e. American Society of Civil Engineers (ASCE)
    - ASCE Subsurface Exploration and Sampling Soils Journal of the Structural Division, paper 5403, "Electrical Transmission Line and Tower Design Guide"
f. United States Department of the Interior, Bureau of Reclamation
    - Design Standard No. 10, Transmission Structures Including Release Number DS-10-8, March 15, 1967
g. International Conference of Building Officials
    - UBC Uniform Building Code, 1982 ed.
h. United States Department of the Interior, Water and Power Resources Services
    - SBR E12 Earth Manual
i. American Concrete Institute
    - ACI 305 R-77, Hot Weather Concreting

j. American Institute of Steel Construction Inc., 8th ed.
   - Specification for the Design, Fabrication and Erection of Structural Steel for Buildings, 1978
   - Code of Standard Practice for Steel Buildings and Bridges, 1976
   - Specifications for Structural Joints, Using ASTM A 325 and A 490 Bolts (1980)
k. American Welding Society (AWS)
   - AWS DI.1-86, Structural Welding Code
l. International Electrotechnical Commission
   - IEC 60056, AC Circuit Breakers above 1 kV
   - IEC 60059, Standard Current Ratings
   - IEC 60060, High-Voltage Test Techniques
   - IEC 60076, Power Transformers
   - IEC 60111, Resistivity of Commercial Hard-Drawn Aluminum Electrical Conductor Wire
   - IEC 60114, Recommendation for Heat-Treated Aluminum Alloy Bus-Bar Material of the Aluminum–Magnesium–Silicone Type
   - IEC 60120, Dimensions for Ball and Socket Coupling of String Insulator Units
   - IEC 60129, Disconnects and Earth Switches
   - IEC 60137, Bushings for Alternating Voltages above 1000 V
   - IEC 60144, Degree of Protection of Enclosures for Low-Voltage Switchgear and Controlgear
   - IEC 60157, Low-Voltage Switchgear and Controlgear
   - IEC 60158, Low Voltage Controlgear
   - IEC 60168, Tests on Indoor and Outdoor Post Insulators of Ceramic Material or Glass for Systems with Nominal Voltage Greater than 1000 V
   - IEC 60044, Instrument (Current and Voltage) Transformers
   - IEC 60207, Aluminium, Stranded
   - IEC 60209, Aluminium Conductors, Steel Reinforced
   - IEC 60214, On-Load Tap Changers
   - IEC 60216, Insulation Thermal Endurance
   - IEC 60233, Tests on Hollow Insulators for Use in Electrical Equipment
   - IEC 60243, Recommended Methods of Test for Electric Strength of Solid Insulating Materials at Power Frequencies
   - IEC 60255, Electric Relays (Protection & Control)
   - IEC 60265, High Voltage Switches
   - IEC 60269, Low Voltage Fuses
   - IEC 60270, Partial Discharge Measurement
   - IEC 60273, Dimensions of Indoor and Outdoor Post Insulators and Post Insulator Units for Systems with Nominal Voltage Greater than 1000 V
   - IEC 60282, High Voltage Fuses
   - IEC 60289, Reactors

- IEC 60292, Low Voltage Motor Starters
- IEC 60296, Specifications for Unused Mineral Insulating Oils for Transformers and Switchgear
- IEC 60298, Metal Enclosed Switchgear
- IEC 60305, Characteristics of String Insulator Units of the Cap and Pin Type
- IEC 60309, Plugs, Socket Outlets and Couplers for Industrial Purposes
- IEC 60337, Control Switches (Low-Voltage Switching Devices for Control and Auxiliary Circuits, Including Contactor Relays)
- IEC 60354, Loading Guide for Oil-Immersed Transformers
- IEC 60364, Electrical Installations of Buildings
- IEC 60372, Locking Devices for Ball and Socket Couplings of String Insulator Units
- IEC 60376, SF_6 Gas
- IEC 60383, Tests on Insulators of Ceramic Material or Glass for Overhead Lines with a Nominal Voltage Greater than 1000 V
- IEC 60408, Low Voltage Air-Break Switches, Air-Break Disconnectors, Air-Break Switch Disconnectors and Fuse-Combination Units
- IEC 60420, HV Switches Combined with Fuses
- IEC 60433, Characteristics of String Insulator Units of the Long Rod Type
- IEC 60437, Radio Interference Test on High-Voltage Insulators
- IEC 60439, Low Voltage Switchgear and Control Gear Assemblies
- IEC 60466, High Voltage Insulation: Enclosed Switchgear and Controlgear
- IEC 60471, Dimensions of Clevis and Tongue Coupling of String Insulator Units
- IEC 60480, Guide for Checking SF_6 Gas
- IEC 60507, Artificial Pollution on HV Insulation
- IEC 60529, Classification of Degrees of Protection Provided by Enclosures
- IEC 60551, Measurement of Transformer and Reactor Sound Levels
- IEC 60567, Guide for Sampling of Gases and of Oil from Oil-Filled Electrical Equipment and for the Analysis of Free and Dissolved Gases
- IEC 60599, Interpretation of the Analysis of Gases in Transformers and Other Oil-Filled Electrical Equipment in Service
- IEC 60644, HV Fuses for Motor Protection
- IEC 60694, Common Clauses for High-Voltage Switchgear and Controlgear Standards

13 Substation Grounding Design Considerations

13.1 INTRODUCTION AND OVERVIEW

Substation grounding design considerations are important to ensure the safety of personnel and the public, to minimize hazard from transferred potential, to protect equipment insulation, to provide a discharge path for lightning strikes, and to provide a low-resistance path to ground. The tolerable limits of body current are:

- Threshold, 1 mA
- Painful (may result in lack of muscular control), 9–25 mA
- Shock energy that can be survived by 99.5% of humans
- Body current:

$$\text{For 50 kg body weight, } I_{B(50)} = \frac{0.116}{\sqrt{t_S}} A$$

$$\text{For 70 kg body weight, } I_{B(70)} = \frac{0.157}{\sqrt{t_S}} A$$

where t_S is the time of exposure in seconds.

The required elements of the grounding grid are as follows:

- Ground grid consisting of bare stranded copper conductor (4/0 AWG or larger), buried 300–500 mm deep in ground and spaced 3–6 m apart, with the conductors bonded at their crossing
- Ground rods, usually 20 mm in diameter, ($\phi \times 3$)-m-long copper alloy rods buried and connected to the grid
- Bare stranded copper wires connecting to the metal frames and the supporting structures
- Gravel or crushed rock, spread in the substation yard to a layer thickness of 75–150 mm

13.2 SOIL-RESISTIVITY MEASUREMENTS

13.2.1 Soil Resistivity

The electrical resistivity or specific resistance of a soil volume is defined by the unit cross-sectional area and the unit length.

13.2.1.1 Rectangular Volume

Ground resistivity is internationally defined as the resistance of a 1-m cube between two opposite faces consisting of the ground material (see fig. 13.1):

$$\rho = r \times \frac{A}{l}$$

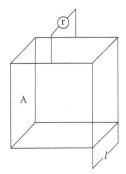

FIGURE 13.1 Soil resistivity for rectangular volume

where
ρ = resistivity, ohm meters
r = resistance, ohms
A = area of cross section, m²
l = length, m

13.2.1.2 Soil Characteristics

Soil characteristics are determined by:

- Effect of salt
- Effect of moisture
- Effect of temperature

Figures 13.2 and 13.3 illustrate the influence of these parameters on ground (earth) resistivity. In the following discussion, the term "earth" denotes the material of the ground.

13.2.2 SOIL MODEL AND ITS IMPACT ON GRID DESIGN

The most accurate representation of a grounding system should be based on the actual variations of earth resistivity present at the substation site. However, it will rarely be economically justifiable or technically feasible to model all variations.

13.2.2.1 Uniform Earth

A soil can be considered uniform if the difference between two extreme values of apparent resistivity is less than 30%.

13.2.2.2 Two-Layer Earth

In most cases, the representation of a ground grid based on an equivalent two-layer earth model (fig. 13.4) is sufficient for designing a safe grounding system. The abrupt changes in resistivity at the boundaries of each layer can be described by means of a reflection factor, k:

$$k = \frac{\rho_2 - \rho_1}{\rho_2 + \rho_1}$$

Substation Grounding Design Considerations

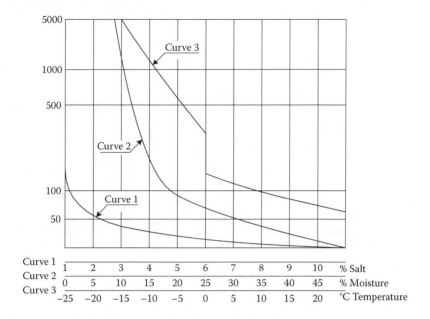

FIGURE 13.2 Effects of temperature, moisture, and salt

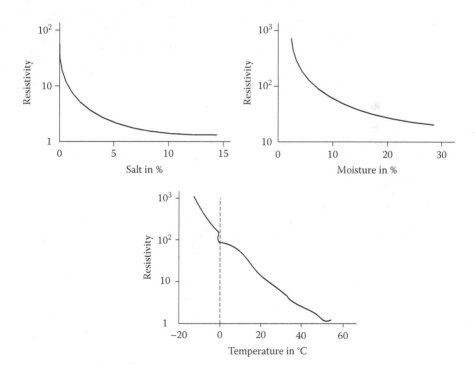

FIGURE 13.3 Earth-resistivity variations

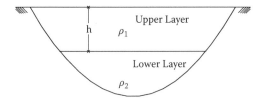

FIGURE 13.4 Two-layer soil

The ground rods have paramount influence on the overall performance of the grounding system, as they penetrate through the upper layer into the lower layer.

13.2.3 Measurement Techniques

The most commonly used measurement method (fig. 13.5) is the Wenner or four-pin method

$$\rho_a = \frac{4 \times \pi \times a \times r}{1 + \left(\dfrac{2 \times a}{\sqrt{a^2 + (4 \times b^2)}}\right) - \left(\dfrac{a}{\sqrt{a^2 + b^2}}\right)}$$

In practice, four rods (pins) are usually placed in a straight line at an interval «a», driven to a depth not to exceed $0.1 \times a$, then $b \approx 0$ can be assumed.

$$\rho_a = 2 \times \pi \times a \times r$$

is the average (approximate) resistivity of the earth at depth «a».

Instrumentation required to measure soil resistivity include:
- Power supply with ammeter and high-impedance voltmeter
- Ratio ohmmeter
- Double-balance bridge
- Single-balance transformer
- Induced-polarization receiver and transmitter

A commercially available instrument, the "Earth Megger," incorporates these instruments.

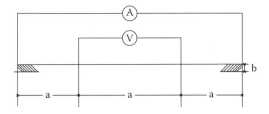

FIGURE 13.5 Measurement techniques

13.3 PERMISSIBLE POTENTIAL DIFFERENCE

The potentials and potential differences that determine the safe design of the grounding grid are illustrated in fig. 13.6.

13.3.1 STEP VOLTAGE

The step voltage (E_{step}) is the difference in surface potential experienced by a person bridging a distance of 1 m between the feet without contacting any other grounded objects. The current path defining the step voltage is illustrated in fig. 13.7.

The step-voltage criterion is defined as follows [S1]:

$$E_{step50} = \left[1000 + 6C_s \times \rho_s\right] \times \frac{0.116}{\sqrt{t_s}}$$

$$E_{step70} = \left[1000 + 6C_s \times \rho_s\right] \times \frac{0.157}{\sqrt{t_s}}$$

where

C_s = reduction factor, a function of reflection factor (k) and the thickness of the crushed rock layer
ρ_s = resistivity of crushed rock
t_s = fault-clearing time, s

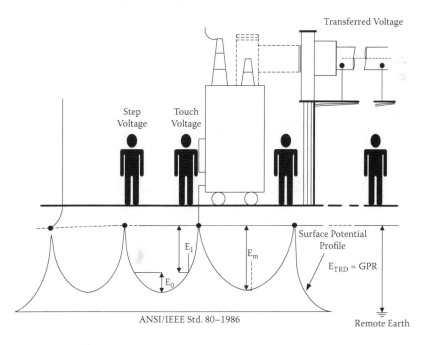

FIGURE 13.6 Basic shock situations

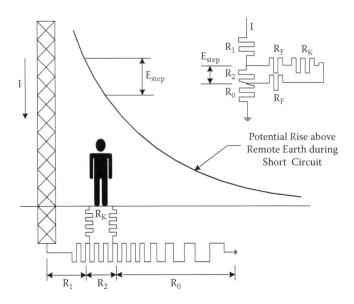

FIGURE 13.7 Step voltage at a grounded structure

Reflection factor $k = \dfrac{\rho - \rho_S}{\rho + \rho_S}$

The actual step voltage (E_S) should be less than the maximum allowable step voltage (E_{step}).

13.3.2 Touch Voltage

Touch voltage (E_{touch}) is the potential difference between the ground-potential rise (GPR) and the surface potential at the point where a person is standing while having a contact with a grounded structure. The current path involving the touch voltage is shown in fig. 13.8. The allowable touch voltage for people weighing 50 or 70 kg, respectively, is defined as follows [S1]:

$$E_{touch\ 50} = \left[1000 + 1.5\ C_S \times \rho_S\right] \times \dfrac{0.116}{\sqrt{t_s}}$$

$$E_{touch\ 70} = \left[1000 + 1.5\ C_S \times \rho_S\right] \times \dfrac{0.157}{\sqrt{t_s}}$$

The actual touch voltage, mesh voltage, or transferred voltage should be less than the maximum allowable touch voltage, E_{touch}, to ensure safety.

Substation Grounding Design Considerations

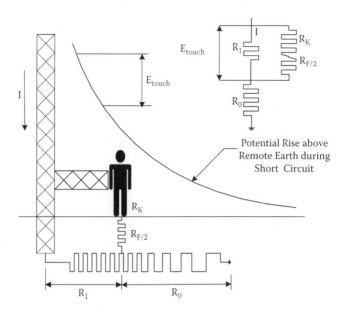

FIGURE 13.8 Touch voltage at a grounded structure

13.3.3 Mesh Voltage

Mesh voltage (E_m) is the maximum surface voltage potential difference between a grid conductor and a point between two grid conductors. It could also be the maximum touch voltage found within the ground grid.

13.3.4 Transferred Voltage

A serious hazard may result during a fault from the transfer of potentials (E_{trans}) between the ground-grid area and outside points by conductors such as communication and signal circuits, low-voltage neutral wires, conduit, pipes, rails, fences, etc. The danger is usually from contacts of the touch type. Induced voltages on unshielded communication circuits, static wires, pipes, etc., may result in transferred potentials exceeding the ground-potential rise (GPR) of both the faulted substation and the source substation. An investigation into possible transferred potential hazards is essential in the design of a safe substation grounding network.

There is a special case of the touch voltage where the voltage is transferred into or out of the substation. Typically, the case of transferred voltage occurs when a person standing within the station area touches a conductor grounded at a remote point or a person standing at a remote point touches a conductor connected to the station grounding grid.

13.3.5 Typical Shock Situation for Gas-Insulated Switchgear

- A fault external to the gas-insulated switchgear (GIS) in which a fault current flows through the GIS bus and induces currents in the enclosures

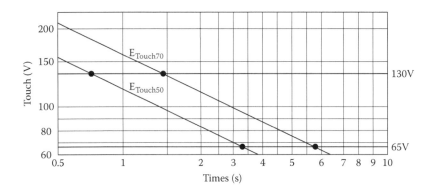

FIGURE 13.9 Touch-voltage limits for metal-to-metal contact and a typical range of enclosure voltages to ground

- An internal fault within the GIS bus system, such as a flashover between the bus conductor and the inner wall of the enclosure

The land area of a GIS is 10–25% of the land area of a conventional substation; it is therefore difficult to obtain the same low grounding resistance as for a conventional substation.

- High-frequency and very-short-duration voltage transients are generated in GIS due to breakdown of gas in short distances and operation of switching devices.
- Enclosures carry induced current similar to the magnitude of the fault current.

The shielding effectiveness of enclosures is determined by the impedance that governs the circulation of the induced currents. With separate enclosures for each phase, the magnitude of circulating currents is influenced by the size and spacing of the enclosures. With the continuous bonded enclosures, different loops of circulating currents are generated under fault conditions, and the metal-to-metal touch voltages are difficult to calculate and control (see fig. 13.9).

13.4 MAXIMUM GROUND CURRENT

13.4.1 Worst-Case Symmetrical Fault Current (I_f or $3I_0$)

- Single line-to-ground fault

$$3I_0 = \frac{3 \times E}{3R_f + R_1 + R_2 + R_0 + j(X_1 + X_2 + X_0)}$$

where:
1. E = phase-to-neutral voltage
2. R_f = fault resistance (usually $R_f = 0$)
3. R_1, R_2, R_0 and X_1, X_2, X_0 = system sequence impedance

Substation Grounding Design Considerations

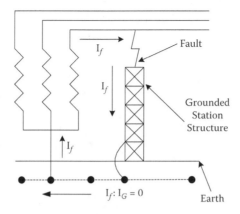

FIGURE 13.10 Fault within local station grounded at local neutral

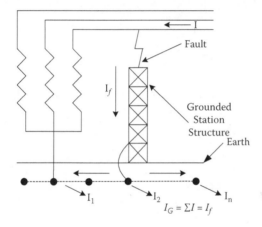

FIGURE 13.11 Fault within local station grounded at remote location

- Line-to-line-ground fault

$$3I_0 = \frac{3 \times E \times (R_2 + jX_2)}{(R_1 + jX_1) \times [3R_f + R_2 + R_0 + j(X_2 + X_0)] + (R_2 + jX_2) \times (3R_f + R_0 + jX_0)}$$

Compute the maximum fault current or obtain the data from the power company. Fault data must include the breakdown of source contributions, such as local, remote, etc.

13.4.2 Symmetrical Grid Current

The current that causes the potential rise of the grounding grid with respect to the remote ground is not the total fault current, but only the portion of the fault current that passes into the earth through the grid. Part of the fault current, depending on the overall grounding configuration, will pass return to the system neutral grounding points directly through the transmission line phase conductors and shield wires (see fig. 13.10, 13.11).

The portion of the symmetrical ground fault current that flows between the grounding grid and the surrounding earth may be expressed as:

$$I_g = S_f \times I_f$$

- I_g = symmetrical grid current
- I_f = rms value of symmetrical ground-fault current
- S_f = current division factor relating the magnitude of fault current and its portion flowing from the grid to the earth

13.4.3 Maximum Grid Current for Design

$$I_G = C_p \times D_f \times I_g$$

- I_G = maximum grid current
- C_p = corrective projection factor for increase of fault current during the station life span for a zero future system growth, $C_p = 1$
- I_g = rms symmetrical grid current, A
- D_f = decrement factor for the entire duration of fault

Figure 13.12 illustrates an example of the fault current division into grid current and the current that returns directly into grounded neutrals for three typical grounding arrangements.

13.4.4 Effect of Station Ground Resistance

In most cases, it is sufficient to derive the ground-fault current by neglecting the system resistance, the station ground resistance, and the resistance at the fault. The error thus introduced is usually small and offers some factor of safety. However, for unusual cases, where the predicted station ground resistance is large compared with system reactance, it is worthwhile to take the resistance into account by including it into the fault calculations.

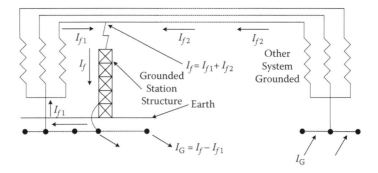

FIGURE 13.12 Fault in station, system grounded at local station and also at other points (From ANSI/IEEE 80-1986, IEEE Guide for Safety in AC Substation Grounding, 1986. With permission.)

Substation Grounding Design Considerations

13.4.5 Effect of Overhead Ground Wires and Neutral Conductors

Where transmission line overhead ground wires or neutral conductors are connected to the station ground, they divert a substantial portion of the ground current away from the station ground grid. Where this situation exists, the overhead ground wire or neutral conductors can be taken into consideration in the design of the ground grid.

It should be realized that connecting the station ground to the overhead ground wire or neutral conductors or both, and through them to the transmission line towers, will usually have the overall effect of increasing the hazard at the tower base while lessening it at the substation, but the probability of exposure to the public is remote.

13.4.6 Effect of Direct Buried Pipes and Cables

Buried cables with their sheaths or armors in effective contact with the ground, and buried metallic pipes, will have a somewhat similar effect when they are bonded to the station ground system but extend beyond its perimeter.

13.4.7 Computation of Current Division Factor

The following methods can be used to compute the current division factor (S_f) and the maximum grid current:

- *Computer programs*: Several programs are available that model the alternative ground-current paths, grounding system, etc., in determining the maximum grid current. Three programs — SMPECC, PATHS, and EMTP — are described in [S4].
- *Graphical analysis*: This method is well covered in [S4], which could be used for estimating S_f. The author and his associates have developed a graphical method for determining the maximum grid current, based on results obtained using the SMPECC program. This method attempts to correlate the substation zero-sequence fault current obtained from a standard short-circuit study to the actual current flowing between the grounding systems and surrounding earth.
- *Estimate reduction factor and flow of current along overhead lines*: Grid current distribution and reduction factor can be estimated if the characteristic impedance of overhead lines is known. Characteristic impedance is given by

$$Z = \frac{Z_C}{S \times 2} + \sqrt{\frac{Z_C \times R_2}{S}}$$

where
Z_C = ground-wire impedance
S = number of ground connections per kilometer
R_2 = footing resistance

$$S_f = \frac{Z}{Z + R_g}$$

S_f = reduction factor,
R_g = grid resistance

The following example illustrates the computation of:

1. Reduction factor (S_f) and symmetrical grid current (I_g):
   - R_g = 0.7 ohm (grid resistance)
   - Max. $3I_0$ = 12.0 kA (200% of present $3I_0$)
   - I_{G1} = 2.0 kA (returning to local transformer)
   - I_g = $3I_0 - I_{G1}$ = 10.0 kA
2. Data on transmission (230 kV) and distribution lines, both having ground wires intact:
   Number of lines entering substation = 4 for transmission = 8 for distribution
   - Z_C = 4 ohm/km for transmission
   -     = 8 ohm/km for distribution
   - S = 3/km for transmission
   -     = 5/km for distribution
   - R_2 = 15 ohm for transmission
   -     = 50 ohm for distribution
3. For 230 kV transmission line:

$$Z_{TL} = \frac{\left[\frac{4}{3\times 2} + \sqrt{4 \times \frac{15}{3}}\right]}{4} = \frac{[0.67 + 4.47]}{4} = 1.28 \text{ ohm}$$

4. For distribution lines, consider only six lines intact:

$$Z_{DF} = \frac{\left[\frac{8}{5\times 2} + \sqrt{8 \times \frac{50}{5}}\right]}{6} = \frac{[0.8 + 8.94]}{6} = 1.62 \text{ ohm}$$

5. Combined impedance ($Z_{TL} + Z_{DF}$) at R_g = 0.7 ohm:

$$Z_{LI} = \frac{1.28 \times 1.62}{1.28 + 1.62} = 0.715 \text{ ohm}$$

$$S_f = \frac{0.715}{0.715 + 0.70} = 0.505$$

13.4.8 Grid Current Returning through the Earth

The grid current returning through the earth (I_G) (rms asymmetrical) is

$$I_G = S_f \times I_g = 5.05 \text{ kA}$$

Note that the division of current will reduce GPR to about 50%.

13.4.9 Decrement Factor

The decrement factor (D_f) is used to determine the effective current during a given time interval after inception of a fault.

$$I_f \times S_f \times D_f$$

This formula gives the equivalent rms asymmetrical fault current flowing between the grounding system and surrounding earth, taking into account the DC component of the initial fault current. Typical values of D_f are given in table 13.1.

TABLE 13.1
Values of D_f

Fault Duration t_f		
s	cycle (60 Hz)	D_f
0.008	0.5	1.65
0.1	6	1.25
0.25	15	1.10
0.5	30	1.0

13.5 SELECTION OF CONDUCTORS AND JOINTS

13.5.1 Basic Requirement

The corrosion resistance retards deterioration in the surrounding environment. The electrical conductivity is such that the element will not contribute substantially to local potential differences in the grounding system. The current-carrying capability is sufficient to withstand the thermal and mechanical stresses during the most adverse conditions of fault-current magnitude and duration. The mechanical strength and ruggedness is sufficient to withstand electromagnetic forces and physical abuse.

13.5.2 Data Required to Design

The magnitude of fault current is the maximum rms single line-to-ground (SLG) fault current, including an allowance for future growth. Use total current for equipment-grounding conductor and 50% current for sizing grid conductor. The duration of fault current use is 3 s for small substations with no redundant relaying and 1 s for large substations with high-speed and redundant relaying.

The material of the conductor and the grounding electrode grid conductors is usually stranded cable made of copper or tinned copper. The most common conductor is hard- or medium-hard-drawn copper wire with a material conductivity of 97%.

The grounding electrodes or rods are usually copper-clad steel, galvanized steel, stainless steel, or concrete-encased electrodes.

13.5.2.1 Type of Connector

The connector must meet the same criteria as the conductor plus the following:

- Large enough to absorb the $I^2 \times R \times t_s$ heating and resist fusing
- Strong enough to withstand the mechanical forces caused by the high current
- Meet the test requirements

13.5.2.2 Maximum Allowable Temperature

- For connectors not qualified per IEEE 837, a 250°C limit is recommended
- 250–350°C for pressure-type connectors
- 450°C for brazed-type connectors
- For exothermic-welded type such as Cadweld, approximately the same fusing temperature as the conductor
- For conductor, 1084°C for copper wire

13.5.3 MINIMUM SIZE (FOR HARD-DRAWN COPPER WIRE)

$$Amm^2 = I \sqrt{\frac{19.7893 \times t_C}{In \times \left[1 + \left(\frac{Tm - Ta}{242 + Ta}\right)\right]}}$$

$$Acmil = 1973.52 \times I \sqrt{\frac{19.7893 \times t_C}{In \times \left[1 + \left(\frac{Tm - Ta}{242 + Ta}\right)\right]}}$$

where
- A = conductor cross-section area
- I = maximum rms current in kA (include for future growth)
- Ta = ambient temperature, °C
- Tm = maximum allowable temperature, °C
- t_C = duration of current flow in s (1 or 3)

The total current flows through the down lead to the grid and then divides into two or more paths at the junction. For sizing the conductor size, use total current for the down lead and 50% current for the grid conductor. The most common grid conductor size is 4/0 AWG copper. More data is given in table 13.2.

13.6 DESIGN OF GROUNDING SYSTEM

For a given area to be grounded:

- Ground resistance, R_g:

Substation Grounding Design Considerations

TABLE 13.2
Data Value

Fault Time (s)	100% Cu Only	97% Cu Only	40% CCS Only	30% CCS Only	97% Cu/Temp. Limits 450°C	97% Cu/Temp. Limits 250°C
30.0	38.4	38.7	75.0	65.8	51.1	64.5
4.0	14.0	14.2	20.8	24.0	18.7	23.5
1.0	7.0	7.1	10.4	12.0	9.3	11.8
0.5	4.9	5.0	7.4	8.5	6.6	8.3

1. Decreases when the number of meshes increases (decrease becomes negligible for a large number of meshes).
2. Gradual decrease with burial depth.
3. Increasing the number or depth of ground rods effectively reduces the resistance until saturation occurs.

- The worst touch and step voltages occur in the outer meshes. Reducing mesh spacing toward the perimeter reduces these voltages.
- Increasing the concentration of deep ground rods around the perimeter will also effectively reduce the excessive step and touch voltages in the outer area.
- For a given length of grid conductor or ground rod, the ground rod discharges much more current into the earth than does the grid conductor.
- Because the current in the ground rod is discharged mainly in the lower portion, the touch and step voltages are reduced significantly compared with that of the grid alone.

13.6.1 EVALUATION OF GROUND RESISTANCE

$$R_g = \rho \times \left[\frac{1}{L} + \frac{1}{20 \times \sqrt{A}} \times \left(1 + \frac{1}{1 + h \times \sqrt{\frac{20}{A}}} \right) \right]$$

where

R_g = station ground resistance, ohm
ρ = average soil resistivity, ohm-meter
A = area occupied by the ground grid, m²
h = depth of the grid, m

For better estimates of the ground resistance of the grids with ground rods, equations such as the Schwarz formula, given below, should be used.

$$R_g = \frac{(R_1 \times R_2) - R_{12}}{R_1 + R_2 - (2 \times R_{12})}$$

where

R_1 = resistance of grid conductors
R_2 = resistance of all ground rods
R_{12} = mutual resistance between the grid conductors and group of ground rods

Schwarz developed a set of convenient formulas, defining R_1, R_2, and R_{12} in terms of basic design parameters, assuming uniform soil. However, in practice, it is often desirable to drive ground rods deep into the ground to reach more conductive soil.

The following site-dependent critical parameters have been found to have substantial impact on the grid design:

- Maximum grid current (I_G)
- Fault duration (t_f) and shock duration (t_s)
- Soil resistivity (ρ)
- Resistivity of surface layer (ρ_s)
- Grid geometry

13.6.2 Design Procedure

Refer to fig. 13.13 (design-procedure block diagram) and the index of design parameters in tables 13.3 and 13.4. These include

- Obtaining utility (power company) standards and guidelines
- Determining substation area (A) and computing soil resistivity from field data
- Determining uniform soil or two-layer model
- Determining maximum expected future fault current ($3I_0$) and maximum possible clearing times (including backup)
- Determining the conductor size for different applications, such as grid, down leads, and equipment-grounding conductors
- Determining tolerable touch and step voltages
- Carrying out initial design (the initial estimates of conductor spacing and ground rod locations should be based on the current I_G and the area being grounded)
- Estimating the grid resistance or resistance of the grounding system (for the final design, more accurate estimates of the resistance may be desired, especially when ground rods are used)
- Determining maximum grid current (I_G) that flows between the ground grid and surrounding earth and calculating the ground-potential rise (GPR)
- Computing the mesh (E_m) and step (E_{step}) voltages for the grid (revise the preliminary design if the computed mesh voltage is greater than the tolerable touch voltage)
- Revising the design if both the computed touch and step voltages exceed the tolerable voltages (these revisions may include smaller conductor spacing and additional ground rods)
- Adding ground rods that may be required at the base of surge arrestors, transformer neutrals, etc.

Substation Grounding Design Considerations

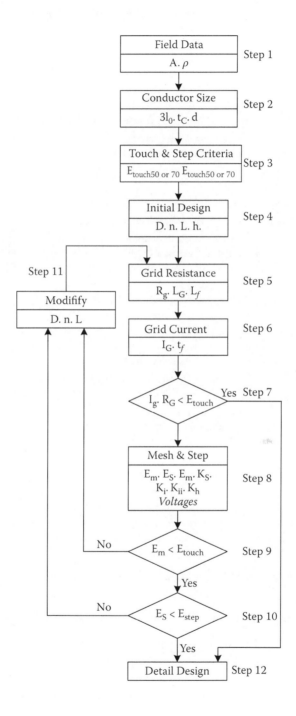

FIGURE 13.13 Design-procedure block diagram

TABLE 13.3
Steps in Substation Grounding Design

Step No.	Description
Step 1	Determine location, carry out soil-resistivity tests; from the test results, determine top-layer resistivity and average resistivity
Step 2	Determine size of substation grid covering the area of the grid
Step 3	Calculate the maximum initial symmetrical value of a SLG fault current at the substation location, considering an SLG fault at all voltage levels in the substation
Step 4	Assume a conductor spacing for the buried conductors and calculate the total length of buried conductors
Step 5	Using the simplified (Laurent) formula, calculate the grid resistance
Step 6	Using the maximum fault current from step 3, calculate the GPR
Step 7	If the GPR exceeds 5 kV (or the utility practice), consider means of lowering the effective value of grid resistance by: Considering the effect of shield wires on lines connected to the grid Improving the physical characteristics of the grid
Step 8	Recalculate GPR and iterate between steps 7 and 8 until the value of the GPR becomes acceptable
Step 9	Determine the acceptable value of touch (mesh) and step voltages based on the time of fault duration for the purpose of grid design; they could be the larger of the backup fault clearing time or 1.0 second (suggested by the author)
Step 10	Calculate the values of acceptable touch and step voltages
Step 11	Calculate the mesh (touch) voltage and step voltage for the grid and compare with the acceptable values from step 10; iterate between steps 10 and 11 if grid mesh voltage is higher than the acceptable touch voltage or if the grid step voltage is higher than acceptable values; note that the increased length of grid conductors will also reduce the GPR
Step 12	Decide on the safety treatment of fences and the spacing of grading wires inside and outside the fence
Step 13	Determine required cross section of grid conductors
Step 14	Examine any potential problem of transferred potentials by communication wires, pipes, rails, etc., and determine protective measures if required

13.6.3 MAXIMUM MESH AND STEP VOLTAGE

- Mesh voltage (E_m)
  For grids with no ground rods or only a few rods away from the perimeter:

$$E_m = \frac{\rho \times I_G \times K_m \times K_i}{L_C + L_r}$$

For grids with ground rods predominantly around the perimeter:

$$E_m = \frac{\rho \times I_G \times K_m \times K_i}{L_C + 1.15\, L_r}$$

TABLE 13.4
Index of Design Parameters

Symbol	Description
$3I_0$	Symmetrical fault current in substation for conductor sizing, A
I_G	Maximum grid current that flows between ground grid and surrounding earth, A
ρ	Soil resistivity, Ω–m
ρ_S	Surface-layer resistivity, Ω–m
h_S	Surface-layer thickness
C_P	Current projection factor for future system growth
C_S	Surface-layer resistivity derating factor
t_C	Duration of fault current for sizing ground conductor, s
t_f	Duration of fault current for determining decrement factor, s
t_S	Duration of shock for determining allowable body current, s
h	Depth of ground grid conductors, m
d	Diameter of conductor, m
A	Total area enclosed by ground grid, m²
D	Spacing between parallel conductors, m
D_f	Decrement factor for determining I_G
n	Number of parallel conductors in one direction
K_m	Spacing factor for mesh voltage, simplified method
K_S	Spacing factor for step voltage, simplified method
K_i	Correction factor for grid geometry, simplified method
K_{ii}	Correction weighting factor that adjusts the effects of inner conductors of the corner mesh, simplified method
K_h	Correction weighting factor that emphasizes the effects of grid depth, simplified method
L	Total length of grounding system conductor, including grid and ground rods, m
R_g	Resistance of ground system, Ω
E_m	Mesh voltage at the center of the corner mesh for simplified method, V
E_s	Step voltage between a point above the outer corner of the grid and a point 1 m diagonally outside the grid for simplified method, V
$E_{touch50}$	Tolerance touch voltage for human with 50 kg body weight, V
$E_{touch70}$	Tolerance touch voltage for human with 70 kg body weight, V
E_{step50}	Tolerable step voltage for human with 50 kg body weight, V
E_{step70}	Tolerable step voltage for human with 70 kg body weight, V

where

K_m = spacing factor for mesh voltage
K_i = correction factor for grid geometry = $0.656 + 0.172\,n$
L_C = total length of the grid conductor
L_r = total length of the ground rod

n = number of parallel conductors in both direction = $\sqrt{n_A \times n_B}$

- Step voltage (E_s)
  For grids with no ground rods or only a few rods away from the perimeter:

$$E_s = \frac{\rho \times I_G \times K_s \times K_i}{L_C + L_r}$$

  For grids with ground rods predominantly around the perimeter:

$$E_s = \frac{\rho \times I_G \times K_s \times K_i}{L_C + 1.15 L_r}$$

  where
  K_s = spacing factor for step voltage
- Limitations of simplified equations for E_m and E_s:
  When using the equations for computing E_m and E_s, the following limits are recommended for square grids or rectangular grids having the same number of conductors in both directions:
    1. $n \leq 25$
    2. $0.25 \text{ m} \leq h \leq 2.5 \text{ m}$
    3. $d < 0.25 \text{ h}$
    4. $D > 2.5$
- Estimate of minimum buried conductor length:
  Minimum length of buried grid conductors to keep the maximum E_{touch} within the grounded area below the safe limit can be computed using equations 77 and 78 in [S1]
- Computer analysis in grid design:
  The use of a more accurate computer algorithm in designing the grounding system may be justified for the following reasons:
    1. One or more of the geometric parameters exceed the limits.
    2. A two-layer soil model is required due to significant variations in soil resistivity.
    3. An unsymmetrical grid makes it impractical to predetermine the location of the worst touch voltage.
    4. Uneven grid conductor or ground rod spacing cannot be analyzed using the approximate methods.

13.7 TREATMENT OF SUBSTATION FENCE

The station fence grounding is of major importance because the outside of the fence is usually accessible to the general public, and the voltage involved is the more dangerous touch voltage. Two different situations may be found in the existing practice for fence grounding:

- Inclusion of the fence within the main ground grid area; this situation is found only when the station fence is located within 3–5 m of grounded station equipment, or when a railway siding is inside the fenced area.

Substation Grounding Design Considerations

- Separation of the fence-grounding system from the main grid and connected to its own ground electrodes.

13.7.1 FENCE-GROUNDING SYSTEM DESIGN

13.7.1.1 Station Fence Connected to the Main Grid and also to Its Own Ground Electrodes

Inclusion of the fence within the ground grid area increases the size of the area and thereby reduces, often substantially, the ground grid resistance and, hence, the maximum ground grid potential rise ($E = I \times R$) as well. To make the grounding installation as effective as possible, particularly during winter months, ground rods equally spaced, not more than 12 m apart, should be installed along the fence and connected to the fence ground electrode. By connecting the fence and its grounding system to the main station ground grid, the maximum touch voltage within the station will be reduced. However, the touch voltage to the public outside the fence will be appreciably greater and is given by

$$E_t = K_i \times \frac{\rho_1}{\pi} \times \frac{I}{L} \ln \left[\frac{\sqrt{h^2 + \frac{x^2}{2 \times h \times a}} \times \left(1 + \frac{x}{D'}\right) \times \left(1 + \frac{x}{D'+D}\right) \times \left(1 + \frac{x}{D'+2 \times D}\right) \cdots}{\left(1 + \frac{x}{D'+(n-1) \times D}\right)} \right]$$

where

I = ultimate ground current, kA
h = depth of buried conductor, m
L = total length of buried conductor, m
x = distance between the fence and point of contact with ground, 1 m
D = spacing of parallel conductors, m
D' = uniform distance between ground grid and fence, m
ρ_1 = upper-layer soil resistivity, Ωm
a = radius of ground rod or grid conductor, m

The touch voltage has to be checked against the maximum tolerable touch voltage as calculated before.

13.7.1.2 Station Fence Is Not Connected to the Station Ground Grid but Has Its Own Grounding System

Isolating the station fence from the main grid (aside from sacrificing improvement of station grid resistance) introduces the possibility of inadvertent electric connection between the grid and the fence areas. Gate-mounted telephones, telephone signal lines, distribution circuits from the station to a gate house, water pipes, rails, etc., could transfer main grid potentials and introduce dangerous local potential differences during faults; remedial protection measures should be considered. If the fence itself is not tightly coupled to the nearby ground by its own adequate ground system, any such inadvertent connections to the main grid could create a hazard along the entire length of the fence under fault conditions. The station fence, therefore, should have its own grounding system.

13.7.2 Station Fence-Grounding Details

Whether the fence is connected to the main grid or not, it should be protected against excessive tough voltage by grading wires, consisting typically of a copper cable buried 150 to 200 mm deep, approximately 1 m (touch distance) outside the fence line and connected to the fence at approximately 12 m intervals by means of copper taps. The grading wire inside the fence is only required if the distance between the fence and the nearest grid conductor is more than approximately 3.0 m.

The size of the copper cable for the ground bus, taps, and jumpers should be 67 mm² (2/0 AWG). If installation of the station fence-ground bus outside the fence is not feasible, the bus should be installed along the fence line. To minimize the touch potential at station gates, a buried grid of 1 m mesh size should be installed encompassing the gate-swing area and connected to the fence-grounding system. The fence's metallic station should not approach closer than 2 m to any metallic component of a privately owned fence.

13.8 CASE STUDY (GROUND GRID DESIGN)

Carry out a preliminary design for a safe grounding system of a 230/44 kV substation, as outlined in fig. 13.14.

Maximum L_G ($3I_0$) current:
- 230 kV = 40 kA for equipment ground
  = 12 kA for grid design
- 44 kV = 8 kA for equipment ground
- 230 kV = single-circuit steel with sky wire
- 44 kV = wooden poles with ground wire
- Substation area = 110 m × 70 m
- Soil – ρ = 130 Ω·m, assuming a uniform soil

13.8.1 Design Parameters

1. Soil resistivity:
   ρ = 130 Ω·m in uniform soil
2. Crushed rock:
   layer thickness = 80 mm (3 in.)
   ρ_s = 3000 Ω·m
3. Maximum ground-fault current ($3I_0$):

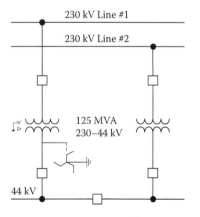

FIGURE 13.14 Case study for a 230/44 kV substation

Substation Grounding Design Considerations

230 kV
  40 kA for equipment ground
  12 kA for grid design
  I_G returning to local transformer = 3.5 kA
4. Ground current distribution:
  I_G returning to remote station = 12 − 3.5 kA
  = 8.5 kA
5. Flow of ground current along transmission and distribution lines:
  Number of lines leaving SS = three for transmission
  = six for distribution
  Ground wire impedance, Z_C = 4 Ω/km for transmission
  = 8 Ω/km for distribution
  Number of ground connections/km, S = 3/km for transmission
  = 5/km for distribution
  Footing resistance, R_2 = 15 Ω for transmission
  = 50 Ω for distribution

$$Z_{TL} = \frac{\left[\frac{4}{3\times 2} + \sqrt{\frac{4\times 15}{3}}\right]}{3} = 1.713\ \Omega$$

$$Z_{DF} = \frac{\left[\frac{8}{5\times 2} + \sqrt{\frac{8\times 50}{5}}\right]}{6} = 1.62\ \Omega$$

Combined impedance of Z_{TL} and Z_{DF}

$$Z_{LI} = \frac{1.713 \times 1.62}{1.713 + 1.62} = 0.8326\ \Omega$$

This impedance will be in parallel with the ground grid resistance
  Fault-clearing time (t_C) = 0.20 s for 230 kV
  = 1.00 s for 44 kV
  Decrement factor (D_f) = 1.1 for 230 kV
  = 1.0 for 44 kV
Allowable potentials

$$E_{step(50)} = \left(1000 + 6\times C_S \times \rho_S\right) \times \frac{0.116}{\sqrt{t}}$$

$$K_n = \frac{\rho - \rho_S}{\rho + \rho_S} = \frac{130 - 3000}{130 + 3000} = -0.9169$$

where C_S = 0.5 and h_S = 0.08

$$E_{step(50)} = (1000 + 6 \times 0.5 \times 3000) \times \frac{0.116}{\sqrt{0.2}} = 2.594 kV$$

$$E_{touch(50)} = (1000 + 1.5 \times 0.5 \times 3000) \times \frac{0.116}{\sqrt{0.2}} = 0.843 kV$$

13.8.2 Grid Design

1. Conductor selection:
   Conductor 97% copper
   $Tm = 450°C$ for Cadweld joints
   $t_C = 1$ s
   For calculation
   $Ta = 30°C$
   $Tm = 450°C$
   Conductor size, cmils = 9.3/A
   For equipment grounding, $A_C = 9.3 \times 40$ kcmil
   $= 372$ kcmil
   $= 400$ kcmil
   For grid conductor $A_{C1} = 9.3 \times 20$ kcmil (current divides into two paths)
   $= 186$ kcmil
   Select 4/0 AWG (107 mm²)

2. Grid data:
   Area $A = 110$ m × 70 m
   Mesh spacing $D = 5$ m (assumed)
   Conductor diameter $d = 0.013$ m
   Depth $h = 0.5$ m
   Parallel n = $\left(\frac{110}{5} + 1\right) \times \left(\frac{70}{5} + 1\right) = 23 \times 15$

   Length $L_C = (23 \times 110) + (15 \times 70) = 3580$ m
   Rod size = 19 mm × 3 m
   Number of rods 80 (assumed) $L_r = 80 \times 3.0 = 240$ m

3. Grid or ground resistance:

$$R_g = \rho \times \left[\frac{1}{L_C + 1.15 \times L_r} + \frac{1}{\sqrt{20 \times A}} \times \left(1 + \frac{1}{1 + h \times \sqrt{20 \times A}}\right)\right]$$

$$R_g = 130 \times \left[\frac{1}{3580 + 276} + \frac{1}{\sqrt{20 \times 110 \times 70}} \times \left(1 + \frac{1}{1 + 0.5 \times \sqrt{20 \times 110 \times 70}}\right)\right] = 0.37\ \Omega$$

4. Reduction factor, S_f:
   $R_g = 0.37\ \Omega$
   $R_{L1} = 0.8326\ \Omega$

Substation Grounding Design Considerations

$$S_f = \frac{R_{L1}}{R_{L1} + R_g} = \frac{0.8326}{0.8326 + 0.37} = 0.6923 \, \Omega$$

5. Maximum grid current I_G through earth:
   $D_f = 1.1$
   $S_f = 0.6923 \, \Omega$
   $I_g = 8.5 \, kA$
   $I_G = S_F \times I_g = 6.923 \times 8.5 = 6.473 \, kA$

6. Ground-potential rise:
   $GPR = I_G \times R_g = 6.473 \times 0.37 = 2.395 \, kV < 3.0 \, kV$

7. Mesh and step voltages:

   Mesh voltage $\quad E_m = \dfrac{\rho \times I_G \times K_m \times K_i}{L_C + (1.15 \times Lr)}$

   $$K_m = \frac{1}{2 \times \pi} \times \left[\ln \times \left[\frac{D^2}{16 \times h \times d} + \frac{(D+2h)}{8 \times D \times d} - \frac{h}{4d} \right] + \frac{K_{ii}}{K_n} \times \ln \times \frac{8}{\pi \times (2n-1)} \right]$$

   where
   $\rho = 130 \, \Omega \cdot m$
   $h = 0.5 \, m$
   $D = 5 \, m$
   $d = 0.013 \, m$
   $K_{ii} = 1.0$
   $Kh = \sqrt{1 + \dfrac{h}{h_0}} = \sqrt{1 + \dfrac{0.5}{1.0}} = 1.225$

   $n = \sqrt{23 x 15} = 18.57$
   $L_C = 3400 \, m$
   $L_r = 240$
   $K_i = 0.656 + 0.172n = 3.85$
   $I_G = 6.473 \, kA$
   $K_m = 0.528$
   $E_m = 443.62 \, V < E_{touch(50)} \, 843 \, V$

   Step voltage, $E_{step(50)} = \quad E_S = \dfrac{\rho \times I_G \times K_S \times K_i}{L_C + (1.15 \times Lr)}$

   $$K_S = \frac{1}{\pi} \times \left[\frac{1}{2h} + \frac{1}{D+h} + \frac{1}{D} \times (1 - 0.5^{n-2}) \right] = 0.3762$$

   $E_S = 331.55 \, V < E_{step(50)} \, 2594 \, V$

13.9 CONCLUSION

All voltages — GPR (I_G, R_G), E_m, E_t, and E_S — are within the safe limits. You may try with a mesh spacing of 6 m. Proceed with a detailed design. Because the GPR > 1000 V (2.395 kV), protection will be required for the communication lines entering the substation/plant. Advise the phone company and give them the values.

REFERENCES

Standards
S1. ANSI/IEEE 80-1986, IEEE Guide for Safety in AC Substation Grounding, 1986.
S2. IEEE 81-1983, IEEE Guide for Measuring Earth Resistivity, Ground Impedance, and Earth Surface Potentials of a Ground System (Part 1), 1983.
S3. IEEE 81.2-1991, IEEE Guide for Measurement of Impedance and Safety Characteristics of Large, Extended or Interconnected Grounding Systems (Part 2), 1991.
S4. IEEE Tutorial Course 86EH0253-5-PWR, Practical Applications of ANSI/IEEE Std. 80-1986, IEEE Guide for Safety in AC Substation Grounding, 1986.
S5. Ontario Hydro Grounding Guide, 1982 ed.
S6. IEEE 487-1992, IEEE Recommended Practice for Protection of Wire Line Communications Facilities Serving Electric Power Stations, 1992.
S7. ANSI/IEEE C62.92.2-1989, IEEE Guide for the Application of Neutral Grounding in Electrical Utility Systems, Part II: Grounding of Synchronous Generator Systems, 1989.
S8. IEEE 32-1972, Requirement, Terminology and Test Procedure for Neutral Grounding Devices, 1972.

BIBLIOGRAPHY

1. Westinghouse, "System Neutral Grounding and Ground Fault Protection," publication PRSC-4B-1979, Westinghouse, 1979.
2. General Electric Co., "Generator Neutral Grounding," GET-1941, Schenectady, NY: General Electric.

14 Electrical Aspects of Power Generation

14.1 INTRODUCTION AND OVERVIEW

A properly selected generator and appropriate excitation system will allow a ride-through during system disturbances and reduce the number of plant shutdowns. This chapter provides some guidelines in the selection and application of a generator and excitation system and related subjects that are important for the overall system operation.

The selection of a generator in a process plant shall therefore be based on the following basic requirements:

- Producing power and energy required for the process and plant services
- Producing power and energy to reduce power cost
- Maintaining plant operations during disturbances on the utility company's transmission system, thus reducing power outages and plant shutdowns

The prime mover (steam or gas turbine or diesel) is sized to meet the requirements under 1 and 2; however, the generator and excitation system must satisfy all three criteria. The detail provided in fig. 14.1 shows a typical single-line diagram. These include:

- Generator parameters
- Salient features of the excitation system
- Generator and system interconnection
- Plant power system and utility interface
- Protection consideration
- Induction generators
- Station auxiliaries

14.2 GENERATOR RATING AND PARAMETERS

Generator rating (output) is given in MVA, which is the vector sum of its MW and MVAR output. The MW rating is determined by the turbine capability, and the MVAR or power factor rating is selected to meet the load reactive power requirements and voltage support. There is a tendency to purchase the generator with a lower power factor, i.e., larger MVA rating than is necessary, to provide enhanced voltage control. Larger generator frames contribute higher short-circuit currents and introduce higher losses such as windage and core losses associated with the bigger

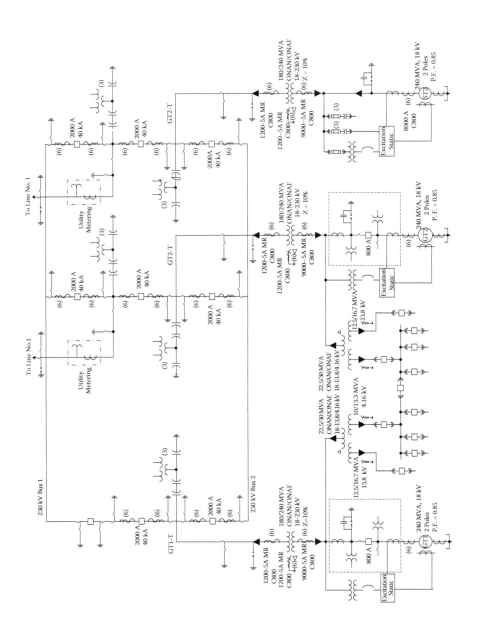

FIGURE 14.1 Typical single-line diagram

Electrical Aspects of Power Generation

physical size. It is better to purchase the generator with an adequate MVAR capability and a high-performance excitation system.

A typical generator reactive capability curve is shown in fig. 14.2 and is based on ANSI C50.12 [S2], which states that a generator shall operate successfully at rated MVA, frequency, and power factor at any voltage not more than 5% above or below rated voltage, but not necessarily in accordance with the standard performance established for operation at rated voltage.

Capability curves for other generator voltages can be obtained from the manufacturer. At lower than rated voltage, the underexcited reactive capability is reduced, whereas the overexcited reactive capability is unaffected for ±5% voltage variation from the rated voltage.

Generator parameters that are of importance for the industrial power system are the short-circuit ratio (SCR), subtransient (X''_d). reactance, and transient (X''_d) reactance. The short-circuit ratio (SCR) is the ratio of the field current at no-load rated voltage to the field current required to produce rated armature current with short-circuited generator terminals. The SCR can be estimated as the inverse of synchronous reactance $(1/X_d)$. A higher value of SCR is an indication of greater stability margin. With the application of a high-speed regulator and exciter, a SCR on the order of 0.5 or higher is considered to be adequate.

Subtransient reactance (X''_d) determines the initial generator current during faults. An approximate value on the order of 12% for 3600 rpm (two-pole) and 17% for 1800 rpm (four-pole) machines can be used. Short-circuit contribution is always a problem

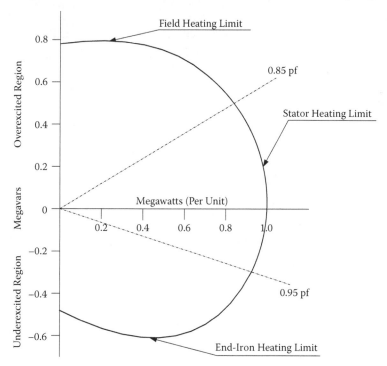

FIGURE 14.2 Generator capability curve

for integrating a generator into an existing power system. A value lower than 15% for two- or four-pole machines is not recommended. Transient reactance (X'_d) is generally in the 25–35% range, and a higher value of (X'_d) compromises on stability.

Another important criterion to be applied to the generator and excitation system is that they must deliver a sustained short-circuit current of 3.0 P.U. (per-unit) for 10 seconds. This feature allows the protective relays to be set to operate within a reasonable time margin during three-phase and phase-to-phase faults.

14.3 EXCITATION SYSTEM

The excitation system consists of an exciter, voltage regulator, and power supply. The system plays an important role in recovery and voltage support during system disturbances. The salient features of the excitation system for a process plant generator shall be:

- High ceiling voltage on the order of 3.0–3.5 P.U. or higher in terms of the rated load field voltage
- Field forcing or field ceiling current of 2.0–2.25 P.U. or higher
- Capability of changing the generator field voltage from the voltage at the rated capacity and voltage to ceiling voltage in less than 150 ms, corresponding to a high-excitation response ratio

The response ratio is defined as the average rate of increase of field voltage during the first 0.5 s divided by the rated field voltage [S8]. This definition is illustrated in fig. 14.3. In general, the higher the response ratio of the exciter, the better is the transient stability of the system. A high-initial-response (HIR) excitation system is

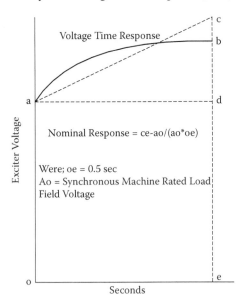

FIGURE 14.3 Excitation system nominal response

Electrical Aspects of Power Generation

capable of attaining 95% of the difference between ceiling voltage and field rated voltage in 100 ms or less. However, for small and medium-sized brushless units used in process plants, a HIR excitation system is costly and difficult to build.

Two types of exciters are available: brushless and static exciters. Brushless exciters are standard with two-pole and four-pole machines rated up to several hundred megawatts.

14.3.1 BRUSHLESS EXCITATION SYSTEM

A typical brushless excitation system is shown in fig. 14.4. The brushless excitation system includes an alternator-rectifier main exciter and a permanent-magnet generator pilot exciter (PMG), both driven directly from the synchronous generator shaft. Field power for the main exciter is supplied from the PMG, which has a set of rotating permanent magnets and a stationary three-phase armature. The high-frequency (420 Hz) output from the PMG is rectified to provide excitation for the stationary armature of the main exciter.

The three-phase output of the rotating armature of the main exciter is fed along the shaft to the rotating rectifier system, whose DC output is conducted to the synchronous generator field. Fuses are connected in series with the diodes to protect the main exciter windings. Redundant diodes must be provided to prevent generator outage upon the loss of a limited number of diodes per phase. Some manufacturers do not use fuses for diode protection.

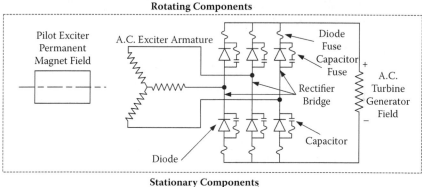

Note: Rotating Components are Mounted on the Same Shaft

FIGURE 14.4 Typical brushless excitation system

A brushless excitation system without PMG offers a faster response to transients. In this system, an AC source, either from the generator terminals or auxiliary power supply, is rectified to provide excitation for the main exciter. A DC source from a station battery must be provided to achieve field forcing when the AC source voltage dips to 80% or lower. This DC source provides power for field flashing when the AC source is taken from the generator terminals.

This system completely eliminates commutators, collector rings, and carbon brushes, and system faults do not affect the source of excitation.

14.3.2 Static Excitation System

The static excitation system is used for large turbogenerators, hydrogenerators, and where fast transient response is required. The system comprises a power transformer, thyristor-controlled rectifier, electronic regulator, and a de-excitation unit. The excitation power is taken from the generator terminals, rectified, and fed to the generator field winding via collector rings. The system can provide a high ceiling voltage (up to 10 PU), extremely short voltage response time, and negative-field forcing. Excitation systems for hydroelectric generators are generally provided with a three-phase full-wave thyristor bridge. This arrangement has the ability to provide a positive or negative voltage to the field and decreases the recovery time during a full-load rejection accompanied by an overspeed condition. The system can provide a higher ceiling voltage than that available with brushless exciters. The design features of a static excitation system are covered in GE publication GET-3007 [5], and the essential elements of the system are shown in fig. 14.5.

14.3.3 Voltage-Regulating System

The function of a voltage regulator is to maintain the output voltage of the generator within a given operating range. With either the brushless or the static excitation

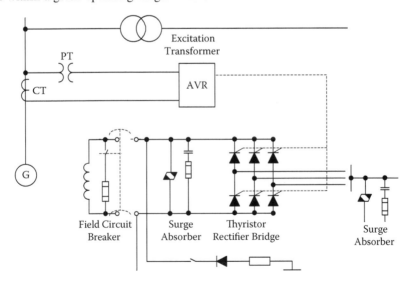

FIGURE 14.5 Generator static exciter

Electrical Aspects of Power Generation 393

system, when a sudden load increase or a system disturbance causes a drop in generator terminal voltage, the regulator senses this change and boosts the exciter field current to a higher level (up to ceiling current) for quick recovery. The voltage-regulating system shall be automatic/manual, high speed, continuously active, static type, complete with the following features:

- Redundant system (two auto/manual systems)
- Three-phase voltage and current-sensing elements responsive to generator output voltage and current, respectively
- Automatic control of the generator terminal voltage to within ±0.50% from no load to full rated load
- DC contactor for positive isolation of the exciter field
- Overexcitation limiter
- Volts per hertz limiter and volts per hertz protection
- Power factor and reactive power MVAR control options in addition to voltage control
- Manual field current regulator and automatic follow-up with the AVR
- Loss-of-potential protection, including transfer of controls from automatic to manual

14.3.4 POWER SYSTEM STABILIZER

Power system stabilizer (PSS) is an element or a group of elements that provides an additional input to the voltage regulator to improve the dynamic performance of the power system. A signal representing the accelerating power is introduced into the automatic voltage regulator algorithm to increase the generator's ability to produce and transmit power in a stable manner by reducing low-frequency rotor oscillations. Power system stabilizers are generally not required for industrial machines; however, this must be checked with the power utility company.

14.4 SYNCHRONIZING

Numerous accidents resulting in machine damage that were associated with human error while synchronizing generators have been reported in a paper by Beckwith [11]. The damage can be cumulative and can be minimized by safeguarding against human error by using check or automatic synchronizing systems. To achieve a successful synchronizing operation, the following parameters must be matched, within a certain tolerance, at the time of breaker closing. Figure 14.6 shows the three cases:

1. Voltage
2. Phase angle
3. Frequency

Excessive phase-angle difference across the synchronizing breaker, just prior to closing, tends to produce a sudden change to the machine torque. A slight frequency difference, coincidental with the phase-angle error, can result in damage to

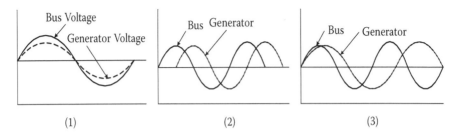

FIGURE 14.6 (1) Voltage difference (generator to bus) (2) Phase difference (3) Frequency difference

the machine. A voltage difference, prior to closing of the breaker, will result in a steady flow of VARs after the breaker is closed. If the generator frequency (speed) and voltage are higher than the system frequency, there could be a substantial sudden change of active and reactive power flow from or into the system.

The synchronizing facility shall include both auto and manual features. Autosynchronizing minimizes the risk factors associated with human error aand shall be used when the machine is brought to speed and paralleled to the system automatically. Manual synchronizing usually becomes a necessity when the generator is supplying isolated system load and then has to be paralleled to the system. Although there are some "system synchronizing" relays available for this duty, they are expensive, and manual matching of frequency and voltage is still required.

Manual closing shall be supervised by a synchronizer or a synchronism check relay that will check phase angle, frequency, and voltage and shall allow closing within a small window. The scheme shall also prevent the operator from holding the breaker control switch in a "close" position and waiting for the synchronism check relay to activate the circuit breaker close signal. The synchronism check relay shall include the following features:

- Adjustable phase-angle settings
- Adjustable slip-frequency settings
- Adjustable voltage settings (undervoltage and differential voltage)
- Adjustable time settings
- Line/bus dead voltage
- Line/bus live voltage

Automatic synchronizers have similar features but also have the ability to raise and lower the governor and exciter set points to match the frequency and voltage automatically. They will also measure the slip speed between the generator and the system and will issue a breaker-close signal in advance, at a suitable time before the generator comes into synchronism, computed from the slip speed and the preset breaker closing time, so that synchronizing occurs with zero slip. A typical synchronizing scheme with manual/auto features is shown in fig. 14.7.

Electrical Aspects of Power Generation

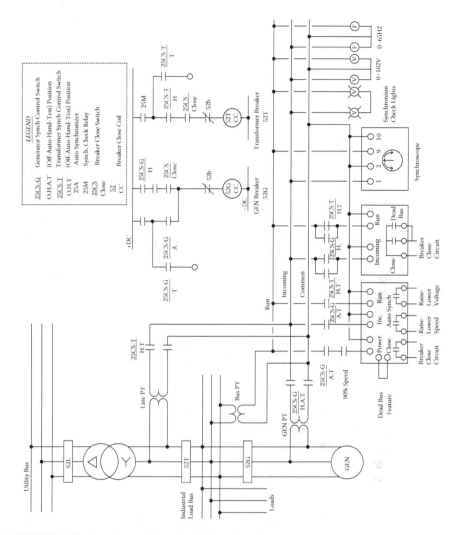

FIGURE 14.7 Auto/manual synchronizing scheme

14.5 INTEGRATION INTO THE POWER SYSTEM

14.5.1 GENERAL

Integration of a synchronous generator into an existing plant power system is a challenge. The impact of the short-circuit current contribution and VAR flow must be resolved to ensure an efficient and reliable operating system. The fault-current contribution from the synchronous generator increases the total fault level and overall system X/R ratio. Due to higher system X/R ratio, the computed fault current as compared against the equipment rating is further increased. The following schemes can be considered to minimize the impact of short-circuit contribution.

14.5.2 Higher Reactance

Higher values of generator subtransient X'_d will reduce the short-circuit contribution from the generator, but this option may incur special design costs.

14.5.3 Current-Limiting Reactor

In applications where the short-circuit level exceeds the switchgear rating by a small margin when the synchronous generator is operating in parallel with the network, a current-limiting reactor with a bypass breaker can be used. The reactor will be bypassed when the generator is not connected to the network. However, VAR flow and voltage regulation must be checked to ensure that the generator terminal voltage does not exceed 1.05 P.U.

14.5.4 Duplex Reactor

In a duplex reactor, the normal working currents in one winding partially erase the voltage drop in the other. With balanced loading, the reactive kVA loss and reactive voltage drop are lower than for two independent reactors. With current flow in one winding section only, the duplex reactor behaves the same as one of the two reactors.

The application of a duplex reactor is an excellent solution to circumvent the problem associated with increased fault level resulting from the integration. However, system engineering must be carried out for the equipment selection. A single-line diagram of an integrated system with a duplex reactor is shown in fig. 14.8.

14.5.5 Unit-Connected System

In this arrangement, the generator delivers the power to the high-voltage (HV) bus through a step-up transformer (fig. 14.9). This system is simple, widely used in utility systems, and has some merits. The disadvantage for industrial systems is that the power has to go through two transformations, thus incurring additional losses. However, for systems with large units or where the short-circuit contribution is large, this arrangement offers the best alternative. The step-up transformer winding connection shall be Delta on the generator side and Wye with neutral grounded on the HV side. Additional benefits from this winding arrangement are:

- Triplen (third) harmonic is blocked.

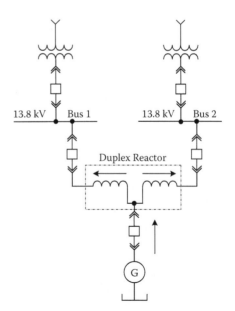

FIGURE 14.8 Integration with a duplex reactor

Electrical Aspects of Power Generation

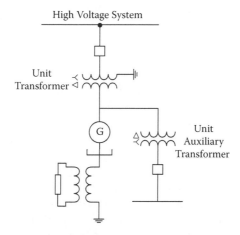

FIGURE 14.9 Unit-connected generator

- There is no transient overvoltage due to ground faults on the HV system.
- High-resistance grounding can be simply applied using a sensitive setting, as the capacitance of the generator connections are kept to a minimum.

14.5.6 Fault-Current Limiters

The fault current limiters are high power, power electronic, or controlled magnetic devices that will pass normal load current while in a low-impedance condition (low regulation), but which automatically transit to a high impedance when a through fault occurs. Such devices can be applied to resolve the problems associated with the increased fault levels. Their application is on the rise in existing installations, and even in some generator circuits.

There are two problems encountered with this approach that must be resolved before the decision is made to apply such a solution. First, the limiter unit must have been tested and certified as a system to ensure that the downstream equipment is adequately protected. The other problem is related to protection coordination, which is almost impractical to achieve because the difference between load and fault current is reduced due to the inherent characteristic of the device.

A fault-current limiter should not be used in a generator circuit directly, as it can cause instability as the synchronizing power, to correct any machine slip, is reduced. In such cases, sequential tripping is easier to apply, as the generator circuit breaker itself is rarely overstressed as it will only be subjected to either its own infeed, for faults on the network, or the system infeed alone, for faults on the generator circuit. It is therefore possible to trip the generator circuit breaker first and then allow the faulted circuit's circuit breaker to open.

Although this concept would require feeder protections to be delayed, the delay to allow the generator circuit breaker to open is short (200–300 ms). For the same reasons as noted above, the generator can be tripped for faults associated with the unit (turbine/generator). Faults within the generator zone are therefore still cleared quickly by the generator protective relays.

14.6 PLANT POWER SYSTEM AND UTILITY INTERFACE

A majority of process plants receive power at voltage levels ranging from 34.5 kV to 230 kV. The IEEE guide for the protection of utility interfaces [S21] has provided some guidelines for the protections at the interface between a utility's network and that of an industrial plant. In North America, two-winding transformers with Delta primary and grounded Wye secondary have traditionally been used to feed industries to prevent the ground fault in the feed from the plant transformers from affecting the utility's ground protections. However, there is a potential problem with having a Delta (ungrounded) high-voltage connection to the system when synchronous generators or large synchronous motors are connected within the plant power.

If a line-to-ground fault occurs on the transmission line feeding the plant, the utility remote (sending end) breaker will trip. However, as there is no ground-fault current fed by the delta to operate any local relays, the high-voltage bus, and hence the utility's line, will remain energized from the plant generator and synchronous motors. This will cause an uncleared ground fault on the HV side of the transformer, and the voltage between the two healthy phases, the phases not involved in the fault, will experience a transient and steady-state overvoltage, with a minimum line-to-ground voltage equal to or higher than the normal phase-to-phase voltage.

This is $\sqrt{3}$ times the normal phase-to-ground rating, thus stressing the insulation of the connected equipment (surge arrestors, voltage transformers, etc.) and, if left so connected for an extended period, can lead to their failure. It also brings up safety aspects for the utility operators and for other loads on the same line, and it can, if auto-reclose is applied on the line, cause an unsynchronized reconnection with the consequences listed in section 14.4, "Synchronizing."

The utility tie transformer primary winding should therefore normally be Wye connected, with the neutral solidly grounded when generation is connected. This connection will prevent overvoltages caused by line-to-ground faults. This connection also facilitates the detection of ground faults, which can then be isolated by tripping the high-voltage circuit breaker. The main transformer can be provided with two different winding configurations to meet these requirements, Wye-Delta or Wye-Wye with tertiary Delta as follows:

- The preferred arrangement is a Wye primary with neutral solidly grounded and Delta secondary. With Wye primary and Delta secondary arrangement, a zig-zag grounding transformer with a neutral resistor is required to achieve a ground source for the plant distribution system. Figure 14.10(a) shows the connections above. This arrangement permits the application of high-resistance grounding for the generator, although with some limitations in settings and higher damage possibilities than with a separate generator transformer.
- The alternative is a Wye-Delta-Wye with the primary neutral solidly grounded, the secondary neutral resistance grounded, and the Delta tertiary with one corner grounded as shown in fig. 14.10(b). This arrangement is not preferred because a low-resistance grounded generator is exposed to greater damage from internal ground faults, particularly during the run-down time

Electrical Aspects of Power Generation

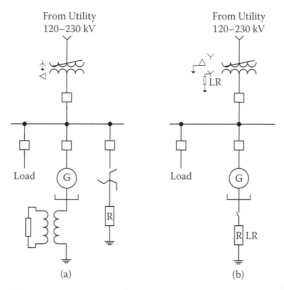

FIGURE 14.10 HV transformer connections

after the generator circuit breaker has opened, and it is more difficult to protect the unloaded tertiary winding. The ground-fault level in the network also changes significantly when the main transformer and the generator are operated separately rather than in parallel.

14.7 SYSTEM DISTURBANCES AND ISLANDING

14.7.1 Faults on HV (Utility) System

Faults on the utility system can cause significant voltage while the fault is cleared, and such dips can interrupt production or shut down the plant. The ability of the system to maintain stability and ride through the disturbance depends on the following:

- Type and duration of faults
- Plant single-line and switching arrangements
- System characteristics
- Motor-load torque and load reacceleration characteristics
- Ability of undervoltage load shedding and motor restart logic

Figure 14.11 illustrates the voltage magnitude on plant buses resulting from faults on the utility transmission system. The voltage magnitudes shown in table 14.1 would be higher if there were voltage support from in-plant generation.

Most of the plant system disturbances result from lightning strikes on the utility (power company) transmission lines. These lightning strikes cause overvoltage and flashover, thus creating line-to-ground faults. The system voltage is depressed until the fault is cleared. The voltage dip will depend on the location, severity, and duration of the fault, noting that, with the common use of extremely high voltage (EHV) transmission, faults on those networks can cause significant disturbances over an extended area.

14.7.2 Performance Criteria for Plant Electrical System and Equipment

To improve the ride-through capability of the system, the generator and excitation system, including the voltage regulator, shall provide the following:

- Sufficient VAR capability
- Sufficiently high initial response in the exciter
- Sufficient capability of plant components (e.g., switchgear, motors, etc.) to survive voltage dips, including:
  1. Higher breakdown torque for induction motor
  2. Higher withstand voltages for control systems, variable-frequency drives, etc.

Typical sag-response curves are shown in figs. 14.12–14.15.

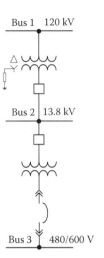

FIGURE 14.11 Percent voltages during a fault on the 120 kV bus (From C. L. Bencel, "Maintaining Process Continuity during Voltage Dips," *IEEE Trans. IAS* 18 (4): (1982). With permission.)

TABLE 14.1
Voltage Magnitudes on Plant Buses Resulting from Faults on the Utility Transmission System

Fault Location	A-B	B-C	C-A	E1	E2
Line-to-Ground (B Phase)					
Bus 1 120 kV	58	58	100	67	33
Bus 2 13.8 kV	88	33	88	67	33
Bus 3 4.16 or 0.6 kV	100	58	58	67	33
Line-to-Line Fault (Phase C-A)					
Bus 1	87	87	0	50	50
Bus 2	50	100	50	50	50
Bus 3	0	87	87	50	50

Electrical Aspects of Power Generation

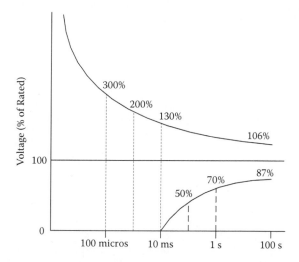

FIGURE 14.12 Computer susceptibility (From Van E. Wagner, Allen A. Andreshak, and J. P. Staniak, "Power Quality and Factory Automation," *IEEE Trans. IAS* 26 (4): (1990). With permission.)

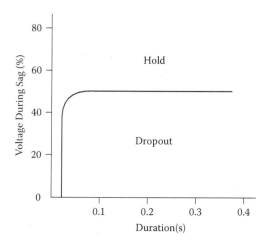

FIGURE 14.13 Starter sag response (From Van E. Wagner, Allen A. Andreshak, and J. P. Staniak, "Power Quality and Factory Automation," *IEEE Trans. IAS* 26 (4): (1990). With permission.)

14.7.3 Steps to Improve Plant Electrical System Ride-Through Capability

The following steps are suggested to determine and improve the plant electrical system ride-through capability:

- Carrying out a system stability study. The study must include line-to-ground faults on the utility transmission lines originating from the plant substation.

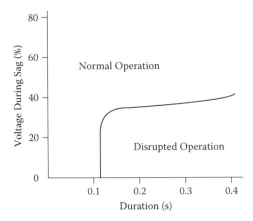

FIGURE 14.14 Sag response for programmable logic controller (PLC) (From Van E. Wagner, Allen A. Andreshak, and J. P. Staniak, "Power Quality and Factory Automation," *IEEE Trans. IAS* 26 (4): (1990). With permission.)

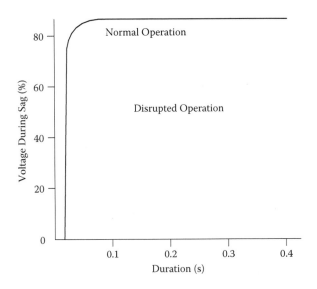

FIGURE 14.15 PLC input/output (120 V AC sag response) [P7] (From Van E. Wagner, Allen A. Andreshak, and J. P. Staniak, "Power Quality and Factory Automation," *IEEE Trans. IAS* 26 (4): (1990). With permission.)

- Lowering the tower-footing resistance within at least 1 km of the plant substation to reduce the probability of lightning exposure.
- Determining the voltage profile at all critical buses, angle for generators and synchronous motors, and slip for induction motors. Induction motor slip shall not exceed 10% following a severe voltage dip.
- Corrective actions, including: (a) transfer trip signal to trip the utility circuit breaker and plant main circuit breaker simultaneously, (b) fast clearing

Electrical Aspects of Power Generation

of faults, (c) uninterruptible power system (UPS) for control system, (d) higher torques for critical motors, (e) high-performance generator excitation system.

14.8 INDUCTION GENERATOR

14.8.1 GENERAL

An induction generator is an induction machine driven above its synchronous speed (negative slip) by a prime mover coupled to its shaft. The machine will generate power back into the system, depending on the magnitude of the slip.

Induction generators do not have a kVA fault rating, since they can produce only kilowatts and take their reactive power requirements from the system. Efficiency is not that important for an induction generator because the generator is usually recovering energy that would otherwise be lost or wasted.

14.8.2 CHARACTERISTICS

The characteristics of an induction machine during motoring and generator operation are shown in fig. 14.16. As a generator, the induction machine has the same speed–torque characteristic as for motor operation, except in reverse.

As the shaft load of an induction motor increases, speed drops and torque increases until the breakdown torque point is reached, where the motor may stall. Speed falls to zero and current rises to the locked-rotor value. Locked-rotor protection will trip the motor, or the motor may experience thermal damage.

14.8.3 TERMINAL VOLTAGE

For an induction motor, the terminal voltage is the vector sum of the internal voltage and the impedance drop. For an induction generator, the internal voltage is the vector

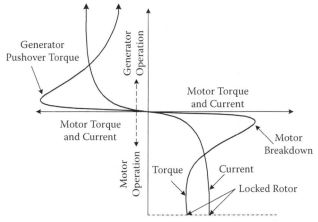

Characteristic Curve – Induction Generator/Motor

FIGURE 14.16 Speed–torque and speed–current for induction machine

sum of the terminal voltage and impedance drop. The generator nameplate voltage shall be the same or higher than the system voltage. For a nominal system voltage of 4.16 kV, the induction generator will be rated at 4.16 kV and the induction motor rated at 4.0 kV.

14.8.4 Excitation

For an induction generator to deliver power to the network, it must receive magnetizing (reactive) VARs from the system. The reactive current consumed by the induction generator varies with the active power delivered to the network. An isolated induction generator excited by a fixed capacitor generates a terminal voltage that varies with the load. To compensate for the reactive power consumed by the induction generator, additional capacitors will be required to meet the utility power factor criteria. No separate DC power source is required for the excitation of induction generators.

14.8.5 Protection

A typical scheme for the protection of an induction generator is shown in fig. 14.17. The protection is almost identical to an induction motor protection, with a few exceptions. The protective features are identified by IEEE device numbers.

A brief description of the protective features follows:

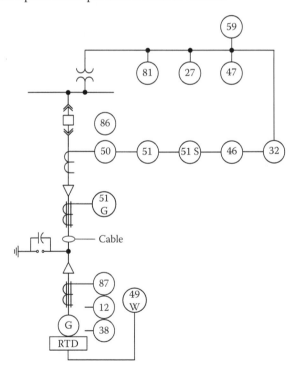

FIGURE 14.17 Typical protection features for an industrial generator

Electrical Aspects of Power Generation 405

- *Reverse Power (32)*: This device disconnects the induction generator when it operates as an induction motor. The direction of power flow will reverse when the prime mover speed falls below the rated synchronous speed.
- *Overspeed (12)*: Upon loss of load, this device will prevent the induction generator from reaching the runaway overspeed level of the prime mover.
- *Phase and ground overcurrent (50/51), (51G)*: These overcurrent devices protect the induction generator against internal and external faults.
- *Negative-sequence current (46)*: This device protects the generator rotor against the negative-sequence overcurrent resulting from system voltage unbalance.
- *Flux balance differential (87)*: This device protects the generator against the internal ground or phase-to-phase faults. This protection is recommended for units rated 1000 kW and above.
- *Winding overtemperature (49W)*: This device is a resistance-temperature detector (RTD)-connected device that protects the generator against sustained overtemperature resulting from overload or blocked ventilation. This protection is recommended for all medium-voltage generators.
- *Underfrequency (81)*: This device protects the generator against uncoordinated reenergization during reclosing of incoming feeders or lines by ensuring that the generator trips before such a reclose occurs. It will also isolate the generator during abnormal frequency operation.
- *Under- and overvoltage (27), (59)*: These devices protect the generator during abnormal system voltages.
- *Bearing overtemperature (38)*: This device disconnects the motor when the bearing oil temperature exceeds the set limit.

14.8.6 Controls

The normal starting sequence for an induction generator is to start the prime mover with the generator off line and close the generator breaker or contactor when the unit has reached synchronous or slightly above synchronous speed. This approach limits the inrush current to a very low value, unlike an induction motor, which draws six to seven times its rated current upon starting.

The output frequency and voltage of a synchronous generator are regulated by the power system frequency, whereas the speed has to be more tightly controlled for an induction generator. The prime mover of an induction generator must be equipped with a speed-governing device to run the unit above the synchronous speed to deliver the desired load and at the same time prevent attaining the prime mover's runaway (push) overspeed upon loss of load.

The torque output of the prime mover must also be limited to prevent overloading. Synchronizing is not required for an induction generator, since the machine cannot generate an emf (output voltage) until it is connected to the system.

14.8.7 Advantages and Disadvantages

The advantages of an induction generator over a synchronous unit are:

- Squirrel-cage rotor: simple, rugged construction
- Voltage regulator, excitation source, and synchronizing equipment not required

- Relatively small effect on system voltage and frequency
- Control and protective devices basically the same as for an induction motor
- Lower capital cost and lower maintenance cost

The disadvantages of an induction generator are:

- Consumes rather than supplies reactive power
- Cannot contribute to maintaining the voltage level of the power system

14.8.8 Differences with an Induction Motor

The upper limit for a commercially available induction generator is 2.5–3.5 MW, although 10 MW-rated units have been built. Although induction motors up to 30,000 hp have been built, there is little demand for induction generators of this size, as most industries that can justify a machine of such size would normally want this to be able to provide standby power upon power system failure, which is not possible with induction machines.

14.9 STATION AUXILIARIES

14.9.1 Auxiliary System Load

The auxiliary load required depends on the type of generator and whether any other essential loads are also fed from the generating system auxiliary system. The thermal plants have the highest loading, as they need to support such loads as boiler feed pumps and induced- and forced-draft fans for the combustion system. Some turbines use process steam, and these loads may be fed from different systems. For the basic generation types, the necessary load requirements in terms of the installed capacity of the generation are as follows:

- Up to 7.5% plant capacity for thermal plants
- Approximately 2% or less for hydroelectric plants and internal combustion (diesel) plants
- Up to 10% plant capacity for a nuclear plant

The last item is included for completeness, as such generation types are not normally used in industrial processes, and there are many more security concerns relative to continuous power, standby generation, and triple-redundant power sources, including the drives themselves, that are appropriate for nuclear power.

14.9.2 Power Supply Arrangement

14.9.2.1 Power Plants

A typical auxiliary distribution system is shown in fig. 14.18. The unit auxiliary bus is powered from the station service system. When the generator is partially loaded and synchronized with the station service system, the bus is transferred from the station service system to the unit auxiliary system. During the transfer, both buses are

Electrical Aspects of Power Generation

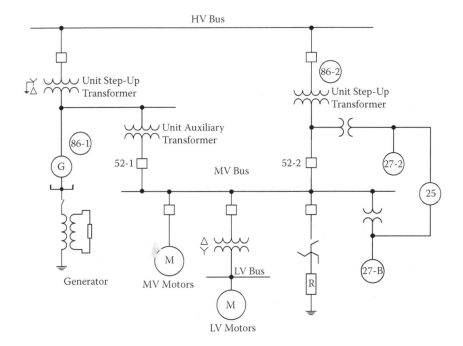

FIGURE 14.18 Station auxiliary system single-line diagram

paralleled momentarily, and the station service breaker is tripped. To shut down the generator, the reverse procedure is applied. During paralleling, the fault current may exceed the equipment rating. However, it is acceptable practice to design the system for a single source (maximum) condition if the duration of parallel operation is short and the transfer system is set up so that the transfer must complete once initiated.

Two types of transfer, manual and automatic, are applied. Manual transfer is initiated by an operator. Transfer during start-up is from the station service to the unit auxiliary system, and during the shutdown, transfer is in the opposite direction.

Hydroelectric stations are usually designed so that the station service can be interrupted for a short time without losing the generation. This makes the transfer system simpler in that dead transfer can be used, and there is no need to check that the systems are in phase before they are paralleled. However, some thermal plants have auxiliaries supplied from unsynchronized or out-of-phase systems that cannot be paralleled. In such cases, dead high-speed transfer, which reaccelerates the running motors after the transfer, is applied.

Automatic transfer can be classified as "residual voltage or delayed," "in phase," and "fast transfer." These schemes are described in IEEE 666 [S22].

14.9.2.2 Industrial Power Systems

A typical auxiliary system is shown in fig. 14.19. Station auxiliaries are powered from the generator/load bus. Auxiliaries are started using the power from the plant power system. The generator is brought to speed and synchronized to the genera-

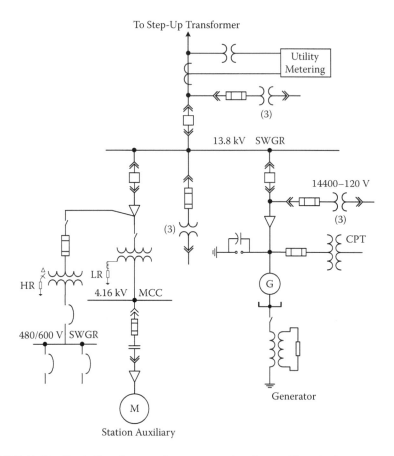

FIGURE 14.19 Single-line diagram of generator and station auxiliary for industrial plants

tor/load bus as required, and the auxiliary loads continue to be fed from the general plant power system.

14.9.3 Design Criteria

14.9.3.1 System Design and Equipment

Any generation system is only as reliable as its auxiliary system, as the generation cannot operate for long without an auxiliary power supply. Although it is possible to run a hydroelectric generator for a few minutes without AC service, thermal units must have their AC auxiliary power restored within milliseconds. All designs have a need for battery-backed DC power for control and protection and critical start-up loads.

Special station service systems need to be designed for generators that need to black start to provide secure supplies during any loss of the utility system. Such

Electrical Aspects of Power Generation

systems usually require separate auxiliary service generators, which can operate on DC battery power alone, to feed the critical loads. These requirements can be summarized as follows:

- Must be reliable
- Needs to be well engineered and coordinated
- Must ride through system disturbances, especially where the generators are needed to secure the process

14.9.3.2 Voltage Considerations

The auxiliary system design and operating voltages must be suitable to feed the auxiliary system equipment, even under severe conditions on the network. The design concepts for voltage are as follows:

- No-load voltage:
  For station auxiliary transformer — limited to 110% of motor-rated voltage
  For unit auxiliary transformer — 112.5% of motor-rated voltage
- Full-load voltage:
  The minimum acceptable voltage at the motor bus — 90% of bus-rated volts (94% of motor-rated voltage)
- Starting voltage:
  The minimum acceptable voltage at the motor bus — 80% of motor-rated voltage, base calculations on the same assumption as for full-load voltage
- Assumptions for voltage calculations:
  Minimum source voltage
  Maximum source impedance
  Positive manufacturing tolerance for impedance values

14.9.3.3 Short-Circuit Considerations

Fault currents should be done to ensure that the equipment ratings are adequate. Calculate maximum fault currents assuming:

- Maximum source voltage
- Minimum source impedance
- Negative manufacturing tolerances for impedance

For protection and voltage-change calculations, the minimum fault-current level must also be calculated, and for that calculation the reverse of the above should be used, as for the voltage calculation.

14.9.3.4 Auxiliary Bus Capacity

Although the basic auxiliary sizing can be used for auxiliary system parametric design, there are many special conditions to be considered in industrial plants. For the generation, much has to be done for the main system, which varies considerably, depending on the particular industrial process.

The bus capacity (voltage and current) should therefore be optimized for the actual situation at the plant under study, during steady-state and transient conditions. This can be investigated using hand calculations but, as discussed in chapter 3, is more conveniently and accurately done — particularly when there are a number of special cases to be investigated — using computer software for load-flow, short-circuit, and motor-starting studies.

REFERENCES
Standards
Generators
S1. ANSI C50.10-1990, General Requirements for Synchronous Machines, 1990.
S2. IEEE C50.12-2005, Requirements for Salient-Pole Synchronous Generators and Generator/Motor for Hydraulic Turbine Applications, 2005.
S3. ANSI C50.13-1977, Requirements for Cylindrical Rotor Synchronous Generators, 1977.
S4. ANSI C50.14-1977 (reaffirmed 1989), Requirements for Combustion Gas Turbine Generators, 1989.
S5. IEEE 115-1995, IEEE Test Procedures for Synchronous Machines, 1995.
S6. IEEE 67-1990, IEEE Guide for Operation and Maintenance of Turbine Generator, 1990.

Excitation System
S7. IEEE 421.1-1986, IEEE Standard Definitions for Excitation Systems for Synchronous Machine, 1986.
S8. IEEE 421.4-2004, IEEE Guide for the Preparation of Excitation System Specifications, 2004.

Neutral Grounding
S9. IEEE 32-1972 (reaffirmed 1991), Standard Requirements, Terminology, and Test Procedures for Neutral Grounding Devices, 1991.
S10. IEEE C62.92.1-2000, IEEE Guide for the Application of Neutral Grounding in Electrical Utility Systems, Part I: Introduction, 2000.
S11. IEEE C62.92.2-1989, IEEE Guide for Application of Neutral Grounding in Electric Utility Systems, Part II: Grounding of Synchronous Generator Systems, 1989.

Protection
S12. IEEE C37.101-1993, IEEE Guide for Generator Ground Protection, 1993.
S13. IEEE C37.102-1995, IEEE Guide for AC Generator Protection, 1995.
S14. IEEE 122-1991, IEEE Recommended Practice for Functional and Performance Characteristics of Control Systems for Steam Turbine Generator Units, 1991.
S15. IEEE 502-1985, IEEE Guide for Protection, Interlocking and Control of Fossil-Fuel Unit Connected Steam Stations, 1985.
S16. IEEE 1046-1991, IEEE Guide for Distributed Digital Control and Monitoring for Power Plants, 1991.
S17. IEEE 1129-1992, IEEE Recommended Practice for Monitoring and Instrumentation of Turbine Generators, 1992.
S18. IEEE C37.91-2000, IEEE Guide for Protective Relay Applications to Power Transformers, 2000.
S19. IEEE C37.110-1996, IEEE Guide for the Application of CTs Used for Protective Relaying Purposes, 1996.

Integration into the Power System and Utility Interface
S20. IEEE C57.116-1989, IEEE Guide for Transformers Directly Connected to Generators, 1989.
S21. IEEE C37.95-2002, IEEE Guide for Protective Relaying of Utility-Consumer Interconnections, 2002.
S22. IEEE 666-1991, IEEE Design Guide for Electric Power Service Systems for Generating Stations, 1991.

Induction Generators
S23. NEMA MG 1, Motors and Generators.

BIBLIOGRAPHY

Generators
1. R. H. McFadden, "Stability Considerations for Industrial Power Systems," *IEEE Trans. IAS* 13 (2): (1977).
2. H. E. Lokay and S. Merry, "Transient and Stability Studies," *Power* Aug. (1969).

Excitation Systems
3. M. Shan Griffith, "Modern AC Generator and Control Systems: Some Plain and Painless Facts," *IEEE Trans. IAS* 12 (6): (1976).
4. T. L. Dillman, F. W. Keay, C. Rackowski, J. W. Skooglund, and W. H. South, "Brushless Excitation," *IEEE Spectrum* Mar. (1972).
5. L. J. Lane, D. F. Rogers, and P. A. Vance, "Design and Test of a Static Excitation for Industrial and Utility Steam Turbine-Generators," publication GET-3007, Schenectady, NY: General Electric.

Neutral Grounding
6. P. G. Brown, "Generator Neutral Grounding," publication GET-1941A, Schenectady, NY: General Electric.
7. D. J. Love and N. Hashemi, "Considerations for Ground Fault Protection in Medium Voltage Industrial and Cogeneration Systems," *IEEE Trans. IAS* 24 (4): (1988).
8. R. H. McFadden, "Grounding of Generators Connected to Industrial Plant Buses," *IEEE Trans. IAS* 17 (6): (1981).

Protection
9. H. K. Clark and J. W. Feltes, "Industrial and Cogeneration Protection Problems Requiring Simulations," *IEEE Trans. IAS* 25 (4): (1989).
10. C. J. Mozina, IEEE Tutorial on Protection of Synchronous Generators, IEEE 95TP102.

Synchronizing
11. Thomas R. Beckwith, "Automatic Synchronizing Considerations and Methods," Beckwith Electric Paper, Largo, FL.

Plant Power System and Utility Interface
12. D. V. Fawcett, "The Tie between a Utility and an Industrial Network When the Utility Has Cogeneration," *AIEE* July (1958).
13. J. Reason, "Cogeneration: Controlling the Industrial/Utility Interface," *Power* Apr. (1984).

System Disturbances and Islanding
14. C. L. Bencel, "Maintaining Process Continuity during Voltage Dips," *IEEE Trans. IAS* 18 (4): (1982).
15. Van E. Wagner, Allen A. Andreshak, and J. P. Staniak, "Power Quality and Factory Automation," *IEEE Trans. IAS* 26 (4): (1990).

Induction Generators

16. R. N. Nailen, "Induction Generators for Process Industries," *IEEE Trans. IAS* 19 (3): (1983).
17. E. L. Owen, "Induction Generator Applications for Petroleum and Chemical Plants," *IEEE Trans. IAS* 19 (6): (1983).
18. H. A. Breadlove, "Protection of Induction Motors/Generators," *IEEE Trans. IAS* 19 (6): (1983).

15 Application of Capacitors

15.1 INTRODUCTION

Capacitors are used in industrial premises for two basic functions: the first, to correct the load power factor, usually to meet the supplying authority's supply requirements and avoid tariff penalties if the required power factor is not met, and the second, to correct voltages within the plant. Individual capacitors can also be attached to circuits feeding loads with inherently low power factors to correct for their low power factor and so improve both the overall plant's net power factor and, at the same time, reduce the voltage drops within the plant.

Capacitor banks and individual capacitors can therefore form either fixed installations, which are always connected when that section of the plant is running, or be switched when required. A common example of the former application is capacitors connected directly in motor circuits that are switched with the motor and keep the net power factor for each motor circuit at an ideal level. Larger fixed units are also used where the process is continuous and the load and power factor of that process are known and vary little over time.

Switched individual units or controlled banks are used where conditions vary, either within the plant or on the in-feeding supply system, and where a variable reactive power level needs to be compensated. Capacitors can only be manufactured as relatively small elements, both in terms of their reactive power rating and their operating voltage. Capacitors used in electrical systems are therefore supplied as capacitor units, also known as capacitor "cans," which are made up of a number of series and parallel internal elements to obtain practical levels of reactive power rating (kVAR) and of service and withstand voltages.

15.2 CAPACITOR APPLICATION

15.2.1 Capacitor Construction

Capacitor units are supplied in three basic configurations: externally fused, internally fused, and fuseless designs. The externally fused design, which is used for most small and individual applications, is designed to rupture the external fuse for the failure of one individual element within a capacitor unit. In the internally fused design, each capacitor element within the capacitor unit is individually fused, and the failure of an individual element ruptures that particular fuse so that the capacitor unit can remain in service. The third type, the fuseless capacitor, is designed so that any element failure produces a stable short circuit in the element that can carry the current of other elements connected with it in series. Again, this design of a capacitor unit can continue to be operated with some individual elements failed.

Externally fused units can be used individually or in self-protected groups, whereas the other two types are usually used within large capacitor banks in series/parallel configurations. Banks made up of these and externally fused units require external circuit protection and overall monitoring to ensure that the individual units that make up the bank are not overstressed when more internal elements fail within the individual units or when additional fuses on the externally fused units rupture.

15.2.2 Individual Capacitors

Individual outdoor capacitors are usually designed for rack mounting on an insulated structure, using individual or group fusing, together with accessories such as switching devices and control facilities. Such capacitors can be provided for operation at voltages less than or equal to 2.4 kV and for reactive power ratings up to 1200 kVAR. Enclosed housings are also available for units in the above voltage range up to 600 kVAR.

15.2.3 Motor Capacitors

Capacitors for individual motor application are available for standard voltages and in the range of 2.5–600 kVAR, three phase. These are used to correct the motor power factor individually, but when such capacitors are coupled with induction motors, there can be overvoltages due to self-excitation, as shown in fig. 15.1. In this figure, the straight lines show the capacitor rating in percentage of the motor rating, and the intercept points A, B, or C on the M curve show the percentage overvoltages expected with that particular capacitor rating.

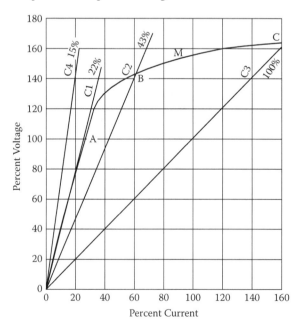

FIGURE 15.1 Motor voltage due to self-excitation influence of capacitor rating

Application of Capacitors   415

15.2.4 Switched Capacitors

For any switched-capacitor scheme, the switching devices must be rated for capacitor switching and have a voltage rating greater than the associated maximum system operating voltage. The device-interrupting rating must exceed the system short-circuit rating and must also interrupt the capacitor current without producing excessive transient overvoltages.

A typical medium-voltage industrial installation is shown in fig. 15.2. In this installation, the incoming cell has an externally operated air-disconnect switch that provides a visible break for maintenance, together with an integral, interlocked ground switch. The switch is located in a separate compartment that allows the main fuses and other components located within the capacitor compartment to be maintained without having to disconnect power to the circuit feeding the bank.

This type of capacitor bank design completely encloses the bank's components within a grounded structure, which eliminates the need for a fenced enclosure and mitigates rodent and pollution problems. Enclosed capacitor banks are more aesthetically pleasing compared with air-insulated open-rack capacitor banks. The capacitor bank assembly is furnished with fused capacitor units, insulators, current-limiting reactors, vacuum switches, main-line fusing, line disconnect switch and grounding switch, surge arrestors, and instrument transformers. A separate section is provided for the protection, monitoring, and control devices together with a separate cable-entrance section.

The doors of the enclosure are key interlocked to prevent entry into a live compartment. Enclosed capacitor banks often use internally fused capacitor units, as these are compact in design and more reliable compared with externally fused capacitor units.

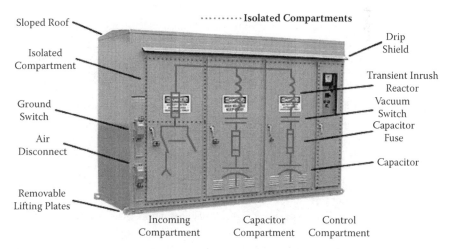

FIGURE 15.2 Two-stage medium-voltage automatic capacitor bank showing location of isolated compartments and bank features

15.2.5 Switching Device Rating

The momentary rating of the switching devices should withstand short-circuit currents for faults and the high-frequency inrush currents associated with switching capacitors. The maximum transient current and net inrush frequency have to be calculated for the more severe of the following two conditions:

- Isolated bank switching

$$I \max. \text{(rms)} = I_C \times \left(1 + \sqrt{\frac{kVA_{SC}}{kVA_C}}\right) = \frac{E_{L-N}}{X_C - X_L} \times \left(1 + \sqrt{\frac{X_C}{X_L}}\right)$$

$$X_C = \frac{kV_{L-L}^2}{3 \text{ ph. MVA}_C}$$

$$f_0 = f_s \times \sqrt{\frac{kVA_{SC}}{kVA_C}} = f_s \times \sqrt{\frac{I_{SC}}{I_C}}$$

- Back-to-back switching

$$I \max. \text{(peak)} = \sqrt{2} \times E_{L-N} \times \sqrt{\frac{C_t}{L_t}}$$

$$f_0 = \frac{10^6}{2 \times \pi \times \sqrt{LC}}$$

where the subscripts "sc" and "c" refer to the short-circuit infeed and capacitor values, respectively, while f_0 is the inrush frequency and f_s is the system frequency. Back-to-back switching refers to switching when other capacitors or a significant level of system capacitance also exists on the bus side of the capacitor bank switch.

The formulae for back-to-back switching given above are simplified versions that apply if the capacitances on both sides of the switch are the same. The formulae to use in other cases are summarized in IEEE C37.012 [S2].

Transient inrush reactors, which are designed to increase the life expectancy of capacitor switches by limiting both the magnitude and frequency of the transient inrush currents associated with back-to-back capacitor bank switching, can be added. These reactors are most commonly applied in multistage medium-voltage capacitor banks or in fixed medium-voltage capacitor banks that are connected to the same switchgear bus as other fixed capacitor banks, as shown in fig. 15.3. In these cases, the switches are usually rated 200–600 A, with a 200 A single-phase group-operated switch being the most common.

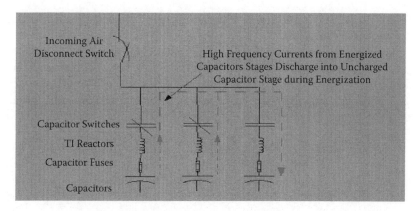

FIGURE 15.3 High-frequency inrush currents that may damage capacitor switches can be reduced by transient inrush reactors.

15.3 LARGE-BANK CONSTRUCTION

15.3.1 Bank Construction

In addition to the small banks of switched capacitors noted above, large fixed capacitor banks are used on the high-voltage buses of major industries to correct the system power factor. Similar banks are also used in the capacitor sections of filter banks. Such filter banks are used to filter the harmonic distortion, caused by certain industrial processes, from the supply system.

To provide capacitors for these higher voltages and to provide the larger capacitive values required, it is necessary to combine capacitor units into banks. A capacitor bank therefore consists of the series connection of lower-rated line-to-neutral voltages capacitor units to achieve the required operating voltage and the parallel connection of the number of units required to achieve the desired reactive power rating.

A capacitor bank of a given size and voltage rating is therefore made up of a number of series and parallel groups. The use of capacitors with the highest possible voltage rating results in a bank design with the smallest number of series groups, and this can provide the most economical capacitor bank design. However, to achieve the greatest sensitivity for the unbalance detection schemes and allow the bank to operate with at least a small number of fuses ruptured, it may be necessary to use a less than optimal mix of series and parallel units.

Capacitor bank design considerations include limiting the permissible capacitor unit overvoltage when other units or elements fail. This is done to avoid cascade failure of the healthy capacitor units and to limit the risk of case rupture for a flashover within the capacitor unit itself. It is also important, in any capacitor installation, to ensure that the maximum operating voltages do not exceed 110% percent of the rated voltage of any individual capacitor. Because of this, the number of parallel capacitor units in each series section is selected, for the externally fused banks, so that the loss of any unit fuse in any series section will not result in such an overvoltage. Additionally, steady-state voltage rises introduced by the flow of harmonic currents into tuned shunt capacitor banks (banks fitted with current-limiting reactors and filter banks) have to be considered.

The IEEE capacitor protection guide [S3] provides recommendations for the maximum and minimum number of parallel capacitor units per series group for externally fused banks. For internally fused and nonfused banks, the ability to detect the failure of individual elements rather than of the unit fuse alone, and hence the performance of these types of capacitor units, requires a different mix of series and parallel groups.

Where only one series section of paralleled capacitors per phase is used and is connected either three-phase grounded Wye or Delta, the unit capacitor external fuse is subjected to the full system short-circuit current available when its associated unit fails. This generally requires that more expensive, high-interrupting-capacity, current-limiting fuses be applied in situations where available fault currents exceed 4 kA.

15.3.2 Capacitor Unit Rating

In terms of capacitor unit voltage ratings, suppliers in the North American market can provide individual capacitor units in voltages ranging from 2.4 to 25 kV, and most industrial users utilize capacitor units rated at or above 7.2 kV.

The desired operating voltage for a capacitor bank is obtained by connecting as many capacitor groups in series as necessary to obtain the required voltage level. Usually, the best engineering choice is to use the lowest number of series groups possible and the use of capacitors with the highest possible rating, as this results in fewer series groups.

One of the advantages of the alternative fuseless design of a capacitor unit is that it usually has higher unit voltage ratings levels available than the other designs. The use of higher voltage units generally provides the simplest and most economical bank design. However, limited inventory or other design considerations may override this rule.

The capacitor unit ratings available have been undergoing frequent changes as the suppliers develop unit configurations to provide the most practical and economical unit sizes for the banks in general use. In general, the trend is toward larger unit sizes, and this has been supported by improvements in manufacturing capabilities. Standard, externally fused capacitor units for shunt capacitor bank applications are 50, 100, 150, 200, 300, and 400 kVAR. No upper limits are defined for internally and nonfused capacitor units.

The capacitor manufacturer's recommendations should be considered in determining the optimum size of a capacitor unit, the number of series sections, the number of units in parallel, and the type of connection used to make up the reactive power requirement for a given application.

In terms of the basic insulation level, the impulse insulation levels of individual capacitor units range from 75 to 200 kV. Table 15.1 summarizes typical basic impulse insulation levels by capacitor unit voltage rating.

The maximum allowable ambient temperature for capacitor equipment installed outdoors with unrestricted ventilation is 40°C (104°F), based on the mathematical

TABLE 15.1
Capacitor Unit Voltage and BIL Rating

Capacitor Voltage Rating (V, rms)	BIL (kV)
2,400–4,800	75
6,640–12,470	95
13,280–14,400	95 and 125
15,125–19,920	125
19,920	125 and 150
19,920–24,940	150 and 200

average of hourly readings during the hottest day expected at the site. Isolated, multiple row and tiers and metal-enclosed or housed units will have maximum ambient ratings of 46°C and 40°C, respectively. Capacitors are designed for continuous operation at −40°C. Where the expected in-service ambient temperatures are lower than −40°C, the manufacturer should be consulted.

15.3.3 Bank Monitoring Criteria

For both the externally fused capacitor units and for the internally fused and fuseless configurations, a number of basic design parameters need to be met. The following are applicable to banks using externally fused capacitors:

- Loss of one (or more if designed for multiple failures) capacitor unit in a group or phase should not produce a voltage across the remaining units in that group or phase that exceeds 110% of the rated capacitor unit voltage.
- In the event of failure of a unit, sufficient current should flow through the failed capacitor unit's individual fuse to ensure clearing in 300 s or less to prevent can rupture.
- The outflow current from the parallel healthy capacitor units must not be high enough to rupture, or to weaken, the fuses on those parallel healthy capacitors.

The last two requirements are based on the normal condition that the energy to rupture the fuse on a faulted unit or element comes mainly from the discharge energy of the total of all the parallel-connected capacitors.

The design criteria above set the minimum number of externally fused units that need to be connected in parallel to ensure effective rupture of the fuse on the faulted unit and hence provides one of the limits in the size and rating of the capacitor units that can be used when configuring a capacitor bank. The other limits are based on ensuring that the bank can continue to operate with one or more capacitor elements failed and that the capacitor unit failure detection systems function correctly. These limits are discussed further in section 15.5.

For internally fused units, the manufacturer makes the same assessment when the series/parallel arrangement of the internal elements inside the capacitor unit is configured. This, in turn, limits the ratings and voltage classes that are practical for the internally fused units.

For the fuseless design there is no need to make any assessment of fuse energy, but the designer has to configure the unit so that when a failed element makes the stable short circuit across the failed element, which is the primary requirement of that design, the voltage over the remaining elements that are in series with the failed element or elements does not exceed the element voltage withstand.

For capacitor banks using the three types of capacitor, it is necessary to have a monitoring system to detect and alarm the initial failures of capacitors within the bank. It is also necessary to detect and then trip the bank once the failures reach a level where the remaining capacitors are reaching their stress limits, thereby removing the bank from service before a cascade failure can occur.

15.3.4 Bus Insulation Systems

Unlike enclosed capacitor banks, open-rack capacitor banks have exposed live parts. This exposure can result in bank outages due to faults originated by birds or animals bridging live parts of the bank to ground. One straightforward method for preventing this type of incident is to use edge-mounted equipment, since the electrical clearances in the racks used in this equipment are sufficiently liberal, making it highly unlikely that any bird or animal will be large enough to bridge any of the live parts.

15.3.5 Interlocking and Safety

Air-insulated capacitor banks are constructed on metal frames and, for indoor units, are often floor mounted and hence are dangerous to approach. Outdoor units may be placed with some level of ground clearance, but their construction makes it relatively easy to climb and approach the live parts. It is also a factor in safe operation that the bank be allowed to discharge before work is done on the bank. It is therefore good practice to provide access control, usually fencing or steel mesh, with the access gate interlocked with the main circuit breaker and a bank ground switch.

This ground switch is also interlocked with the circuit switch, but also has a time-delay control so that it is not possible to close the ground switch or enter the capacitor enclosure until sufficient time has elapsed for the capacitors to have discharged. In this respect, it is always a factor in the design of individual capacitor units to have internal discharge resistors to ensure that the unit discharges in a reasonable time.

15.4 CONTROL AND MONITORING SYSTEMS

Capacitors are used within industrial plants to control power factor or voltage or both. Switched capacitors need to be switched to correct these system conditions. However, the control logic may be based on the direct measurement of the voltage levels and reactive power flow, or it may be based on a number of different indirect parameters within an industrial plant such as current flow, ambient temperature, time of day, or some combination of these factors.

A variety of physical options are also available in terms of price, features, and quality. These range from high-priced panel-mounted relay-based monitoring systems to low-priced socket-mounted controls. In recent years, capacitor controls based on electromechanical relays have been superseded by more modern electronic types, including solid-state and microprocessor-based controllers.

The control and application of capacitors are closely related subjects, and a discussion of one has to necessarily involve the other. In fact, in the typical situation, both the type of control and its adjustment are dictated by the type of the capacitor installation.

Capacitor switching controls measure the system parameters selected to reflect system reactive power requirements and provide outputs to close or open the capacitor switches when the appropriate set points are at their set levels. Theoretically, at least, any intelligence that changes only when a change in the reactive power levels needed can be utilized to switch capacitors automatically. However, in practice, selecting a signal that accurately reflects the requirements of the system often turns out to be the most difficult part of the problem.

The single-input types of control, such as voltage and time controls, are generally less expensive in initial cost and installation, less complex, and easier to adjust, but they are less flexible in application than the dual-input types, such as reactive power and current-based voltage controls. Typical schemes are shown in figs. 15.4, 15.5, and 15.6 for the various capacitor bank control arrangements.

The optimum choice of control should be the least expensive type that will switch the capacitor bank to meet the operating requirements of the network. To facilitate comparison, the operating characteristics and relative cost of the more common controls used are summarized in table 15.2.

15.5 CAPACITOR PROTECTION AND MONITORING

The protection and monitoring systems for capacitor banks consist of three parts: (a) the fault protection on the circuit feeding the bank, (b) the individual capacitor protection, and (c) the systems that monitor the health of the individual capacitors within the bank.

The protection of the circuit feeding the capacitor banks is basically the same as that provided for other industrial feeders, except for making an allowance in the settings to handle the capacitive inrush, which is discussed in more detail in chapter 11.

15.5.1 CAPACITOR FUSING

Devices or circuit protections are generally used for one or both of the following two basic reasons:

- To protect the device from overloads
- To protect the system from failure within the device

The failure rate of externally fused capacitor units is relatively high, hence there is a need to provide protection by detecting such failures. The requirements for these fuses are discussed in more detail in section 15.3 and are summarized again here:

- Voltage rating ≥ 110% of capacitor voltage rating

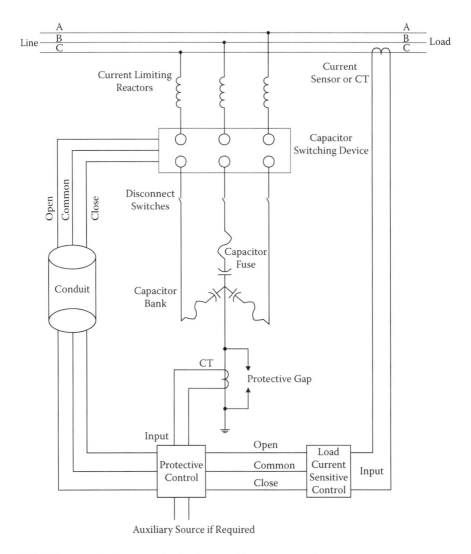

FIGURE 15.4 Typical capacitor bank ground Wye connected

- Short circuit rating ≥ maximum available fault current
- Continuous current rating ≥ 165% of the capacitor current rating, 150% for ungrounded Wye connections
- Must clear in less than 300 s (preferably 30 s) for a capacitor fault
- Must withstand the energy that a healthy capacitor contributes to a failed parallel unit

Internally fused capacitors are designed with their internal fusing based on the same criteria, with the internal configuration and fuse characteristics chosen by the manufacturer to ensure reliable fuse operation. As the internal fuses cannot be

Application of Capacitors

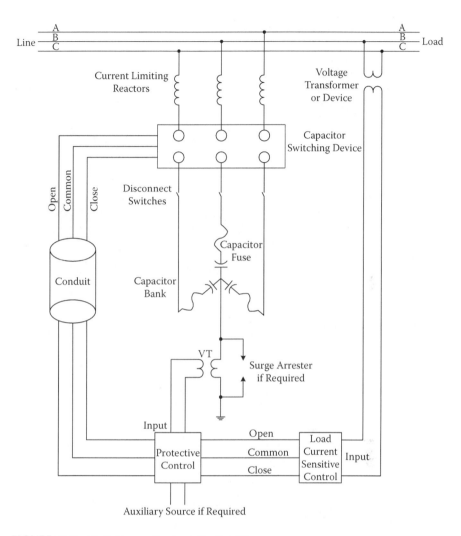

FIGURE 15.5 Typical capacitor bank floating Wye

accessed, their reliability is critical for the long-term reliability of the internally fused units. Internally fused units are usually rated to withstand the rupture of four to five internal element fuses before the capacitor must be removed from service. The associated monitoring system is set up so that warning alarms are issued when the first or second element fails, with tripping initiated for subsequent failures.

Fuseless capacitors do not have internal fuses and are designed to form stable short circuits when elements fail so that they can operate continuously with a number of elements in that condition. The internal configuration usually allows a similar number of internal failures (four to five) before the capacitor must be removed from service.

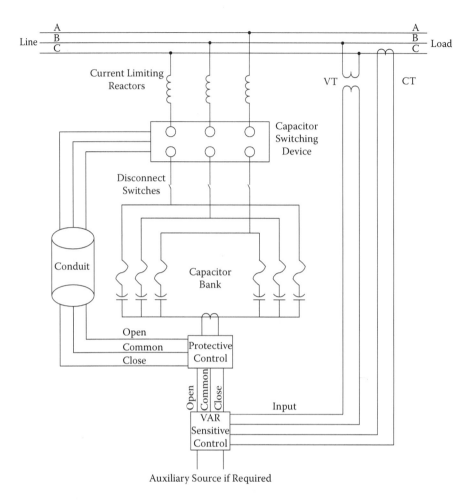

FIGURE 15.6 Typical capacitor bank Wye-Wye connected

15.5.2 Protective Control and Monitoring

The purpose of a capacitor bank's protective control is to remove the bank from service before any units or any of the elements that make up a capacitor unit are exposed to more than 110% of their voltage rating. When capacitor units in a capacitor bank fail, the amount of increase in voltage across the remaining units depends on the connection of the bank, the number of series groups of capacitors per phase, the number of units in each series group, and the number of units removed from one series group. A similar effect occurs on the internal elements that make up a capacitor unit.

Such monitoring is desirable for both externally and internally fused units to prevent a cascade failure of the remaining units and their associated fuses. For the fuseless designs, such a cascade failure would tend to become a short circuit and require the main circuit protection to operate, so it is somewhat more important to detect the failure before it cascades in fuseless banks.

TABLE 15.2
Comparison of Operating Characteristics of Capacitor Controls

Type of Control	Advantages	Disadvantages
Manual	No control device necessary, since the bank's switching device is operated by substation personnel	Requires attendants at the substation
Time	Nonelectrical control input allows application at any point on the circuit	Can only be applied on feeders where power factor and demand have a regular daily variation that is repeated weekly; use is limited to locations where established switching schedule will not cause high voltage on holidays or during other abnormally light load periods; insensitive to abnormal voltage conditions
Temperature	Nonelectrical input allows application at any point on the circuit; connects capacitors when the outside ambient temperature exceeds a temperature set point	Can only be applied where direct correlation with load increases and temperature can be established, such as air conditioning loads
Voltage	Applies and removes shunt capacitors when system voltages operate outside of the allowable voltage tolerances	Can only be applied where voltage drops appreciably under load; more difficult to coordinate with voltage regulators and other switched capacitor banks; requires a suitably located potential transformer
Current	Can be applied at any point on the circuit where the load current can be monitored; not directional; responds to current changes	Requires current transformer; adjustment slightly more complex than other controls
kVAR	Most effective in minimizing losses or power factor because it senses fundamental quantity being corrected (kVAR); current and voltage sources available for general testing on feeder	Most expensive control method; insensitive to abnormal voltage conditions; requires current transformer and potential transformer

Protective monitoring controls are available for capacitor banks connected Wye-Wye, grounded-neutral capacitor banks, and ungrounded-neutral capacitor banks, as shown in figs. 15.7 and 15.8. This topic is discussed further in chapter 11, section 11.7.

The scheme applicable to double Wye-configured banks is shown in fig. 15.7. This scheme detects the current that flows when an unbalance exists between the neutrals of two ungrounded banks. This scheme has the advantage that it is insensitive to system voltage unbalance and any standing third-harmonic current or voltages, and is therefore the preferred arrangement if the bank is of sufficient size to make this practical. For Delta banks, a similar principle can be adopted using an "H" configuration of capacitors on each phase.

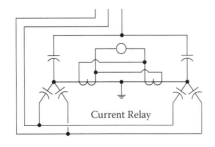

FIGURE 15.7 Different neutral current

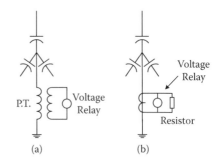

FIGURE 15.8 (a) Neutral voltage; (b) neutral current

For single Wye-grounded neutral capacitor banks, the most straightforward protective control is neutral-current-type relaying. This scheme operates on the neutral current generated because of the unbalance caused by capacitor failures in any phase. The major advantage of the neutral protection scheme is that it is relatively inexpensive, but it needs to be immune to third harmonics and to be set to withstand the standing unbalance voltages and outflow current on external ground faults as seen at the bank neutral.

If there are not too many series sections, this monitoring control can be set to alarm and trip at two different levels of neutral current to provide (a) early detection that a problem exists in the bank and (b) adequate protection should additional capacitor units fail.

The floating-neutral protective control is similar to that for an ungrounded neutral bank, except that a voltage transformer (VT) is used in the neutral (usually rated 15 kV) to indicate neutral voltage shift upon loss of units. The control voltage is sensitive and subject to the same limitations and advantages as the protective control for grounded-neutral capacitor banks.

15.6 CAPACITOR-BANK RESONANCES

15.6.1 Introduction

With any application of capacitor banks, there is always the risk of resonance. This is due to the interaction of the bank's capacitance with the inductive reactance characteristics of the network. Damping is also low due to the design objective of keeping system losses, and hence the net resistance, down to a minimum. Any application of capacitors should be checked for resonance using a simple method. If that calculation indicates a resonant frequency close to a multiple harmonic of the system frequency, a full harmonic analysis should be done.

The method illustrated below is only for approximation and will not be valid if the capacitors are installed at other buses in the system. It will therefore be necessary to carry out a harmonic analysis using a full-model harmonic load-flow program if:

- Capacitors are installed at multiple locations
- System fault levels, configuration, and load vary significantly

Application of Capacitors

15.6.2 EXAMPLE OF A RESONANCE RISK CALCULATION

When a bank of a particular size is proposed, then it is necessary to calculate the resonance risk frequency, f_r, for maximum and minimum fault levels that could occur on that bank. The simplified approach to check resonance at the capacitor bus uses the following formula:

$$\text{Resonance frequency} = f_r = f_{60} \times \sqrt{\frac{kVA_{SC}}{kVA_C}}$$

where
- kVA_{SC} = three-phase short-circuit kVA
- kVA_C = capacitor bank kVA

The design arrangement for this example is illustrated in fig. 15.9 for the following parameters:

- Design load = 1.5 MVA
- Actual power factor of the drive(s) = 85%
- Desired power factor on the bus = 96%

First calculate the required size (kVA_C) of the capacitor bank as follows:

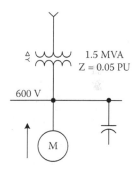

FIGURE 15.9 Example of a resonance risk calculation

$$kVA_C = 1500 \times \left(\sin(\cos^{-1} 0.85) - \sin(\cos^{-1} 0.96)\right)$$

$$kVA_C = 1500 \times (0.527 - 0.28)$$

$$kVA_C = 370.5 kVA \sim 400 kVA \text{ bank}$$

Then calculate the fault levels based, in this case, on an infinite source in-feed on the high side of the transformer:

- In-feed from the system at the 600 V level: 30 MVA
- Assumed in-feed from the parallel connected motor(s): 8 MVA
- Maximum three-phase 600 V bus fault level: 38 MVA
- Minimum three-phase 600 V bus fault level: 30 MVA

Then calculate the resonance risk frequency, in this case expressed as the ratio to the system frequency or the harmonic level, for the maximum fault level as follows:

$$f_r = \sqrt{\frac{38}{0.4}} = 9.7$$

Then calculate for the minimum fault level:

$$f_r = \sqrt{\frac{30}{0.4}} = 8.66$$

The bank rating above is safe if there is no triplen harmonic, but it would be problematic if such harmonics exist, as the harmonic range of resonant risk frequencies, between minimum and maximum system fault level conditions, includes a multiple of three.

REFERENCES

Standards
S1. ANSI C37.06, "Preferred Ratings and Related Required Capabilities for AC High-Voltage Circuit Breakers Rated on a Symmetrical Current Basis."
S2. ANSI/IEEE C37.012, "IEEE Application Guide for Capacitance Current Switching for AC High-Voltage Generator Circuit Breakers Rated on a Symmetrical Current Basis."
S3. ANSI/IEEE C37.99-2000, "IEEE Guide for the Protection of Shunt Capacitor Banks," 2000.
S4. ANSI/IEEE C37.04, "IEEE Standard Rating Structure for AC High-Voltage Circuit Breakers Rated on a Symmetrical Current Basis."

BIBLIOGRAPHY

1. Dow Chemical Co., "Environmental Impact Statement, Dow XFS-41G9L Capacitor Fluid," Midland, MI: Dow Chemical.
2. McGraw-Edison Co., "An Environmentally Acceptable Liquid Dielectric for Power Capacitors," McGraw Edison.
3. Sangamo Electric Co., Power Capacitor Application, Measurement and Control (papers), Workshop Seminar, Springfield, IL.
4. Northeast Power Systems Inc.; available on-line at http://www.nepsi.com, accessed 14 July 2007.
5. Canadian Electricity Association, "Capacitor Can Protection Review," CEA report 388 T 893, Montreal: SNC-Shawinigan Inc., 1995

16 Impact of Nonlinear Loads on Power System and Equipment

16.1 INTRODUCTION

Harmonics are a convenient way of expressing the waveform distortion caused by nonlinear loads in electrical power systems. Before the introduction of power electronics, most nonlinear loads were the result of static rectifier systems or the operation of transformers of shunt reactors at voltages beyond their magnetizing knee point, i.e., beyond their linear region.

Since the introduction of power electronics into industrial systems, waveform distortion has become more common, and the need for analysis of the resultant harmonics has increased. Other waveform distortions occur when transformers are energized, but these harmonics are transient in nature. Harmonics can also cause resonance with, and overloading of, devices designed to operate at fundamental frequencies only.

To control the level of harmonics in any system, it is important to know both the source of harmonics and the devices that are sensitive to such harmonics to limit their effect in any network.

16.2 HARMONICS AND RESONANCE

16.2.1 What Are Harmonics?

Harmonics are voltages or currents in the electrical system at some multiple of fundamental frequency (60 Hz in North America).

Any periodic wave can be described mathematically as a sum of sinusoidal wave forms in a Fourier series, named after the French mathematician Jean Fourier (1768–1830). The frequency of each sinusoid is an integer multiple of the frequency represented by the fundamental periodic cycle or frequency. Each term in the series is referred to as a harmonic of the fundamental frequency. A square wave, as shown in fig. 16.1 (common for a full-wave static rectifier), is the sum of an infinite series of odd harmonic sinusoidal wave forms. This can be expressed as follows:

$$E(\omega t) = \frac{4}{\pi} \times E_{max} \times \left(\cos \omega t + \frac{1}{3}\cos(3\omega t) + \frac{1}{5}\cos(5\omega t) + ... + \frac{1}{n}\cos(n\omega t) \right)$$

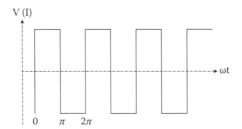

FIGURE 16.1 A periodic wave is a sum sinusoidal in a Fourier series

where
- $\dfrac{4}{\pi} \times E_{max}$ = magnitude of the fundamental
- third harmonic = 1/3 of the fundamental
- fifth harmonic = 1/5 of the fundamental, and so on to the nth odd harmonic

A similar series of even harmonics is an omnidirectional square wave such as that produced by a half-wave rectifier.

On the same basis, any complex waveform can be converted into a series of appropriately valued harmonics. As it is far easier to analyze system problems using the standard fundamental AC network formulae, it has become standard practice to analyze the effects of waveform distortion by first analyzing the harmonic makeup of any complex waveform.

As an example, for a switched converter, the harmonic current generated by the converter is given by the formula:

$$\text{Harmonic order } h = n \times P \pm 1$$

where
P = pulse number
n = an integer

Therefore, the harmonics produced by various types of converter, where h is the harmonic order, i.e., a list of the harmonics that are being produced, can be deduced as:

- For a 6-pulse system, h = 5, 7, 11, 13, 17, 19
- For a 12-pulse system, h = 11, 13, 23, 25, 35, 37
- For a 16-pulse system, h = 17, 19, 35, 37, 53, 55

There are three major harmonics indices used in the analysis of waveform distortion. These are:

- Harmonic voltage: $V_h = \sqrt{\sum\limits_{h=2}^{n} V_h^2}$

- Harmonic current: $I_h = \sqrt{\sum_{h=2}^{n} I_h^2}$
- Total harmonic distortion (THD)

There are two THDs, the first for the voltage and the second for the current:

$$THD_V = \frac{\sqrt{\sum_{h=2}^{50} V_h^2}}{V_1} \times 100\ \%$$

$$THD_I = \frac{\sqrt{\sum_{h=2}^{50} I_h^2}}{I_1} \times 100\ \%$$

These THD values are an indicator of the total level of harmonic distortion and hence the severity of the waveform distortion. These factors are often used to express the limits of distortion that a utility will accept from an industrial user or for the harmonics generated by any power electronics device.

16.2.2 Harmonic Current Generated by Nonlinear Loads

All nonlinear loads cause some level of harmonic distortion, but with the introduction of power electronics to the control of motors, and their use to provide controllable rectifier systems and self-adjusting switching power supplies, the number of systems that can cause such distortion has increased significantly.

Typical values of harmonic current generated by such nonlinear loads are given in table 16.1. It is worth noting that the fifth harmonic current generated by a six-pulse, pulse-width-modulated (PWM) drive can be much greater than the theoretical value of 20% due to the variation in switch-off points. PWM drives not only generate excessive harmonic currents, but they can also inject even as well as triplen harmonic currents.

16.2.3 Effects of Harmonics

The distortion in the waveform in the load current of any nonlinear device causes similar changes in the voltage waveform relative to the harmonic impedance of the source network. This voltage distortion affects both the current and voltage for all other loads connected to that system. The common effects of such harmonic distortion are as follows:

TABLE 16.1
Harmonic Current Generated by Nonlinear Loads (% of Fundamental)

Harmonic Order	Rectifier 6/12-Pulse	VFD PWM6-Pulse	SMPS$_a$	PC	Fluorescent Lamp
3	...	...	81	57	28.7
5	20/...	25–47	60.6	38	2.6
7	14.3/...	16	37.5	23	2.2
9	...	...	15.7	13	...
11	9.1/9.1	8.7	2.4	23	...
13	7.7/7.7	4.5	6.3	5	...
15	...	...	7.9	3	...
17	5.9/...	3.7	...	...	...
19	5.3/...	1.9	...	...	...
THD	28/14.3	51	116	75	29

a SMPS: switch mode power supply.

16.2.3.1 Motors and Generators

Generators and motors are adversely affected by harmonics in the networks to which they are connected. Typical effects are:

- Increased heating due to iron and copper losses at the harmonic frequencies
- Higher audible noise emission as compared with sinusoidal excitation
- Harmonic currents in the rotor

The harmonic currents noted above are caused by harmonics in the stator winding, which will produce harmonic currents in the rotor, e.g., 5th- and 7th-order stator harmonics will produce 6th-order rotor harmonics, while 11th- and 13th-order stator harmonics will produce 12th-order rotor harmonics. These rotor harmonic currents will result in increased rotor heating and pulsating or reduced torque.

It should also be noted that system unbalance (standing unbalance or ground faults), expressed as negative-sequence currents, can also reflect into the rotor as harmonic currents, which add to those noted above. Generators can also produce harmonics and, in particular, triplen harmonics that can circulate through adjacent Wye-grounded transformers when generators are directly connected to a load bus. The use of the Delta-connected generator transformers can control this, as discussed in chapter 14.

16.2.3.2 Transformers

The stray-loss factor for copper conductors varies as the square of the load current and the square of the frequency, and will therefore vary with the harmonic mix in the power supply. Although the percentage contribution to distortion by higher harmonics decreases as the harmonic frequency rises, its heating effect, even if the

Impact of Nonlinear Loads on Power System and Equipment

harmonic percentages are low, could rise substantially. These aspects are covered in more detail in section 5.5 of chapter 5, "Power Transformers and Reactors."

16.2.3.3 Capacitors

Any capacitance in an AC network can produce a risk of resonance with the inductive parts of the network. Although electrical networks are designed not to have any resonances at fundamental frequencies, when the multiple frequency effects of harmonic distortions are considered, there is always the possible risk of system resonance. These and other effects of harmonics on capacitors and capacitor banks are as follows:

- Resonance imposes considerably higher voltages and currents in capacitors.
- The capacitor bank acts as a sink for higher harmonic currents, which increases the heating and dielectric stresses.
- The losses in a capacitor are proportional to the reactive output (kVAR), which, in turn, is proportional to the frequency. These losses are increased, and the overall capacitor life is shortened with increasing harmonics.

To avoid or minimize such problems, capacitor banks can be tuned to reject certain harmonics by adding reactance, as discussed in section 16.5 of this chapter and in chapter 15.

16.2.3.4 Power Cables

Power cables are inherently capacitive and, as noted above for capacitor banks, their capacitance can produce a risk of resonance with the inductive parts of the network. These resonance risks and the harmonics themselves can produce the following problems for cable systems:

- Cables involved in system resonance may be subjected to voltage stress and corona.
- Increased heating due to higher rms current, skin effect, and proximity effect. The skin effect will vary with the frequency and conductor size.

Power cable conductors commonly lie very close to one another, and therefore the high-frequency currents in the outer skin of one conductor influence the spread and behavior of high-frequency currents in the skin of the adjoining conductors, giving rise to a "proximity effect." The skin effect and proximity effect are proportional to the square of a frequency. Cables therefore have to be derated if there is significant harmonic distortion, particularly if I_{THD} is greater than 10% [4, 5]. These aspects are discussed in more detail in chapter 10.

16.2.3.5 Electronic Equipment

Power electronic equipment is susceptible to misoperation if there are significant levels of harmonic distortion. Some of the control systems for power electronic devices use zero crossing detection to control switching. Harmonic distortion can result in shifting of the voltage zero crossing points, and these changes can be critical for many types of electronic control circuits. Also, if incorrect switching occurs, more harmonics can be produced, compounding the problem.

16.2.3.6 Switchgear and Relaying

Harmonic currents in switchgear will increase heating and losses in switchgear in the same way as has been discussed for power cables above. Similarly, voltage distortions can cause problems for voltage transformers (VT) and connected relays, while current distortion can do the same for current transformers (CT) and current-operated relays. In particular, electromechanical overcurrent relays exhibit a tendency to operate more slowly [8], while static underfrequency relays that use zero crossing for frequency measurement are susceptible to substantial changes in operating characteristics [9] in the same way as noted above for electronic control circuits. Harmonics can also impair the speed of operation of electromagnetic-type differential relays.

Modern digital relays use filtering techniques to produce the fundamental frequency only so that the fault current and voltage measurement circuits are not significantly affected by harmonics. Correct filtering requires the relay algorithm to track the system frequency, and most relays have a limited frequency range over which they are designed to operate. Although the fault-measuring capability may not be compromised, its ability to correctly measure and detect overloading conditions depends on its ability to measure the heating effect accurately, i.e., the full rms values. Although corrections can be made in the settings when the harmonic mix is known, it is important to know the cut-off frequency of the relay to be able to fully compensate and protect devices, such as high-frequency filters.

16.2.3.7 Fuses

Fuses suffer a derating factor because of the heat generated by harmonics. Fuses can therefore mal-function under the influence of harmonics. These effects must be considered so that the fuses can be derated correctly. Also, the unwarranted and frequent blowing of fuses is an indication of the presence of unexpected harmonics or changes in the harmonic mix in that system.

16.2.3.8 Telephone Interference

Communication systems are susceptible to interference from harmonics, particularly those that result in frequencies in the audible range. Any increase in audible telephone interference needs to be investigated, since this is often the first warning that harmonics are present or are increasing. When calculating the influence of harmonics and their associated frequencies on adjacent communication circuits, two factors are used, the weightage factor, W, and the telephone interface factor, TIF. The weighting in the W factor is based on the sensitivity of the human ear and is shown in fig. 16.2. The communication interference due to inductive coupling between power and communication circuits is calculated using the TIF, which is defined as:

Impact of Nonlinear Loads on Power System and Equipment

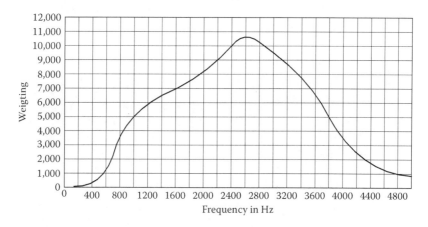

FIGURE 16.2 TIF weightage values

$$TIF = \frac{\sqrt{\sum (V_f \times W_f)^2}}{V_t}$$

where

$$W_f = 5 \times P_f \times f$$

where
- V_f = rms voltage at frequency f
- W_f = TIF weightage factor at frequency f
- V_t = rms fundamental
- P_f = C message weighting
- 5 = constant to simulate captive coupling
- f = frequency

16.2.4 Harmonic Resonance

Harmonics, being a multiple set of frequencies over and above the fundamental system frequency, expose the combination of inductance and capacitance in a power system to multiple risks of resonance. Typically, there are two types of resonance, parallel and series, as discussed below.

16.2.4.1 Parallel Resonance

This type of resonance can occur when the source or interface is paralleled by a branch with inductive and capacitive components, as illustrated in fig. 16.3. If Z_S is the equivalent system impedance for the power system as a whole, then:

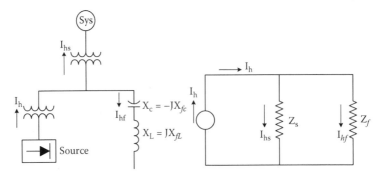

FIGURE 16.3 Parallel resonance circuit

$$Z_f = Z_{fc} + Z_{fl}$$

$$I_h = I_{hs} + I_{hf}$$

$$I_{hf} = \frac{Z_S}{Z_f + Z_S} \times I_h$$

$$I_{hs} = \frac{Z_f}{Z_f + Z_S} \times I_h$$

Z_f is usually large compared with Z_S at the fundamental frequency of the system. However, for an increase in frequency, Z_f will decrease where

$$X_c = \frac{1}{2 \times \pi \times f \times c}$$

Z_S will decrease where

$$Z_S \approx X_S = 2 \times \pi \times f \times L_S$$

(the resistive component is usually so small as to be negligible). If $(Z_f + Z_S)$ becomes small at a particular frequency, then I_{hS} and I_{hf} will increase. If

$$Z_f \times (Xc + X_L) \;\rightarrow\; 0, \text{ then } I_{hs} = 0, I_{hf} = I_h \ .$$

The capacitor and the reactor will then become a filter at that frequency. At the resonance frequency f_r, $Z_L = -X_c$, Z is very high when excited from a source at this frequency. A high circulating current will then flow in the inductance-capacitance loop, as shown in fig. 16.4.

As an example, if the system shown in fig. 16.4 has the following impedance values at 60 Hz:

Impact of Nonlinear Loads on Power System and Equipment

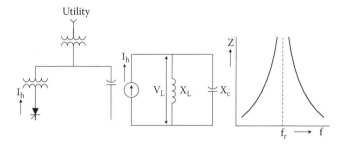

FIGURE 16.4 Example of parallel resonance

$X_{fL} = 0.6\ \Omega$
$X_{fc} = -30.23\ \Omega$
$X_L + X_c = 0.61$
Inject 1.0 A ($I_h = 1.0$ A)

then

$V_L = 0.61$ V
$I_L = 1.02$ A
$I_c = 0.02$ A

However, reducing the balance for the seventh harmonic (420 Hz) $I_7 = 1/7$ A, the following values are then obtained:

$X_L = 4.2\ \Omega,\ X_c = -4.22\ \Omega$
$X_L + X_c = 92.4\ \Omega,\ V_L = 13.2$ V
$I_L = 3.13$ A, $I_c = 3$ A

The system is therefore closer to resonance at the seventh harmonic, and the noted current amplification may occur.

16.2.4.2 Series Resonance

This type of resonance can occur if a resonance loop occurs as illustrated in fig. 16.5. Using the same concepts developed below in section 16.2.4.1, at the resonance frequency, f_r, then:

$$+JX_L = -JX_c$$

The only limitation on the current is then the net resistance R, which is usually very small for a high-efficiency distribution system.

An example of series-resonance-type situations is illustrated in fig. 16.6, where I_h is the source of the harmonic distribution and hence the source of each particular harmonic.

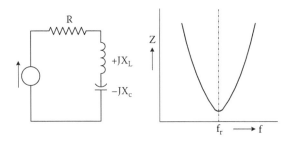

FIGURE 16.5 Series resonance circuits

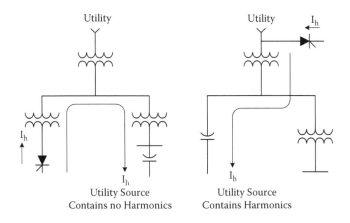

FIGURE 16.6 Example of series resonance

In both examples it is critical to know the harmonic impedance of the main system, as this may influence both the risk of resonance and the effectiveness of any filtering design to suppress the resonant conditions.

16.2.5 IEEE Standards for Harmonic Distribution Levels

IEEE 519 [S1] provides some, but not all, requirements to limit the harmonic injection from the variable frequency drives (VFDs) into the system. The philosophy of IEEE 519 is that the customer is responsible for limiting the harmonic currents injected into the power system and the utility is responsible for maintaining the quality of the voltage waveform.

Harmonic current distortion limits as extracted from IEEE 519 [S1] are given in tables 16.2, 16.3, and 16.4 for distribution systems, subtransmission systems, and transmission systems, respectively. Harmonic current limits for low- and medium-voltage systems rated 120 V through 69 kV are provided in table 16.2. These limits are based on the size of the load (I_L) with respect to the system short-circuit level (I_{SC}). All power generation equipment is limited to these values of current distortion, regardless of actual I_{SC}/I_L.

TABLE 16.2
Current Distortion Limits for General Distribution Systems (120 V through 69 kV), Maximum Harmonic Current Distortion in Percent of I_L

	Individual Harmonic Order (Odd Harmonics)					
I_{SC}/I_L	h < 11	11≤ h < 17	17 ≤ h < 23	23 ≤ h < 35	35 ≤ h	TDD
<20[a]	4.0	2.0	1.5	0.6	0.3	5.0
20 < 50	7.0	3.5	2.5	1.0	0.5	8.0
50 < 100	10.0	4.5	4.0	1.5	0.7	12.0
100 < 1000	12.0	5.5	5.0	2.0	1.0	15.0
>1000	15.0	7.0	6.0	2.5	1.4	20.0

[a] Even harmonics are limited to 25% of the odd harmonic limits above.

Source: IEEE 519-1992, IEEE Recommended Practice and Requirements for Harmonic Control in Electrical Power Systems, 1992.

TABLE 16.3
Current Distortion Limits for General Subtransmission Systems (69 kV through 161 kV), Maximum Harmonic Current Distortion in Percent of I_L

	Individual Harmonic Order (Odd Harmonics)					
I_{SC}/I_L	h < 11	11≤ h < 17	17 ≤ h < 23	23 ≤ h < 35	35 ≤ h	TDD
<20	2.0	1.0	0.75	0.3	0.15	2.5
20 < 50	3.5	1.75	1.25	0.5	0.25	4.0
50 < 100	5.0	2.25	2.0	0.75	0.35	6.0
100 < 1000	6.0	2.75	2.5	1.0	0.5	7.5
>1000	7.5	3.5	3.0	1.25	0.7	10.0

Source: IEEE 519-1992, IEEE Recommended Practice and Requirements for Harmonic Control in Electrical Power Systems, 1992.

TABLE 16.4
Current Distortion Limits for General Transmission Systems (>161 kV) Dispersed Generation and Cogeneration

	Individual Harmonic Order (Odd Harmonics)					
I_{SC}/I_L	h < 11	11≤ h < 17	17 ≤ h < 23	23 ≤ h < 35	35 ≤ h	TDD
<50	2.0	1.0	0.75	0.3	0.15	2.5
≥50	3.0	1.5	1.15	0.45	0.22	3.75

Source: IEEE 519-1992, IEEE Recommended Practice and Requirements for Harmonic Control in Electrical Power Systems, 1992.

- I_{SC} = maximum short-circuit current at PCC
- I_L = maximum demand load current (fundamental frequency component) at PCC

The voltage distortion limits at the connection point for any consumer are the responsibility of the utility. The required levels as expressed in IEEE 519 [S1] are presented in table 16.5. However, individual utilities may set more stringent values.

16.3 VARIABLE-FREQUENCY DRIVES

The application of variable-frequency drives (VFDs), especially pulse-width-modulated (PWM) drives, is growing faster than the technology is able to remedy its undesirable effects on power systems and equipment, especially where additional drives of this type are added into an existing system or when multiple drives of this type exist. Unfortunately, many users purchase VFDs mainly on the basis of their first cost, with no consideration for additional costs that can be required to remedy the problems associated with that drive type or the combined effects of multiple drives using this technology.

As experience has been gained by the suppliers of such drives in their operation in typical situations, VFD technology has made substantial improvements. These improvements minimize the adverse effects of the resultant internal harmonics on power systems and adjacent equipment. These improvements include the addition of active front-end filters to minimize harmonic injection into the power system and sine wave filters on the inverter output to limit the impact on the motors themselves. The active front-end filter (AFE) will minimize or eliminate several issues, including:

- The need for "K"-rated transformers (standard transformer design can handle up to 5% waveform distortion)
- The need for external harmonic filters
- Resonance problem associated with adjacent power factor correction capacitors

TABLE 16.5
Voltage Distortion Limits

Bus Voltage	Maximum Individual Harmonic Component (%)	Maximum THD (%)
69 kV and below	3.0	5.0
115 kV to 161 kV	1.5	2.5
Above 161 kV	1.0	1.5

Source: IEEE 519-1992, IEEE Recommended Practice and Requirements for Harmonic Control in Electrical Power Systems, 1992.

Active filters can provide reactive compensation themselves, and therefore the need for additional capacitors for power factor correction may not be necessary. Although the AFE (active filter) technology has existed for a while, the equipment for application with VFDs is only available through a few manufacturers. Some adverse effects of their use, such as noise emission, still need resolution.

The application of sine wave filters on the inverter output will eliminate the need for definite purpose inverter duty motors and bulky symmetrical cables for the motor feeders. The design aspects of cabling VFDs are dealt with in more detail in chapter 10.

16.4 SYSTEM NEUTRAL GROUNDING

The design of system neutral grounding is discussed in detail in chapter 4. To summarize, the most common methods of system neutral grounding [6, 7] in an industrial plant are as follows:

- High-resistance (HR) grounding for three-phase three-wire systems rated at 600 V and below. The neutral resistor is rated up to 5 A, and it is permissible to operate the system with a line-to-ground fault on one phase. The grounding system rating is based on the system phase-to-neutral voltage and current returning to the neutral.
- Low-resistance (LR) grounding for medium-voltage systems. The neutral resistor is rated 100–400 A, and a ground fault is cleared in the shortest possible time.
- High-resistance (HR) grounding for medium-voltage systems where power interruption will be detrimental for the process. HR grounding is limited to systems not exceeding 5 kV where the changing current is less than 5.5 A. The neutral resistor is rated 5.0 A on the primary of the grounding bank, and continuous operation with a line-to-ground fault in one phase is permitted.

For large drives and rectifiers with dedicated transformers, the system neutral is not grounded. Ground-fault detection equipment comprising AC and DC relays are utilized for alarm and shutdown.

When inverters are applied to motors, the motor winding can be exposed to higher than normal voltages due to the neutral shift (voltage difference between the source neutral and the motor neutral). According to NEMA MG1, Part 31 [S5]: "The magnitude of the neutral voltage can be reduced if the inverter is connected to an ungrounded power source or, if this is not possible, by isolating it from the source ground by using an isolation transformer or by using separate reactors in both the positive and the negative direct current link."

Many VFD installations are connected from a common HR grounded bus, and the following must be considered when connecting such a drive to a HR grounded system:

- The resistor rating is based on system phase-to-neutral voltage and the fundamental undistorted alternating current level. For a 600 V system, the resistor will be rated 347 V, 69 ohm for a 5.0 A continuous rating.

- A line-to-ground fault on the DC bus will result in a current through the resistor. For a six-pulse 600 V drive, the line-to-ground voltage will be 234 VlN/2 (or 405 V). The resistor will be subjected to 17% higher voltage, and the ohmic value must be increased to limit the current to 5.0 A. The ground-fault detection equipment needs to be suitable for both AC and DC.
- A line-to-ground fault on the inverter output will cause fault current to flow from the AC system into the fault on the inverter output. The current distribution and its impact on the supply are not easy to predict.

The following recommendations are therefore made:

- Group the VFDs and feed them through a common step-down transformer with an ungrounded secondary.

Use isolating transformers for individual drives. Use half of the transformers with Delta-Wye winding and the other half Delta-Delta. This will tend to cancel fifth and seventh harmonic currents, although the amount of cancellation will depend on the loading.

16.5 DESIGN RECOMMENDATIONS FOR MITIGATION OF HARMONICS

16.5.1 Capacitor Bank Applications

In most industrial harmonics power systems, the primary objective for installing capacitors is to meet the utility power factor requirements as expressed in its tariff rates. Additional benefits are better voltage regulation and lower losses. Commonly used locations are shown in fig. 16.7.

Any capacitor bank can be a source of parallel resonance with the system inductance. The best approach to avoid resonance problems is to install large capacitor banks at the main bus. This solution offers the following advantages:

- More available reactive power to the system as a whole
- Easier control of harmonic voltages and currents
- Lower capital costs, as large banks are more economical in terms of purchase cost
- Reactors can be added to shift the resonant frequency away from the characteristic harmonic frequency of the plant

Capacitors can also be combined with reactors to develop harmonic filters at the troublesome resonance harmonic frequencies. The resonant frequency at the capacitor bus can be calculated by:

$$f_r = f_s \sqrt{\frac{kVA_{sc}}{kVA_c}}$$

Impact of Nonlinear Loads on Power System and Equipment

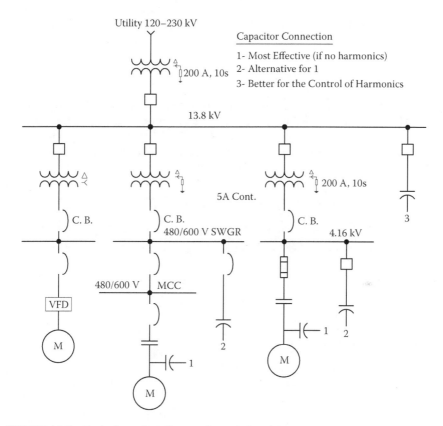

FIGURE 16.7 Typical one-line diagram for an industrial system

where
- f_r = resonant frequency
- f_s = system frequency, 60 Hz
- kVA_{sc} = three-phase system fault level in kVA
- kVA_c = three-phase capacitor-bank rating in kVA

16.5.2 Variable-Frequency-Drive Application

Based on the discussion of the problems associated with VFDs in section 16.3, the following recommendations are put forward:

- Study the harmonic impact before making a commitment or purchasing the drive.
- Preference shall be given to the power system friendly drives. These drives shall include features such as minimum harmonic injection into the power system and having a sine wave filter on the inverter output.

- When it is feasible, group the drives and connect them to a common bus. Feed the bus through a K-rated transformer [S3] with a Delta primary and an ungrounded Wye secondary winding. A common active front end (AFE) filter for the group should be configured to limit the harmonic current distortion to less than 2%.
- Feed the individual or isolated drives through a K-rated transformer.
- Motors used in VFD applications shall be definite purpose inverter duty, conforming to NEMA MG1 Part 31 [S5]. Also follow the other recommendations provided in this document for the application of VFDs.
- Motor feeders shall use three conductors with three symmetrically placed ground conductors and heavy-gauge continuous aluminum sheath. Do not use Teck-type cable, as noted in chapter 10, section 10.4.

The suggested arrangement for the power supply to low-voltage and medium-voltage drives is shown in figs. 16.8 and 16.9, respectively.

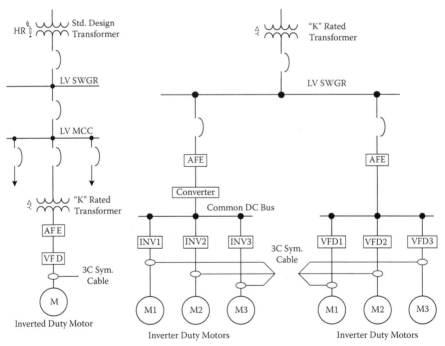

FIGURE 16.8 Suggested arrangements for power supply to low-voltage (<600 V) drives

Impact of Nonlinear Loads on Power System and Equipment

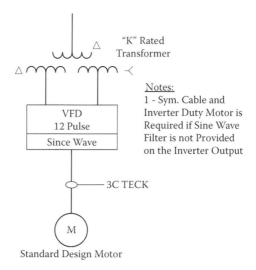

FIGURE 16.9 Suggested arrangements for power supply to medium-voltage drives

REFERENCES

Standards

S1. IEEE 519-1992, IEEE Recommended Practice and Requirements for Harmonic Control in Electrical Power Systems, 1992.
S2. IEEE 1100-1992, IEEE Recommended Practice for Powering and Grounding Sensitive Electronic Equipment, 1992.
S3. ANSI/IEEE C57.110-1986, IEEE Recommended Practice for Establishing Transformer Capability When Supplying Non-Sinusoidal Load Current, 1986.
S4. IEEE 141-1993, IEEE Recommended Practice for Electric Distribution System for Industrial Plants, 1993.
S5 NEMA MG 1-1993, Motors and Generators, 1993.

BIBLIOGRAPHY

1. Austin Bonnett, "Analysis of the Impact of Pulse Width Modulated Inverter Voltage Waveforms on AC Induction Motors," *IEEE Trans. IAS* 32 (2): 386–392 (1996).
2. Dave Busse et al., "Characteristics of Shaft Voltage and Bearing Currents," *IEEE Ind. Appl. Mag.* 3 (6): 21–32 (1997).
3. Mauri Peltola, "ABB Drives: Harmonic Distortion," publication ST-312 (6/1/93), 1993.
4. Ajit Hiranandani, "Calculation of Cable Ampacities Including the Effects of Harmonics," *IEEE Ind. Appl. Mag.* 4 (2): 42–51 (1998).
5. J. M. Bentley and P. J. Link, "Evaluation of Motor Power Cables for PWM AC Drives," *IEEE Trans. IAS* 33 (2): 342–358 (1997).
6. J. C. Das and R. H. Osman, "Grounding of AC and DC Low Voltage and Medium Voltage Drive Systems," *IEEE Trans. IAS* 34 (1): 205–216 (1998).
7. J. R. Dunki-Jacobs, "The Reality of High Resistance Grounding," *IEEE Trans. IAS* 469–475.

8. W. A. Elmore et al., "Effect of Waveform Distortion on Protective Relays," *IEEE Trans. IAS* 29 (2): 404–411 (1993).
9. A. A. Girgis et al., "Effect of Voltage Harmonics on the Operation of Solid State Relays in Industrial Applications," *IEEE Trans. IAS* 28 (5): 1166–1173 (1992).

Index

A

Absorptive glass-microfiber (AGM), 187
Active front-end filter (AFE), 440, 441
AEIC, *see* Association of Edison Illuminating Companies (AEIC)
AFE, *see* Active front-end filter (AFE)
AGM, *see* Absorptive glass-microfiber (AGM)
Air-blast breakers, 171
Air-cooled generators, 96
Air-core mutual reactors, 136
Air-magnetic air quenching breakers, 170
Alstom guides, 135
Altitude
   and battery sizing, 189
   dielectric correction factors, 101
   effect on conductor size, 340–341
   effect on insulation, 203, 340
   and substation design, 340–342
   zones, 341–342
Arc-flash hazard, 55–57
Areva, 135
Askarel-immersed transformers, 118–119
Association of Edison Illuminating Companies (AEIC), 246
   testing cable insulation, 250
Auxiliary current transformers, 136, 146
Auxiliary systems, 407–408
   bus capacity, 409–410
   design and equipment, 408–409
   distribution, 406–407
   load, 406
   short circuits, 409
   sizing, 409
   voltage considerations, 409

B

Bar-type CT, 135
Basic impulse (BIL), 112
Basic insulation level (BIL), 325, 348
Batteries, 181
   application voltage levels, 189–190
   correction factor, 193
   duty cycle, 193
   float charge, 192
   lead-acid, 187, 192
   load classification, 190
   nickel-cadmium, 187, 192
   protection fuses, 184
   rating, 190
   sizing, 191

   system voltage, 191–192
   types of battery, 185–188
Battery chargers, 188–189
   capacity, 188
Bearings, *see* Motor bearings
BIL, *see* Basic impulse (BIL); Basic insulation level (BIL)
Breakaway torque, 212
Brushless excitation system, 213–214, 391–392
Bus capacity
   for auxiliary systems, 409–410
Bus design
   considerations, 342–343
   rigid bus, 345–346
   ring bus, 332
   strain bus, 346–347
Bus-differential schemes, 318–320
Bushing current transformers (CTs), 114, 135–136, 144
Bus insulation
   capacitor banks, 420
Bus protection, 317
   current-transformer saturation, 317
   high-impedance bus-differential scheme, 320
   high-impedance voltage differential, 320
   linear coupler bus-differential scheme, 320
   nondirectional overcurrent relays, 318
   stabilizing resistor, 319

C

Cable ampacity calculations, 63
Cable application
   short-circuit withstand, 243–244
   for variable-frequency drives (VFD), 243–246
Cable insulation
   characteristics, 247
   EPR testing, 249–250
   high-voltage DC testing, 249
   levels, 247–248
   reliability, 248–249
   standards, 246
   testing, 249
   types, 246
Cables
   control, 250–251
   harmonics effect, 433
   system capacitance data, 94
Cable selection, 237–238
   fault-current criteria, 239
   insulation levels and system grounding, 239

447

448 Index

load current criteria, 238
voltage-drop limitations, 238
Cable shielding
    grounding, 241
    inner shield, 239
    outer shield, 240
    overall size, 240–241
    recommended practice, 241–243
Canadian Electrical Code (CEC), 3, 247, 292–293
Canadian Standards Authority (CSA), 247
Capacitor banks
    application, 413
    bus insulation, 420
    construction, 417–418
    for harmonics mitigation, 442–443
    monitoring, 419–420
    protection, 315–316, 317, 421
    protective control, 424, 426
    resonances, 426–428
    safety, 420
Capacitors; *see also* System capacitance data
    construction, 413–414
    control, 420–421
    fusing, 421–423
    individual, 414
    motor, 414
    protection, 314–315
    and resonance, 433
    switched, 415
    unit rating, 418–419
Capacitor voltage transformers, 147–148
CEC, Canadian Electrical Code (CEC)
Circuit breakers, low-voltage, 31–34, 156
    features, 155
    interrupting duty, 35
    metal-enclosed, 35, 36, 155
    molded-case, 36
Circuit breakers, medium- and high-voltage
    classification, 170
    close-and-latch duty, 33–34
    indoor ratings, 158–159, 160–161
    interrupting duty, 32–33
    outdoor ratings, 162–163, 164–165
    performance requirements, 169–170
    rating by E/X method, 173–174
    selection, 167–168
    standards, 156, 168–169
    types, 157, 170–173
Class C current transformers, 138, 139
Class K current transformers, 138, 139
Class P current transformers, 152
CL/L period, *see* Close-and-latch (CL/L) period
Close-and-latch (CL/L) period, 31
Conservator design, 111–112
Conservator diaphragm, 112
Control cables, 250–251

Conveyor drives, 209
Coordination curves, 295
Cranes, 209
Cross-linked polyethylene (XLPE), 246
Crushers, 209
CSA, *see* Canadian Standards Authority (CSA)
CT, *see* Current transformers (CTs)
Current distortion, 438, 439
Current-limiting fuse, 262
Current-limiting reactor, 396
Current transformers (CTs), 54
    accuracy class and burden, 139–141
    accuracy voltage, 144
    auxiliary, 136, 146
    bushing type, 114, 135–136
    connections, 254
    equivalent circuit, 137, 138
    European standards, 151, 152
    excitation characteristics, 138–139
    metering, 146, 152
    polarity, 145
    remnant flux, 144
    saturation, 141–144
    in series, 145

D

DC power supply, 63
    direct, 181, 182
    distribution system, 181, 182–185
    for protection and control, 9
Delta-Wye transformers, 54
Diesel generators, 7, 8, 9, 312, 352
Digital current transformers, 136–137
Direct-stroke shielding, 349
Distribution types
    expanded radial, 9–10
    primary selective, 10
    secondary selective, 10, 11
    simple radial, 9
    sparing transformer, 10, 11
Dornenberg ratio, 118
Dry-type transformers, 100, 102
    accessories, 122
    impedance, 121
    insulation class and temperature rise, 120
    insulation level, 121
    primary system fault level, 120
    surge arresters, 122
    tap changer or taps, 121
    types and rating, 119–120
Duplex reactor, 396

E

EHV, *see* Extremely high voltage (EHV)
Electrical codes, 3
Electric motors, 217–218, 219

Index

Electromagnetic interference (EMI), 253, 255
Electromagnetic transient program (EMTP), 27, 63
Emergency power supply, 7
EMI, *see* Electromagnetic interference (EMI)
EMTP, *see* Electromagnetic transient program (EMTP)
EPR, *see* Ethylene propylene rubber (EPR) insulation
Equivalent circuit
   of a current transformer, 137, 138
Ethylene propylene rubber (EPR) insulation, 246
European standards, 151–153
Excitation systems
   brushless, 391–392
   high-initial response, 390–391
   nominal response, 390
   performance criteria, 400
   power system stabilizer (PSS), 393
   static, 392
   for synchronous motors, 213–214
   voltage-regulating system, 392–393
Extremely high voltage (EHV), 137, 172

F

Feeder protection, 313
Fence grounding, 380–382
Formulae
   per-unit values, 27–28
   using, 27–28
Fused starter, 179
Fuse let-through charts, 167
Fuses, 166, 260
   advantages/disadvantages, 166
   for capacitor protection
   harmonics effect, 434
   minimum rating, 260, 262
   size, 167
   time-current, 262
   transformer protection, 262

G

Gas-insulated switchgear (GIS), 325
   configurations, 174
   design concepts, 175, 176
   main features, 175
   shock situation, 367–368
   specifications, 176, 177
Gas-oil seal system, 110–111
Gauss-Siedel algorithm, 29
Gelled electrolyte (gel-cell) battery, 187
Generator neutral grounding, 80
   equipment sizing, 82–84
   high-resistance and system low-resistance grounding, 85
   high-resistance grounding, 81–82
   low-resistance grounding, 84
   other methods, 85
   solidly grounded neutral, 80–81
Generator protective devices
   accidental energizing, 301
   breaker fail, 310
   differential, ground, 308
   differential phase, 307
   field ground, 310
   frequency under/over, 307
   loss of field, 304
   negative-sequence current, 305
   out-of-step, 307
   overcurrent, 306
   overfluxing, 301
   overvoltage, 307
   reverse power, 303
   stator ground-fault, 309
   thermal overload, 305
   winding overtemperature, 305
Generators; *see also* Induction generators
   air-cooled, 96
   capability curve, 389
   diesel, 312
   harmonic effects, 432
   integration, 395–397
   parameters, 389
   performance criteria, 400
   rating, 387, 389
   recommended protection, 298–301
   relay selection, 296
   selection, 387
   synchronizing, 393–394, 395
   tripping schemes, 312–313
   in unit-connected system, 396–397
Generator step-up transformers, 124–126
GIS, *see* Gas-insulated switchgear (GIS)
GPR, *see* Ground-potential rise (GPR)
Ground-fault device, 287
Ground-fault relay, 10, 68, 262, 269, 287
Grounding, *see* Generator neutral grounding; Substation grounding; System neutral grounding
Grounding, mines, 86
   distribution system, 87
   ground-fault relaying, 87
   for low-voltage systems, 87–88
   neutral grounding resistor, 87
   safety, 86
   shock hazard, 86
Grounding pads, 208
Grounding transformers, 90–93
   protection of, 271, 272
Ground-potential rise (GPR), 366, 367
Ground resistance, 370

H

Harmonic current, 431, 432
Harmonic resonance
  parallel, 435–437
  series, 437–438
Harmonics, 429–431
  analysis, 57–58
  effects of, 431–435
  IEEE standards, 438–440
  mitigation, 442–445
  switchgear and relaying, 434
  telephone interference, 434
  and variable-frequency drives (VFD), 440–441, 443–445
Harmonic voltage, 430
High-initial-response (HIR) excitation, 390–391
High-resistance (HR) grounding, 75–76, 90
  calculation, 72, 74
  for delta-connected systems, 74–75, 77
  design, 73
  directly connected resistor, 81
  of generator neutral, 81–82
  ground fault, 71–72
  hybrid, 85
  for low-voltage systems, 75, 76, 89
  normal operation, 71
  recommendations, 76
  and system charging current, 72, 73, 75
  transformer-resistor combination, 82
  using, 67, 68
  for wye-connected generator neutral, 90
HR grounding, *see* High resistance (HR) grounding
Hybrid high-resistance grounding, 85

I

ICEA, *see* Insulated Cable Engineers Association (ICEA)
IEC, *see* International Electrotechnical Commission (IEC)
IEC-rated contactors, 179
IEEE, *see* Institute of Electrical and Electronics Engineers (IEEE)
IEEE Gold Book, 4
IGBT, *see* Insulated-gate bipolar transistors (IGBT)
Immersed transformers, *see* Oil-filled transformers
Induction generators, 403
  advantages/disadvantages, 405–406
  characteristics, 403
  controls, 405
  excitation, 404
  limitations, 406
  protection, 404–405
  terminal voltage, 403–404

Induction motor equivalent network, 199
Induction motors, 209–210; *see also* Squirrel-cage induction motors
  accessories, 208
  slip-ring, 209–210
  zero-sequence capacitance, 96
Inductive voltage transformers, 146–147
Institute of Electrical and Electronics Engineers (IEEE), 3
Insulated Cable Engineers Association (ICEA), 246
Insulated-gate bipolar transistors (IGBT), 210
Interlocking, 4, 5, 184, 332, 420
International Electrotechnical Commission (IEC), 247
Inverters, 181

K

K-rated transformers, 123, 124, 440

L

Lifting devices, 208
Lightning protection, *see* Shielding
Linear couplers, 136
Load flow, 20, 28–29
  studies, 28–29
Load-flow runs, 29
Load tap changers (LTC), 106
Low-reactance grounding, 85
Low resistance (LR) grounding, 77–78
  for delta-connected system, 77
  recommendations, 78
  for wye-connected systems, 76–77, 90
LR grounding, *see* Low resistance (LR) grounding
LTC, *see* Load tap changers (LTC)

M

Magnetizing inrush current, 262
Maintenance, 5, 354
Masts, 351
Maximum continuous operating voltage (MCOV), 341
MCCBs, 55
MCCs, *see* Motor control centers (MCCs)
MCOV, *see* Maximum continuous operating voltage (MCOV)
Mesh voltage, 367
Metering current transformers, 146
Motor bearings
  antifriction, 207
  protection against failures, 207
  shaft current, 207–208
  sleeve, 206
  supply voltage unbalance, 207

Index

thrust, 207
Motor capacitors, 414
Motor control centers (MCCs), 67, 181
  bus bars, 179
  IEC-rated contactors, 179
  low-voltage, 177, 178
  medium-voltage, 178
  protection, 179
  short-circuit capacity, 178–179
  size, 179
Motor controllers
  fused starter, 228
  NEMA E2, 278
  starting methods, 230–234
  switchgear-type controllers, 230
Motor protection; *see also* Motor-protection schemes
  recommended, 279–279
  relay selection, 274
Motor-protection schemes
  acceleration torque, 276
  current transformer selection, 276
  electrical code, 278
  high-torque load, 278
  NEMA E2 Motor controllers, 278
  recommended, 280
  redundancy, 277
  relay characteristics, 276
  RTD bias, 278
Motor protective devices
  differential, 288
  ground fault, 287
  locked rotor, 285
  loss of field protection, 288
  out-of-phase reenergizing, 290
  out-of-step, 288
  overload, 285
  rotor protection, 289
  short circuit, 287
  slip-ring flashover protection, 289
  surge protection, 291
  unbalanced voltage/current, 282
  underfrequency, 285
  undervoltage, 281
  variable-frequency drive (VFD), 291
  winding overtemperature, 285
Motors, classified areas, 205–206
Motors, medium-voltage; *see also* Motor bearings; Motor controllers; Squirrel-cage induction motors; Voltage drop
  acceleration torque, 197–198
  constant-torque-drives, 195–196
  friction, 198
  harmonics effect, 432
  induction-motor equipment, 199
  inertia, 198
  load categories, 195

speed-torque curve, 197
starting frequency, 198–199
steady-state, 196–197
torque as function of speed, 196
variable frequency drives, 217–218
voltage rating and rise time, 218
Motor starting, 63, 231
  autotransformer, 230–231, 232
  capacitor, 232
  captive transformer, 233
  direct-on line, 230
  part winding, 234
  reactor, 230
  reactor-capacitor, 232
  variable frequency, 233–234
  variable voltage, 233
Motor terminal box, 208–209

N

National Electrical Code (NEC), 3, 294
National Electrical Manufacturer's Association (NEMA), 3, 246
  code letters, 201
  design letters, 200
  of normal load inertia, 213
National Electrical Safety Code (NESC), 3, 345
NEC, *see* National Electrical Code (NEC)
NEMA, *see* National Electrical Manufacturer's Association (NEMA)
NEMA E2 Motor controllers, 278
NESC, *see* National Electrical Safety Code (NESC)
Network protection and application Guide, 135
Neutral conductors, 371
Neutral grounding equipment
  generator neutral, 90
  for low-voltage systems, 89
  for medium-voltage systems, 89, 90
  Wye-Delta grounding transformer, 92–93
  Zig-Zag grounding transformer, 74, 77, 85, 89, 91, 271
Newton-Raphson algorithm, 29
NFPA 70, 55, 63

O

Occupational Safety and Health Administration (OSHA), 56
OCT, *see* Optical current transformers (OCT)
Off-circuit taps, 106
Oil circuit breakers, 170–171
Oil-filled transformers, 101, 102
  accessories, 115–116
  bushings, 114
  classification, 107
  cooling designation, 108–109
  cooling equipment, 116–117

core grounding, 117
gas analysis, 117–118
gas-oil seal, 110–111
impedance, 108
installation, 126–127
liquid preservation, 110–112
nameplate, 117
tap changers, 113–114
testing, 129–130
voltage, power rating and temperature rise, 107–108
winding insulation levels, 112
On-load tap changers, 105, 106
Optical current transformers (OCT), 136–137
OSHA, see Occupational Safety and Health Administration (OSHA)
Overcurrent, 54
coordination, 55, 56
primary protection, 259
secondary protection, 259–260
Overexcitation, 125, 214, 393
protection, 272–273
Overhead ground wires, 371
Overvoltage, 69

P

PCB (polychlorinated biphenyl), 118
P-class current transformers, 152
Personnel protective equipment (PPE), 4
PFW, see Power frequency withstand (PFW)
Plant distribution systems; see also Distribution types
in-plant generation, 7
main substation, 5
primary, 6–7
reliability, 4
secondary, 7
single (one)-line diagram form, 6
Plant main substation, 5
PlC, see Programmable logic controllers (PLC)
Post-type CT, 136
Power cables, see Cables
Power demand, 11
demand for utilization equipment, 12, 15
energy consumption, 11–12, 13, 14
from equipment list, 15
estimates of power demand, 14
estimates of power requirements, 13
factors used, 15
Power frequency withstand (PFW), 84, 112
Power fuses, see Fuses
Power plants, 406–407
Power system device function numbers, 256
Power system stability, 59
stability study for an ammonia plant, 60–62
transient behavior of synchronous generators and motors, 59–60

Power system stabilizer (PSS), 393
PPE, see Personnel protective equipment (PPE)
Programmable logic controllers (PLC), 181
Protection; see also Generator protective devices; Motor protection; Transformer protection
relay-setting coordination, 255, 257
schemes and relay selection, 254–255
Protection-class current transformers, 152
Protective device coordination, 53–54
data required for a study, 54
time interval for coordination of overcurrent elements, 54–55
Protective devices; see also Generator protective devices; Motor protective devices
Pull-in torque, 212
Pullout torque, 212
Pull-up torque, 212
Pulse-width-modulated (PWM), 210, 431, 440
PWM, see Pulse-width-modulated (PWM)

R

Radio frequency interferences (RFI), 255
Reactors, 127–128
current-limiting
installation, 126–127
protection, 317
testing, 128
Rectifier transformers, 126
Relaying
building location, 338
for capacitor-bank protection, 315
ground-fault, 88
harmonic effect, 434
time-delayed devices, 36
Resistance-temperature detectors (RTD), 208
Resonance risk, 427–428
Resonant grounding, 85
RFI, see Radio frequency interferences (RFI)
Rigid bus, 345–346
Rogers ratio, 118
RTD, see Resistance-temperature detectors (RTD)

S

Safety; see also National Electrical Safety Code (NESC)
codes, 3, 17, 55, 126, 187, 254
in HV substation, 86
of life, 3–4
procedures, 353–354
of SF gas-insulated switchgear, 175
SCADA, see Supervisory control and data acquisition (SCADA)
Schwarz formula, 375
SCR, see Short-circuit ratio (SCR)

Index

Sealed-tank system, 110
SF_6 gas circuit breakers, 171–172
SF_6 gas-insulated switchgear, *see* Gas-insulated switchgear (GIS)
Shielding
    direct-stroke protection, 349
    IEEE guides, 352
    lightning protection, 351–352
    masts, 351
    overhead wires, 349–351
Shock hazard, 86, 88, 240
Short-circuit ratio (SCR), 389
Short circuits, 27, 29; *see also* Circuit breakers
    contribution from different sources, 29–31
    current multiplying factors, 37–38, 156
    impedance, 103
    motor protective devices, 287
    rating calculations, 45–52, 155, 173–174
    rotating machine impedance multipliers, 36–37
    steps required for study, 31
    system and equipment data, 39–45
    for time-delayed relaying devices, 36
    X/R ratio of system component, 38
Silicone-immersed transformers, 119
Sleeve bearings, 206
Slip-ring induction motors, 209–210
Soil resistivity, 361–364
Solidly grounded neutral, 79–80
    for generators, 80–81
    recommendations, 80
Space heaters, 208
Spare transformer, 10
Squirrel-cage induction motors, 199; *see also* Motor bearings
    for classified areas, 205–206
    enclosures, 204–205
    inertia, 201–202
    insulation systems, 203
    magnetizing consumption, 214
    NEMA code letter, 201
    NEMA design letters, 200
    power supply voltage and frequency, 203–204
    service factor, 202–203
    starting power, 204
    surge-withstand capability, 203
    torque characteristics, 199–200
Starting torque, 212
Static excitation system, 392
Static trip units, 55
Station battery, *see* Batteries
Steady-state power, 196–197
Step voltage, 365–366, 380
Strain bus, 346–347
Substation design; *see also* Substation grounding
    altitude, 340–342
    bus design, 342–347
    commissioning, 352–354
    conductor spacing, 342
    configuration by type, 331
    control and metering, 336, 338
    information needed, 327
    location, 325–327
    measurement unit standards, 327
    medium- and low-voltage, 352
    protection, 335–336
    recommended configurations, 334–335
    relay building location, 338
    shielding, 349–352
    site testing, 352–354
    standards, 328
    surge protection, 348–349
    switching configuration, 330–331
    system phasing, 328–329
    transformer connections, 328–329
    transformer requirements, 339
    and utilities, 338–339
Substation grounding, 347–348
    and buried cables, 371
    case study, 382–386
    conductors and joints, 373–374
    current division factor, 371–372
    decrement factor, 372
    design steps, 378
    fence grounding, 380–382
    grid current return through earth, 372
    ground resistance, 370, 374, 375–376
    maximum grid current, 370, 371–372
    and overhead wires, 371
    permissible potential difference, 365–368
    required grid elements, 361
    soil resistivity, 361–364
    symmetrical grid current, 369–370
    worst-case symmetrical fault current, 368–369
Supervisory control and data acquisition (SCADA), 28
Surge arrestors, 349
Surge capacitors, 93, 208, 279, 291
Surge protection, 208
    substations, 348–349
    transformers, 273–274
Switched capacitors, 415
Switchgear; *see also* Gas-insulated switchgear (GIS)
    current transformer accuracy voltage, 144, 145
    harmonics effect, 434
    load-interrupter, 157, 166
    low-voltage, 155–156
    medium-voltage, 156–157
Switchgear-type controllers, 30
Switching
    configuration, 330–331

device rating, 416
harmonics effect, 433–434
transients, 62–63
Synchronous motors, 211
accelerating, 215–216
accessories, 208
application, 217
current pulsations, 216–217
excitation system, 213–214
inertia (wk2), 213
insulation system and temperature rise, 212
loss of field protection, 288
out-of-step protection, 288
power factor improvement, 214
power supply voltage and frequency, 211
rotor protection, 214, 289
starting, 215–216
torque characteristics, 212
torque pulsations, 215–216
zero sequence capacitance data, 96
Synchronous torque, 212
System capacitance data
air-cooled generators, 96
bus, 95
induction motors, 96
outdoor apparatus bushings, 95
overhead lines, 93–94
power cable, 94
power transformers, 95
surge capacitors for rotating machines, 93
synchronous motors, 95
System charging current, 72, 73, 75, 93
System disturbances, 399–403
System neutral grounding, 67–68, 441–442; *see also* High-resistance (HR) grounding; Neutral grounding equipment; System capacitance data
low-resistance grounding, 76–78
solidly grounded neutral, 79–80
ungrounded system, 68–70
System planning, 3–5

T

Tap changers, 106
Taps, 106
TCC, *see* Time-current characteristic curves (TCC)
T-class current transformers, 139, 140
Telephone interference, 434–435
THD, *see* Total harmonic distortion (THD)
Through-fault current, 266–268
Time-current characteristic curves (TCC), 54
TNA, *see* Transient network analyzer (TNA)
Torque, 212
Total harmonic distortion (THD), 431
Touch voltage, 366, 367
Transferred voltage, 367

Transfer system, 407
Transformer protection, 149–150
devices, 257–260
device settings, 261–262
differential protection, 260, 268–271
electrical codes, 264–265
grounding transformers, 271, 272
low-voltage unit substation transformers, 263–264
for overexcitation, 272–273
with a primary fuse, 260, 262
surge protection, 273–274
Transformers; *see also* Current transformers (CTs); Type; Voltage transformers
altitude correction factors, 100, 101
basic construction, 99
categories, 102–103
European standards, 151
generator step-up, 124
grounding of secondary, 150–151
harmonics effect, 432–433
inspection, 131–132
K-factor, 123, 124
load loss, 122
nonflammable liquid-filled, 118–119
nonlinear loads, 122–123
normal service conditions, 99–100
parallel operation, 106–107
properties of liquids, 119
short circuit impedance, 103, 104
short-circuit withstand, 102–103
taps or tap changer, 105
testing, 128–131
through-fault current damage curves, 266–268
in unusual air and water temperatures, 101, 102
vector group, angular displacement, and terminal markings, 103, 105
Transformers, dry-type, 119
accessories, 122
impedance, 121
insulation class and temperature rise, 120
insulation level, 121
primary system fault level, 120
surge arresters, 122
tap changer or taps, 121
type and rating, 119
Transformers, oil-filled, 107
accessories, 115–116
bushing current transformers, 114
classification of mineral-oil-immersed transformers, 107
cooling designation, 108–109
cooling equipment, 116–117
core grounding, 117
gas analysis, 117–118